AEROSOL INHALATION:
RECENT RESEARCH FRONTIERS

AEROSOL INHALATION: RECENT RESEARCH FRONTIERS

Proceedings of the International Workshop on Aerosol Inhalation, Lung Transport, Deposition and the Relation to the Environment: Recent Research Frontiers, Warsaw, Poland, September 14–16, 1995

Edited by

J. C. M. MARIJNISSEN

Delft University of Technology, Delft, The Netherlands

and

L. GRADOŃ

Warsaw University of Technology, Warsaw, Poland

KLUWER ACADEMIC PUBLISHERS

DORDRECHT / BOSTON / LONDON

A C.I.P. Catalogue record for this book is available from the Library of Congress

ISBN-13: 978-94-010-7259-5 e-ISBN-13: 978-94-009-1694-4
DOI: 10.1007/978-94-009-1694-4

Published by Kluwer Academic Publishers,
P.O. Box 17, 3300 AA Dordrecht, The Netherlands.

Kluwer Academic Publishers incorporates
the publishing programmes of
D. Reidel, Martinus Nijhoff, Dr W. Junk and MTP Press.

Sold and distributed in the U.S.A. and Canada
by Kluwer Academic Publishers,
101 Philip Drive, Norwell, MA 02061, U.S.A.

In all other countries, sold and distributed
by Kluwer Academic Publishers Group,
P.O. Box 322, 3300 AH Dordrecht, The Netherlands.

Photograph cover first appearing in:

G.C. Ferron: Deposition of Polydisperse Aerosols in two Glass Models Representing the Upper Human Airways, 1977, Vol. 8, pp. 409–427.
Reprinted with kind permission from Elsevier Science Ltd, The Boulevard, Langford Lane, Kidlington OX5 1GB, UK.

Printed on acid-free paper

TABLE OF CONTENTS

FOREWORD

Different groups in the world are working on the measurement and modelling of air flow and aerosol particle deposition in the lung. Others are working on the characterization of airborne particles and still other groups are investigating the effects of particles on health and environment.

To stimulate an interdisciplinary discussion, Warsaw University of Technology and Delft University of Technology organized a workshop on "Aerosol Inhalation, Lung Transport & Deposition and the Relation to the Environment, Recent Research and Frontiers." The workshop was held in the beautiful Jablonna Palace near Warsaw, Poland, September 14-16, 1995, just before the '95 European Aerosol Conference in Helsinki.

This Workshop tried to bring together the different disciplines to get a complete view of the aerosol inhalation, lung transport & deposition in relation to the current environment. Prominent researchers from all the involved subjects were invited to give state-of-the-art overviews including a few hot-item detailed topics. Beside these speakers, other involved scientists were invited to come to a fruitful discussion.

This book contains the proceedings of the Workshop, and it includes the worked-out panel discussion on the important issue of the social and economical consequences of the latest views on the effects of small particle inhalation.

Leon Gradoń, TU Warsaw
Jan Marijnissen, TU Delft

INTRODUCTION

The *Workshop Aerosol Inhalation, Lung Transport and Deposition and the Relation to the Environment*, held at the Jablonna Palace, Warsaw, Poland during September 14 to 16, 1995, was sponsored by Warsaw University of Technology (TU Warsaw), Delft University of Technology (TU Delft), Gesellschaft für Aerosolforschung (GAeF), TSI Incorporated and Astra Draco AB.

Host and Chairman was Professor Leon Gradoń of the TU Warsaw; Co-Chairman was Dr. Jan Marijnissen of the TU Delft.

Preface - Introduction

During the 14th to 16th September, 1995 a group of highly esteemed aerosol scientists - physicists, chemists, biologists, toxicologists, etc. - met in Warsaw at a workshop dealing with the newest developments in the field of aerosol health subjects, including inhalation, lung transport and deposition as well as with relation of these topics to the environment. In important review contributions and interdisciplinary discussions this very contemporary field was reevaluated and suggestions for future research needs were made and summarized. The Workshop complemented an earlier International Workshop held in Delft in May 1993 on "Transfrontier Air Pollution and the Law of the Atmosphere". This earlier Workshop, supported by The Rand Corporation, also dealt with scientific issues related to the setting of air pollution standards.

Genius Loci

The choice of Poland and Warsaw for this workshop was a very good decision. It is also symbolic and tangible evidence that the science and the scientists of the Eastern European Countries are considered to be an integral part of the scientific community of the free democratic world. This is an extremely gratifying development, especially in light of the fact that Polish scientists were directly involved in aerosol physics already in very early days. It was Marian Smoluchowski who first recognized and theoretically described the phenomenon of fine particle coagulation. It was Marie Curie-Sklodowska who, already at the beginning of our century, showed that radon decay products exist as very fine particles which form natural radioactive aerosols in air.

Atmospheric Aerosols and Environmental Health

Recent epidemiological studies indicate health effects on the general population at air particulate mass concentrations which lie considerably below existing air quality standards (AQS). This new data indicates that increases in human mortality and morbidity are associated with levels of air particulate pollution (anthropogenic atmospheric aerosols: AAA) significantly lower than those previously thought to adversely affect human health (Morton Lippmann). The causative mechanisms of these observed adverse effects on the general population, mainly on children, the elderly and vulnerable groups (e.g. the chronically ill) are, however, practically unknown. Numerous controlled toxicological investigations of individual chemical species have shown that specific constituents of ambient air particulate matter are associated with adverse biological effects, including carcinogenicity (Roger McClellan).

Anthropogenic atmospheric aerosols exist as an aerodispersed system of solid and liquid particles with different sizes, particle shapes and chemical composition. Additionally, the chemical composition of single particles can be anisotropic, which means that the chemical composition on the surface of the particle is different from that inside the particles. Anthropogenic atmospheric aerosols are produced in the atmosphere, and during their residence time they undergo many physical, physicochemical and chemical processes. The resulting product is a polydispersed system of chemically heterogeneous particles with complex toxic and carcinogenic potential. Biochemically active components may be present at the particle surface or inside the particles. AAA can contain short-lived chemical species, including free radicals and other metastable components. AAA are formed by two basic mechanisms: Dispersion and condensation (including chemical reactions). The resulting total AAA is a mixture of the primary and secondary aerosol and will therefore have a bimodal size distribution*. As both components are of different origin, they have also different chemical composition. Chemical composition is size dependent. The fine fraction of the AAA (particle sizes less than about 2 μm) is associated with the most toxic air pollutants, both inorganic and organic. This fine fraction can be designated as the "Lung Toxic Aerosol Mode" and could serve as a possible marker for the dosage of the toxic and carcinogenic components of the AAA. New toxicological as well as epidemiological studies underline the importance of highly dispersed AAA with particle sizes less or even much less than 1 μm. The measurement of such aerosols requires new physical and instrumental approaches (Wladislaw Szymanski).

Aerosols and the Lung

The mechanisms by which inhaled particulates (AAA) induce adverse health effects depend, initially, on their depth of penetration, deposition and retention of these particles in the lung. The response to deposited material has been shown to be markedly greater when the tracheobronchial and alveolar regions of the respiratory tract, as opposed to the extrathoracic region, are involved.

For understanding these separation, transportation and retention processes in the human respiratory system, experimental studies on humans and laboratory animals as well as theoretical computational studies are very necessary and useful.

In several new experimental studies the behaviour of aerosols in human nasal and oral airways, including the ultrafine size range, could be clearly established (David Swift). The same is valid for the behaviour of hygroscopic aerosol particles in the lung (George Ferron). Existing lung models were also substantially improved; it was shown that the improved models are able to satisfactorily describe particle deposition in the different branches of the lung as well as particle transport within lung structures (Werner Hofmann). Particle shape can also strongly influence separation and deposition of aerosols in the airways (Jan

---

* The third, course, mode is not considered here

Marijnissen). The total aerosol separation in the "lung filter" is determined by a superposition of several partial physical mechanisms. The final result is a total size-dependent particle collection in the lung tree (James Gentry). Besides the prediction of the size-dependent aerosol particle separation in the airways, very advanced mathematical models enable prediction of the entire dynamics of inhaled aerosols. Biological and physiological mechanisms influencing the transport and clearance of inhaled aerosols in the human respiratory system can be clearly described and well predicted (Werner Stöber, Leon Gradoń). Nevertheless, the experimental animal model remains still a very useful object for the verification of theoretical studies and for inhalation studies at special conditions (Magnus Dahlbäck, Ronald Wolff).

In the majority of cases aerosols in the occupational as well as in the ambient atmosphere are electrically charged. Charge influences the particle separation efficiency in the airways. In some cases aerosol particle deposition can be significantly enhanced, for example by 20% or more, by electrical charge (Beverly Cohen). The primary and secondary air flows, mixing and expanding conditions as well as normal and pathological inhomogenities in the lung tissue play a very important role in the whole particle dynamics inside the lung (Yoshio Otani, Dariusz Orlicki). The toxic and carcinogenic effects of particles deposited in the different lung compartments and the biochemical reactions occurring with these particles can be correlated to several physical and chemical properties of single particles: their size, shape, porosity, surface, solubility, chemical composition, etc. The correlations previously mentioned concerning the concentration of AAA and increases in mortality/morbidity of the general population seem to be also related to these particle properties. New observations, experimental results and hypotheses underline the role of very fine insoluble particles with dimensions less and much less than 1 μm; mainly particles in the size range between 10 and 50 nm are now of interest for toxicologists. Although these particles have very low mass, there are extremely many of them, and their sheer numbers mean that they present a large surface and interface when they reach the alveoli in the lung. Once in the alveoli they may induce oxidant production, lung inflammation, and biological hyper-activity (Günter Oberdörster).

A particular biological role of the atmosphere can be attributed to relatively coarse particles, namely to pollen grains, which produce serious allergic effects in the general population. These effects seem to be considerably aggravated by gaseous and particulate air pollutants (Gabriel Peltre). Last but certainly not least, one useful protection measure against toxic aerosols is formed by new types of masks and respirators. The modern air filters are satisfactorily efficient for retention of all respirable particles. But they need to be thoroughly tested by appropriate and scientific methods (Klaus Willeke).

Panel Discussion

The panel discussion dealt with the newest results in the field of deposition and retention of inhaled particulate matter and with social, economic-regulatory and research implications of the latest developments in relation to atmospheric particles and the human respiratory system. In the first part of this discussion, chaired by Ronald Wolff, the role of the "ultrafine" atmospheric aerosol particles was evaluated. Dozens of studies conducted in the past five years suggest that the ultrafine insoluble particles are very probably responsible for increased risks of respiratory-, cardiovascular-, and cancer-related death, as well as pneumonia, reduced lung function, extra hospital admissions, asthma, and other respiratory problems. In spite of the huge social and financial costs of these problems, there is a great deficiency in the existing knowledge. To rectify this hiatus in knowledge, comprehensive biological and toxicological studies are necessary. Much more understanding is needed, e.g. concerning the effects of ultrafine particles on susceptible groups within the general population.

In the second part of this discussion, chaired by Kvetoslav Spurny, more practical aspects of these environmental health problems were considered. Firstly, the role of the aerosol physics and chemistry was mentioned. Since it has been demonstrated that particles with an aerodynamic diameter of 10 μm or less penetrate into the thoracic region (the bronchus and bronchioles), the term "Thoracic Particles" was proposed and used to describe such particulate fractions. This fraction is measured as a mass concentration, in $\mu g/m^3$, and is designated as "PM-1O". This definition, unfortunately, neglects the fact that PM-10 particles are composed of a mixture of two different particle fractions. Neglecting this fact causes important aspects of particulate chemistry and toxicity to be neglected. For this reason several aerosol scientists proposed that the PM-10-mass should be sampled as two fractions: the fine fraction, PM-2.5 (particles with aerodynamic diameters less than 2.5 μm) and the coarse fraction, PM-2.5-10 (particles with aerodynamic diameters between 2.5 and 10 μm). The necessary know-how and the equipment to sample both fractions are available. In addition to the total mass of both these fractions, a continuous, in situ and an on-line real time monitoring of carbon black and PAH-aerosol could e.g. be a complementary option. Some of the latest epidemiological studies showed a better correlation of the increased mortality with PM-2.5 than with PM-10 (Kvetoslav Spurny). The results of the recent epidemiological studies support the discussions proposing a revision of existing air quality standards (AQS) for particulate air pollutants. The existing AQS relating to the PM-10 particulate fraction (annual mean: 50 $\mu g/m^3$) will probably be decreased in the United States. And an additional AQS for the PM-2.5 fraction should be proposed. Of course, such revisions and changes can not be done without considering the social and economic-regulatory impacts (Morton Lippmann, Roger McClellan). The European point of view, presented by Othmar Preining, is that the existing definition of PM-10 fraction can sufficiently evaluate and represent the hazard from urban particulate air pollution. The coarse fraction of the PM-10 contains about 40 % natural dispersed aerosols. The fine fraction of PM-10 consists for about 60 % of AAA. Soluble ions, elemental and organic carbon (including toxic organic micropollutants) are in the fine fraction.

Results presented in the following lectures, as well as opinions formed in plenary and panel discussions, have contributed greatly to the understanding and more precise description and definition of this important environmental, interdisciplinary problem. The necessary research needs were also pointed out. Furthermore, the effect of being together socially, the excellent Polish hospitality, the spirit of the barock Poniatowskis' Jablonna Palace as well as the wonderful musical evening realized by the Dutch pianists Mme's Jetje van Wijk and Marlies van Gent contributed to the very pleasant and familiar atmosphere of this Workshop.

Kvetoslav R. Spurny
Leon Gradoń
Saul Lemkowitz
Jan Marijnissen

The Challenge of the Epidemiologic Evidence for Excess Mortality and Morbidity Associated with Atmospheric Aerosols*

Morton Lippmann, Ph.D.

Nelson Institute of Environmental Medicine, New York University Medical Center, Tuxedo, NY 10987, USA

Keywords

Acid aerosols; air pollution; bronchitis; coal smoke; hospital admissions; morbidity; mortality; particulate matter; sulfate; sulfuric acid.

Abstract

A rapidly growing body of epidemiological data has shown statistically significant associations between the concentrations of particles in the ambient air and rates of mortality and morbidity in populations exposed to concentrations below current USEPA standards and WHO-EURO air quality guidelines. When other known risk factors have been taken into account in regression analyses, it has been possible to show that daily changes of ambient particulate matter (PM) are associated with: 1) changes in daily mortality; 2) hospital admissions for respiratory diseases; 3) emergency room visits for respiratory diseases; 4) respiratory symptoms; and 5) lung function. For these various responses, there is sometimes a lag of one to a few days between exposure and response.

Differences in long-term average levels of PM pollution have been associated with differences in: 1) annual mortality rates; 2) the prevalence of chronic disease and symptoms; 3) lost time from work or school; and 4) reduced lung function baselines. When multiple regressions are performed which include pollutant vapors as well as PM, the PM variable tends to be most influential. The only exceptions appear to be hospital admissions and symptoms, where ozone (O_3) is sometimes as, or more, influential on the regressions than PM.

Despite the consistency and coherence of these findings in human populations, there has been a reluctance by many to accept the associations as causal in the absence of more positive findings in controlled inhalation exposure studies in humans and laboratory animals. Only one PM component, i.e., hydrogen ion (H^+) and one other ambient air pollutant, i.e., ozone (O_3), have produced any comparable responses at peak ambient concentrations.

This paper reviews the available epidemiologic evidence for PM-associated health effects and possible causes for such effects.

---

* Prepared for presentation at: Workshop on "Aerosol Inhalation, Lung Transport and Deposition, and the Relation to the Environment, Recent Research Frontiers," Warsaw, Poland, 14-16 September 1995.

J. C. M. Marijnissen and L. Gradoń (eds.), Aerosol Inhalation: Recent Research Frontiers, 1–25.

Introduction

For the purpose of this review, the epidemiologic literature will be discussed in terms of a number of specific temporal periods, i.e.:

- The highly polluted coal-smoke era when descriptive analyses seemed adequate (from 1873 to the early 1960's)
- The less-polluted intermediate era of relatively crude statistical analyses (from ~ 1960 to ~ 1985)
- The recent past and the application of increasingly sophisticated analytical techniques (since ~ 1985)

In reviewing the historic literature, we have the benefit of current knowledge about the relative strengths of associations between pollutants within ambient air mixtures and population-based response rates. Thus, we can examine and intercompare the published data in ways that the original authors could not, and thereby gain new insights from past experience.

Specific components of PM mixtures, such as the H^+ content of acidic aerosols, have seldom been measured directly, and inferences about exposures have generally relied on surrogate measures. In most cases, airborne PM concentrations have been expressed in terms of mass concentration per unit volume (e.g., $\mu g/m^3$), as determined by dividing the mass accumulated by a sampling substrate (such as a filter disc or sheet) by the volume of air from which the particles were extracted. The sampled mass is often dominated by the largest particles that reach the collector. Thus, the concentration of total suspended particulate matter (TSP) is numerically larger than that for PM_{10} (thoracic particulate matter with aerodynamic diameters below 10 μm), which, in turn, is larger than that for $PM_{2.5}$ (fine particulate matter with aerodynamic diameters below 2.5 μm). The $PM_{2.5}$ contains the accumulation mode of the atmospheric aerosol (which extends from 2.5 μm down to ~ 0.1 μm). This mode includes most of the mass of the secondary aerosols formed in the atmosphere from gaseous pollutant precursers by chemical reactions, including: sulfuric acid and its ammonium salts; ammonium nitrate; and photochemically formed organics. The $PM_{2.5}$ fraction also includes the transient, rapidly coagulating ultrafine aerosol mode, which extends from ~ 0.1 μ down to ~ 0.01 μm, but the mass of this mode makes little contribution to the overall mass of the $PM_{2.5}$. On the other hand, in terms of the particle number concentration of the ambient aerosol, the ultrafine mode is dominant.

Some epidemiological studies have relied on indirect measures of PM concentration. These include black smoke, a.k.a. British Smoke (BS) (Schwartz and Marcus, 1990; Lippmann and Ito, 1995), and a similar coefficient known as KM (Kinney et al., 1995), both of which are related to the reduction in optical reflectance from a filter surface as dust is accumulated, and coefficient of haze (CoH) (Fairley, 1990) which represents light transmittance through a filter. All three (BS, KM, CoH) are measures of unburned carbon, and are generally converted to $\mu g/m^3$ on the basis of a local calibration that compares the optical index to the mass concentration determined by means of a collocated manual gravimetric measurement. Airport visual range measures, and light scattering in an optical sensing zone, can also be calibrated to equivalent mass concentration of fine particles by local calibration (Schwartz, 1994; Burnett et al., 1994), although some components of $PM_{2.5}$, especially

$SO_4^=$, are much more influential on light-scatter than are others (Lippmann et al., 1979; U.S. EPA, 1979).

While it is unlikely that all components of PM are equally toxic, the finding of quite similar exposure-response coefficients for mortality and morbidity indices across communities that are widely divergent in climate and local PM sources (Pope et al., 1995a) suggests that no one component is solely responsible for the effects. On the other hand, in the relatively few studies that have utilized multiple PM metrics in their regression analyses, the strength of association improves as one goes from TSP to PM_{10} to $PM_{2.5}$ to $SO_4^=$, the latter being largely a component of the $PM_{2.5}$ fraction of the aerosol (Ozkaynak and Thurston, 1987; Thurston et al., 1994).

The $SO_4^=$ in the samples is the molar sum of the H_2SO_4, NH_4HSO_4, and $(NH_4)_2SO_4$ captured on the filter. The first two, with free H^+, are strong acids, while the $(NH_4)_2SO_4$ is a weak acid. Controlled exposure studies in both humans and laboratory animals have shown alterations in lung function and particle clearance rates can be produced by low concentrations the strong acid aerosols (Koenig et al., 1983; Bauer et al., 1988; Leikauf et al., 1981, 1984; Schlesinger et al., 1979, 1983, 1990; Lippmann et al., 1982, 1987). Thus, the $SO_4^=$ associations with health effects seen in the epidemiological studies suggest that $SO_4^=$ was serving as a surrogate of H^+ (Lippmann, 1989).

COAL SMOKE ERA

Coal smoke in the classic U.K. episodes was strongly acidic (Commins and Waller, 1963). While the H^+ concentrations prior to the December 1962 episode were not known, the studies of the earlier episodes are informative in terms of the nature and extent of the responses recorded and of the population groups affected. The Ministry of Health (1954) report on mortality and morbidity during the London Fog of December 1952 also reviewed data on excess bronchitis mortality during six prior fog episodes extending back to December 1873. Of these, December 1952 was third highest in excess mortality. The report attributed an excess of ~ 4,000 total deaths to the exposures during the 1952 episode, and Table 1 shows the deaths, by week, divided according to cause. For all causes, deaths peaked in the first full week, and were still above baseline levels two weeks after that. The specific cause with the greatest number of excess deaths over the four weeks was bronchitis (1,156 excess deaths) and it had the greatest relative risk (RR=6.67). The next greatest increase, for heart disease (737 excess deaths), had an RR of only 1.82. The all cause relative risk was somewhat higher (1.96). Table 2 shows that most of the excess deaths occurred in individuals over 55 years of age (2,616 excess deaths over the four weeks), but there was an excess for all age groups beyond 4 weeks of age. Overall, the excess mortality was concentrated among the elderly with pre-existing disease.

As shown in Figure 1, the daily death rate rose rapidly with the onset of the fog, and peaked one day after the peak of pollution by the measured pollutant indices, i.e., black smoke (BS) and sulfur dioxide (SO_2). There was a coherent rise in hospital emergency bed admissions, which peaked two days after the pollutant peaks. Both the deaths and hospital bed admissions remained elevated for many days after the fog lifted. Note also that hospital admissions exhibit dips on Sundays, a finding consistent with normal hospital admissions and staffing.

TABLE 1. GREATER LONDON DEATHS DIVIDED ACCORDING TO CAUSE - NOV. AND DEC. 1952

Cause	Av. Number of deaths in weeks ending 8th, 15th, 22nd, 29th Nov.	Number of deaths registered in week ending				For weeks ending 6th, 13th, 20th, 27th Dec.	
		6th Dec.	13th Dec.	20th Dec.	27th Dec.	Av. No. of deaths	RR
Pulmonary Tuberculosis	17	14	77	37	21	37.25	2.19
Lung Cancer	34	45	69	32	36	45.50	1.34
Heart Disease	226	273	707	389	272	410.25	1.82
High Blood Pressure	14	19	47	36	21	30.75	2.20
Other Circulatory	22	26	46	31	32	33.75	1.53
Influenza	2	2	24	9	6	10.25	5.13
Pneumonia	31	45	168	125	91	107.25	3.46
Bronchitis	51	76	704	396	184	340.00	6.67
Other Respiratory	6	9	52	21	13	23.75	3.96
Ill-defined Causes	20	25	79	35	37	36.50	1.83
All Other Causes	340	411	511	412	316	412.50	1.21
All causes	763	945	2,484	1,523	1,029	1,495	1.96

From: Comm. on Air Pollution: Interim Report, Cmd 9011, London. H.M. Stationery Office (Dec. 1953).

TABLE 2. GREATER LONDON DEATHS DIVIDED ACCORDING TO AGE - NOV. AND DEC. 1952

Age	Av. Number of deaths in weeks ending 8th, 15th, 22nd, 29th Nov.	Number of deaths registered in week ending				For weeks ending 6th, 13th, 20th, 27th Dec.	
		6th Dec.	13th Dec.	20th Dec.	27th Dec.	Av. No. of deaths	RR
Weeks:							
0-4	20	16	28	19	12	18.75	0.94
4-52	8	12	26	15	11	16.00	2.00
Years:							
1-4	7	6	7	13	7	8.25	1.18
5-14	4	4	6	6	2	4.50	1.18
15-24	7	9	7	14	7	9.25	1.32
25-34	11	16	28	17	11	18.00	1.64
35-44	26	36	64	29	34	40.75	1.57
45-54	70	80	204	96	83	115.75	1.65
55-64	133	157	448	251	167	255.75	1.92
65-74	211	254	717	444	258	418.25	1.98
75 and over	266	355	949	619	437	590.00	2.22
All ages	763	945	2,484	1,523	1,029	1,495.00	1.96

From: Comm. on Air Pollution: Interim Report, Cmd 9011, London, H.M. Stationery Office (Dec. 1953).

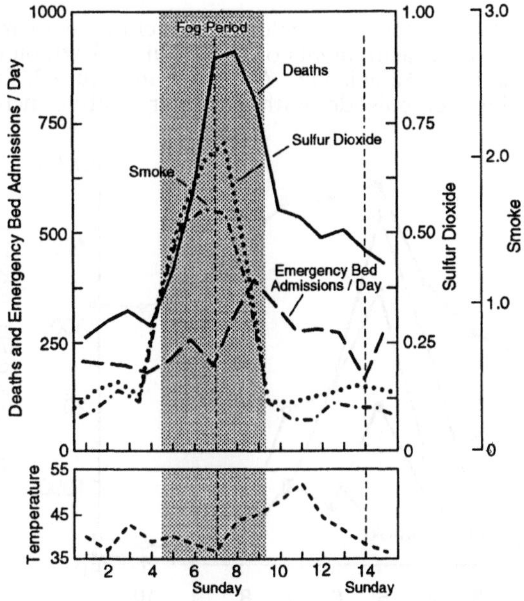

Figure 1. Metropolitan London total mortality and emergency bed admissions during the 1952 pollution episode.

The Ministry of Health Report (1954), also noted that there was a clear association between chronic air pollution and the incidence of bronchitis and other respiratory diseases. The death rate from bronchitis in England and Wales (where coal smoke pollution was very high) was much higher than in other European countries (with much lower levels of coal smoke pollution).

The very high chronic coal smoke exposure in the U.K., associated with a high prevalence of chronic bronchitis, appears to have created a large pool of individuals susceptible to "harvesting" by an acute pollution episode.

Both the December 1873 and the December 1952 London fog episodes happened to coincide with the annual Smithfield Club cattle shows in London, and it is interesting to note that there was excess mortality and morbidity among the prize cattle, but not among pigs and sheep, on exhibit on both occasions (Anon, 1874; Ministry of Health, 1954).

There was speculation at the time of the 1952 episode that the causal factor, within the pollution mixture, for the excess mortality and morbidity was the sulfuric acid that was formed in the fog droplets due to oxidation of dissolved SO_2. Such a role for acid droplets would be consistent with the anecdotal evidence that ammonia vapor was protective. Attending physicians in some of the hospitals found that placement of ammonia air wicks in the hospital wards led to symptomatic improvements among the newly admitted respiratory patients (Bates, D.V., 1987, personal communication). Also, the greater ammonia levels in the pens housing the pigs and sheep may have provided them with more protection than that available to the prize cattle in their well-cleaned stalls at the Smithfield shows.

Commins and Waller (1963) developed a technique to measure H_2SO_4 in urban air, and made daily measurements of H_2SO_4 at St. Bartholomew's Hospital in Central London during the 1962 episode. As shown in Figure 2, the airborne H_2SO_4 rose rapidly during the 1962 episode, with a greater relative increase than that for black smoke (BS).

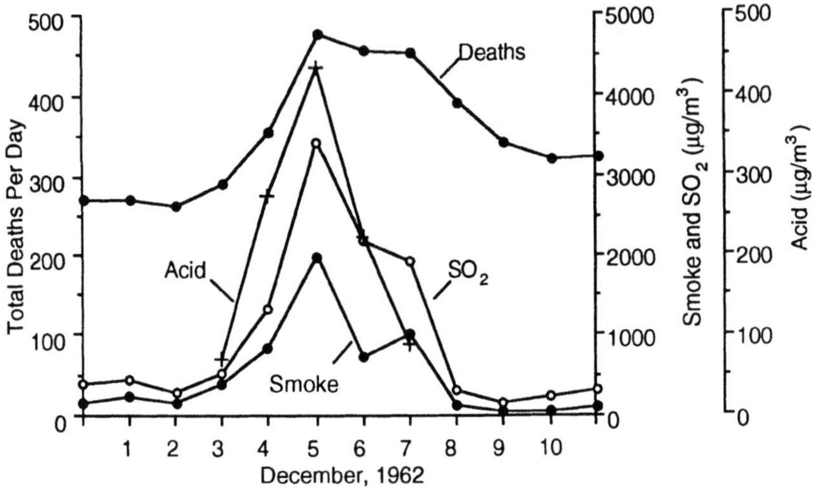

Figure 2. December 1962, London pollution episode.

The December 1962 London fog episode was the last to produce a clearly evident acute harvest of excess deaths, albeit a much smaller one than December 1952. The U.K. Clean Air Act of 1954 had led to the mandated use of smokeless fuels and, as shown in Figure 3, annual mean smoke levels had declined by 1962, to about one-half of the 1958 level. The annual average SO_2 concentrations had not declined by 1962, but dropped off markedly thereafter, along with a further marked decline in BS levels. For the period between 1964 and 1972, the measured levels of H_2SO_4 followed a similar pattern of decline.

During the later part of the coal smoke era in the U.K., researchers begin to study the associations between long-term daily records of mortality and morbidity and ambient air pollution. In the first major time-series analysis of daily London mortality for the winter of 1958-1959, Martin and Bradley (1960) and Lawther (1963) used the readily available BS and SO_2 data. They estimated that both pollutants were associated with excess daily mortality when their concentrations exceeded about 750 $\mu g/m^3$. However, additional analyses of this data set led to different conclusions. For example, Ware et al. (1981) concluded that there was no demonstrable lower threshold for excess mortality down to the lowest range of observation (BS ≈ 150 $\mu g/m^3$), as illustrated in Figure 4.

In terms of time-series analyses of morbidity, a study by Lawther (1970) reported the daily symptom scores of a panel of patients with chronic bronchitis in relation to the daily concentrations of BS and SO_2. As shown in Figure 5, there was a close correspondence between symptom scores and both pollutant indices.

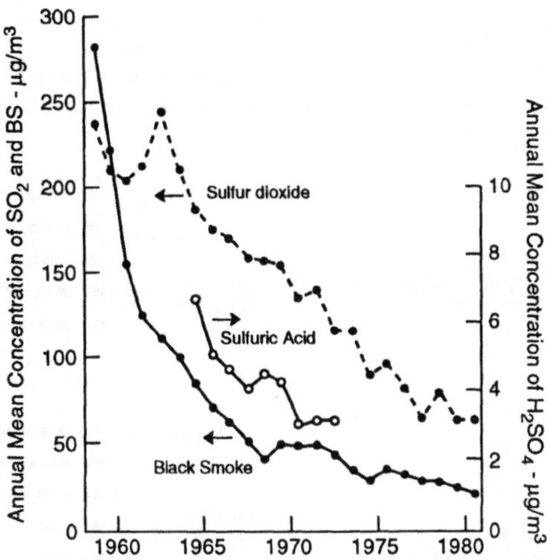

Figure 3. Long-term trends in annual mean atmospheric concentrations of black smoke (BS) and sulfur dioxide (SO_2) at seven stations in Greater London, and annual mean concentration of sulfuric acid (H_2SO_4) at St. Bartholomew's Hospital in Central London.

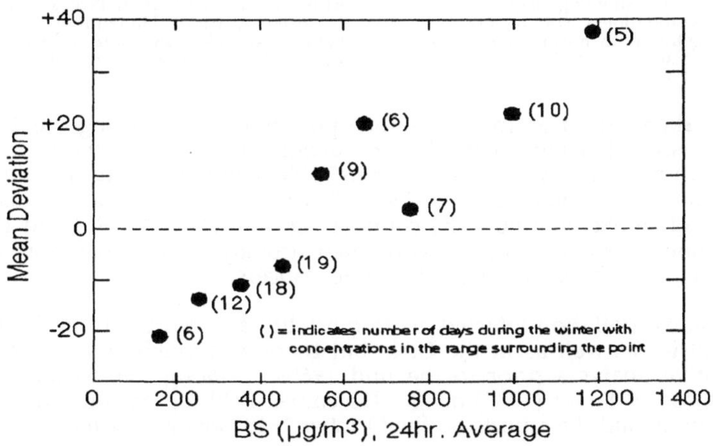

Figure 4. Martin and Bradley (1960) data for winter of 1958-9 in London as summarized by Ware et al. (1981), showing average deviations of daily mortality from 15-day moving average by concentration of black smoke (BS).

Chronic coal smoke exposure clearly had effects on lung function. Holland and Reid (1965) analyzed spirometric data collected on British postal workers in 1965. By that time, pollution levels were well below their peaks, but the postal workers had been exposed out-of-doors for many years when pollution levels were higher. As shown in Figure 6, the London postal workers had lower forced expiratory volumes in one second (FEV_1) and peak expiratory flow rates (PEFR) than their country town counterparts. As indicated in the figure, the deleterious effects of smoking were accounted for in these analyses. Within each smoking category, the differences between the London and country town means were attributed to pollution on the basis that pollution levels were, on average, twice as high in London as in the country towns.

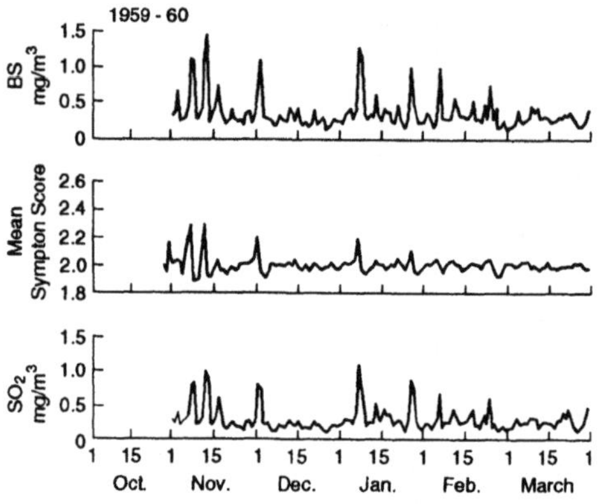

Figure 5. Results from 'diary' studies in London, winter 1959-60, showing day-to-day variations in the illness score for bronchitic subjects together with mean daily concentrations of black smoke (BS) and sulfur dioxide (SO_2).

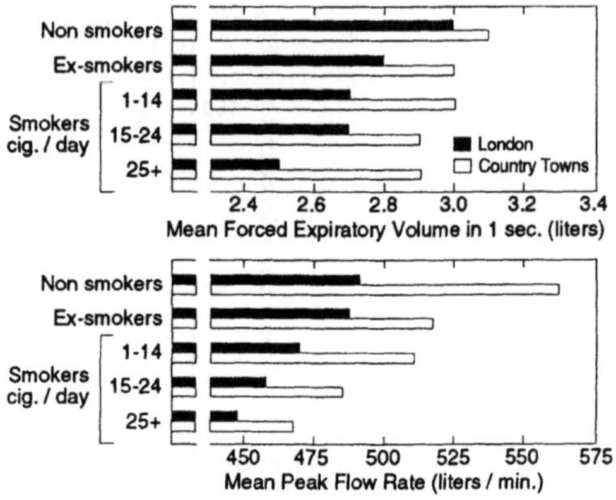

Figure 6. Cross-sectional study of lung function in British Postal Workers in 1965 standardized to age 40. Adapted from: Holland, W.W. and Reid, D.D. (Lancet 1:445-448, 1965) SO2 and smoke levels in the country towns were about half those in London.

The marked reduction in U.K. smoke pollution levels during the 1960s was shown to be associated with a marked reduction in annual mortality in County Boroughs by Chinn et al. (1981). As shown in Table 3, mortality rates in middle-aged and elderly men and women for the 1969-1973 period were no longer associated with an index of smoke pollution. By contrast, for both the 1948-1954 and 1958-1964 periods, the index of smoke exposure correlated strongly with annual mortality rates for both chronic bronchitis and respiratory tract cancers.

Coal smoke pollution affected acute mortality and morbidity in the U.S. and in other countries as well as in the U.K., but there was much less documentation and quantitative analyses prior to the mid 1960s. Among the notable non-U.K. reports are those of Firket (1936) on the December 1930 fog episode in the Meuse Valley in Belgium, and the reports on the October 1948 Donora, PA fog episode.

The December 1930 fog in the Meuse Valley was associated with 60 deaths from a population of ~ 6,000, but the pollutant concentrations were not measured.

In Donora, a valley town of about 10,000 people at a bend in the MonongahEla River south of Pittsburgh, there were steel mills, wire mills, zinc works and a sulfuric acid plant along the river bank for the entire length of the town. As reported by Schrenk et al. (1949), a persistent valley fog was associated with 20 excess deaths as well as acute morbidity among 43% of the population. About 10% were reported to have severe effects requiring medical attention. In a ten-year follow-up of the affected population, Ciocco and Thompson (1961) reported increased mortality

TABLE 3. STANDARDIZED ANNUAL MORTALITY RATE REGRESSION COEFFICIENTS ON SMOKE* FOR 64 UK COUNTY BOROUGHS
(from: Chinn, S. et al., J. Epid. Comm. Health 35: 174-179, 1981)

Sex	Ages	Mortality in	Cancer of Trachea, Bronchus & Lung	Chronic Bronchitis
Males	45-64	1969-1973	0.07	0.02
		1958-1964	0.53++	0.32+
		1948-1954	0.71+++	0.48+++
	65-74	1969-1973	0.15	-0.06
		1958-1964	0.68+++	0.31
		1948-1954	0.87+++	0.37+
Females	45-64	1969-1973	-0.02	-0.02
		1958-1964	-0.64++	0.33+
		1948-1954	0.49+	0.49++
	65-74	1969-1973	0.07	0.03
		1958-1964	0.25	0.40+
		1948-1954	0.61++	0.31

* Based on index of black smoke pollution 20 years before death of Daly (Br. J. Prev. Soc. Med. 13: 14-27, 1959).
\+ $p < 0.05$
++ $p < 0.01$
+++ $p < 0.001$

rates and incidence of heart disease and chronic bronchitis among the residents who had reported acute illness in 1948 in comparison to residents who did not. Some of the findings are illustrated in Figures 7 and 8.

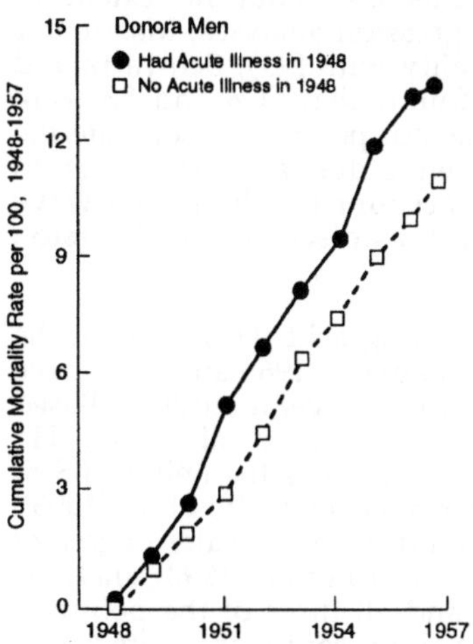

Figure 7. Cumulative mortality in Donora men by year since acute fog exposure in 1948.

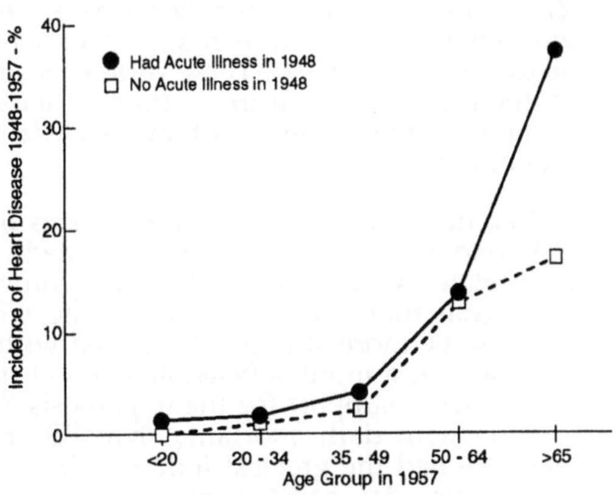

Figure 8. Ten year incidence of heart disease in Donora males with exposures to 1948 air pollution episode. Data from: Ciocco, A. and Thompson, D.J., Am. J. Public Health 51:155-164 (1961).

THE INTERMEDIATE ERA

As noted earlier, there were marked reductions in BS and SO_2 in London in the 1960s and 1970s (Figure 3), and corresponding reductions in U.K. mortality rates (Table 3). The public health authorities in the U.K. congratulated themselves on their impressive accomplishments and declared victory. They did, however, maintain the seven-station air monitoring network in Greater London that continued to generate daily concentrations of BS and SO_2. The locations of these sites are shown in Figure 9, along with that of the research station at St. Bartholomew's Hospital in central London, where daily H^+, SO_2, and BS measurements were made after 1963. Daily mortality data for the winter seasons (November, December, January and February) became available in the fall of 1958, and year-round daily measurements of BS and SO_2 became available at all sites in the spring of 1965.

Many analyses of the influence of air pollution on daily mortality in humans have taken advantage of the availability of the very large set of daily mortality data for Greater London and the temporally coincident data on temperature (T), humidity, BS and SO_2. Most analyses of associations between the environmental variables and morality have utilized the average values for the seven sites in Greater London. All of the time-series studies using these London data sets reported associations between daily mortality and temperature and, after adjustment (for temperature, day-of-week, and season) residual associations of daily mortality with BS (Mazumdar et al., 1982; Shumway et al., 1983; Roth et al., 1986; and Schwartz and Marcus, 1990). As shown in Figure 10 for winter data between 1958 and 1972, the Schwartz and Marcus analysis indicated a steep slope at the lower BS concentrations, and a flattening of the response at higher BS concentrations.

The various investigators differed in their conclusions about the extent of association of daily mortality with SO_2. In multiple regression analyses, Mazumdar et al. found no independent association of daily mortality with SO_2; Schwartz and Marcus found only a small role for SO_2; Shumway found that SO_2 and BS were equally predictive; and Roth et al. concluded that it was not possible to separate the influence of BS and SO_2. These differences in results were due, at least in part, to the differences in the nature of the methods used to account for the longer-wave phenomena in the time-series that affect daily mortality, as discussed by Thurston and Kinney (1995).

The data on the daily measurements of aerosol strong acid (H^+) made at the central London site (St. Bartholomew's Medical School) between 1963 and 1972 have been correlated with concurrent daily records of mortality in several studies. Based on the initial time-series analysis of the winter data (Thurston et al., 1989), H^+ appeared to be more strongly associated with total daily mortality than either BS or SO_2. However, a more detailed analysis of the full-year data set by Ito et al. (1993), did not provide support for the hypothesis that H^+ would have a greater degree of association with daily mortality than BS or SO_2. In the Ito et al. (1993) analysis, temperature had the greatest influence in all seasons, and all three of the pollution variables (same day and lagged one or two days) were significantly associated with daily mortality. There were limitations imposed on the analysis by the use of only one monitoring site, the limited precision of the measurement of H^+, and the selection of statistical filters for controlling the confounding long-wave influences.

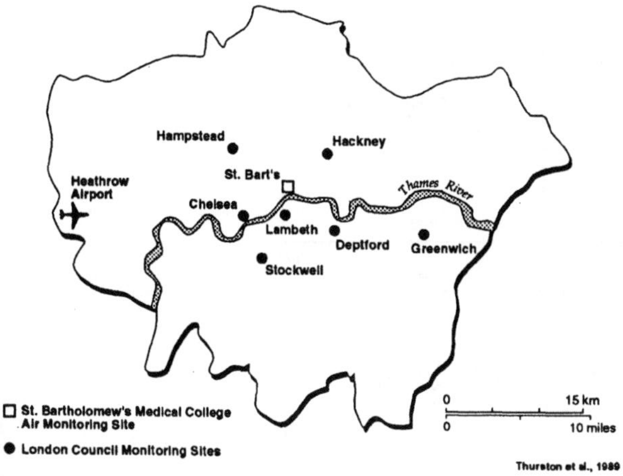

Figure 9. Historical London, England, air pollution monitoring sites.

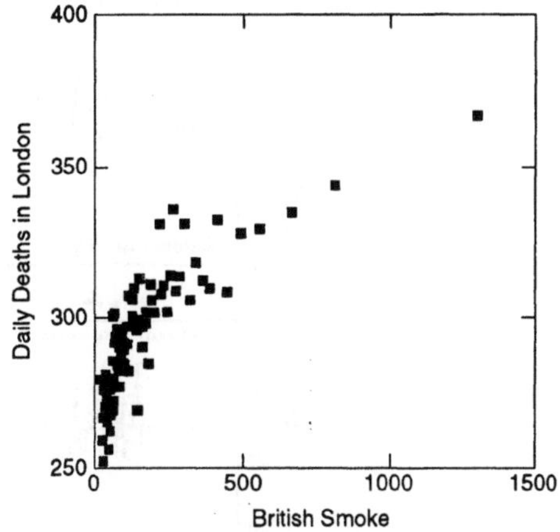

Figure 10. Daily mortality in London, by Britsh Smoke level, for the winters of 1958-1972. Each point represents the mean total mortality and mean British Smoke (in µg/m^3) for 20 adjacent values of British Smoke, after sorting by British Smoke. From: Schwartz and Marcus, Am. J. Epidemiol. 131:185-194 (1990).

All of these previous time-series analyses of the London data relied on mathematical models to account for other known influences on daily mortality rates. In an exploratory further analysis of the Central London data set, Lippmann and Ito (1995) developed an alternate approach for separating the effects of season and temperature on daily mortality from those of pollution. They analyzed the data for each season separately, and within each season, they restricted analyses to those days in which ambient temperature had little, if any, influence on mortality. Regressions were performed for BS, SO_2, and aerosol H^+. For the winter period (November-February) mortality was most closely correlated with H^+, while for the rest of the year, it was most closely correlated with SO_2. In all seasons, the correlation was poorest for BS. The results for winter and summer are illustrated in Figures 11 and 12.

One current hypothesis, based upon available evidence in the literature, is that daily mortality is most closely related to exposures to fine particles of combustion origin, and that this accounts for the closer association of daily mortality with PM_{10}, $PM_{2.5}$, $SO_4^=$, and H^+ than with TSP, which is often dominated by coarse particles, or with BS, which is usually dominated by carbon content. The greater association of daily mortality with SO_2 than with BS (Lippmann and Ito, 1995) may simply indicate that SO_2 is a better surrogate for fine particles than BS.

Sulfate ion concentrations were used as the prime PM pollutant metric in many early U.S. population based studies of acute respiratory morbidity and mortality. An early example of a morbidity study is the one of Dohan et al. (1962) of the incidence of respiratory disease lasting more than seven days in women workers at assembly plants in five U.S. cities in relation to $SO_4^=$ concentrations in those cities (Figure 13). $SO_4^=$ has also served as a surrogate measure of exposure to combustion-related particulate in mortality studies in the U.S.

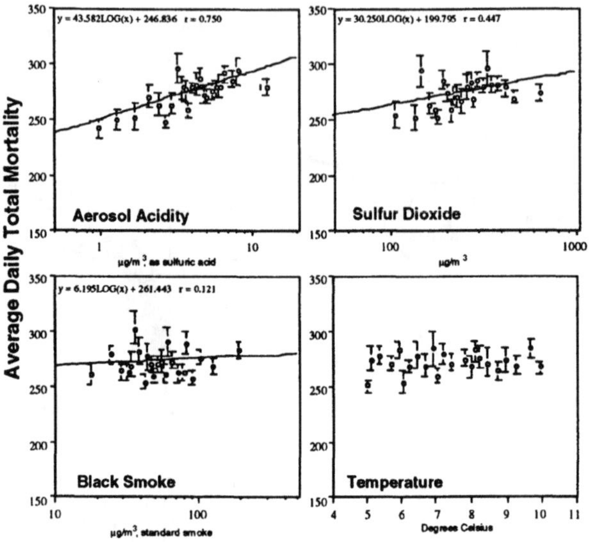

All Plots for winter data where 5°C ≤ temperature ≤10°C (n=461) were produced by taking averages of 20 adjacent points that were sorted by pollution or temperature.

Figure 11. Mortality in London: 1965-1972, winter days with temperatures between 5 and 10°C.

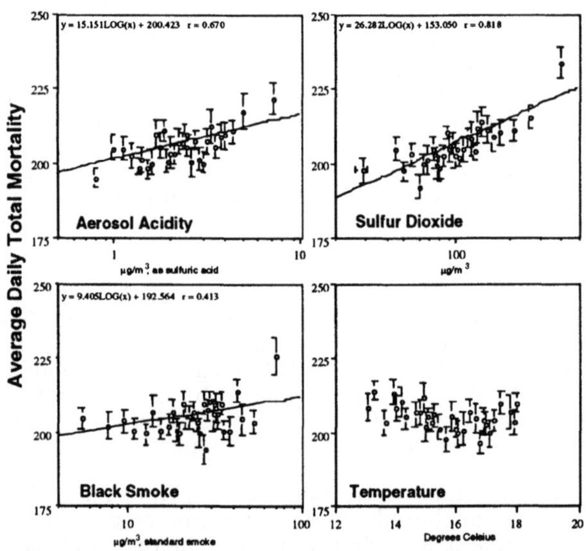

All Plots for summer data where 13°C ≤ temperature ≤18°C (n=610) were produced by taking averages of 20 adjacent points that were sorted by pollution or temperature.

Figure 12. Mortality in London: 1965-1972, summer days with temperatures between 13 and 18°C.

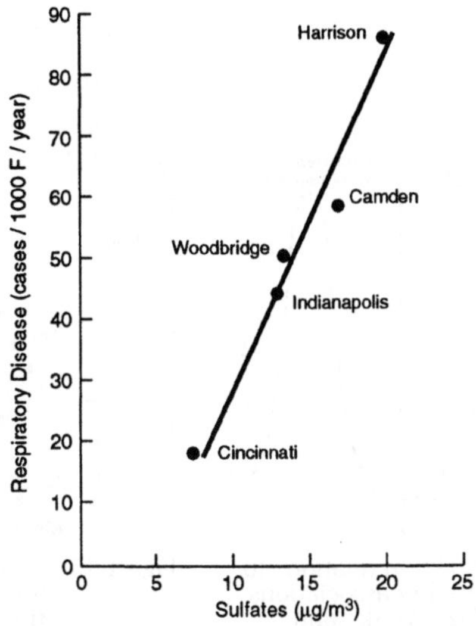

Figure 13. Incidence of respiratory disease lasting more than seven days in women (mean of three years) versus concentration of suspended particulate sulfates in the city air at test sites. From: Dohan, F.C. and Taylor, E.W. Am. J. Med. Sc., 337-339 (1960).

Sulfate is often a relatively large fraction of $PM_{2.5}$, it is non-volatile, it is stable on filters used for air sampling, it can be easily extracted from the filters, and it can be accurately analyzed with relatively simple and inexpensive procedures. Furthermore, it generally correlates better with mortality and indices of morbidity in populations than do other frequently measured PM indices, such as TSP, BS, CoH, and PM_{10}. Lippmann (1989) proposed that $SO_4^=$ is the best surrogate for H^+ exposure, the latter being the most likely causal factor for the observed associations between PM and chronic mortality. The hypothesis, illustrated in Figure 14, was based on the Ozkaynak and Thurston (1987) annual mortality analysis. Their actual data, for $SO_4^=$ and mortality, are shown in Figure 15. There has been a reluctance on the part of many to accept the $SO_4^=$-mortality association as likely to be causal on the basis of ecological analyses such as those of Ozkaynak and Thurston (1987), or the earlier analyses of Lave and Seskin (1977). Many skeptics felt that the results could have been due to confounding by differences among the communities in smoking, occupations, ethnicity, etc. However, later prospective cohort studies, to be discussed in the next section, have reported similar associations.

THE RECENT PAST

In the past five years there has been a great increase in the number of time-series studies of the associations between daily ambient air pollutant concentrations and daily rates of mortality and hospital admissions for respiratory diseases. Also, there is important new information from two prospective cohort studies of annual mortality rates. In terms of morbidity, there has been a rapid growth of the literature showing associations between airborne particle concentrations and exacerbation of asthma, increased symptom rates, decreased respiratory function and restricted activities.

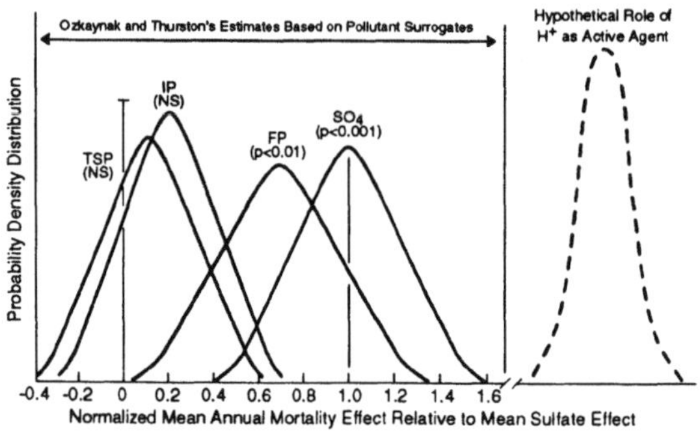

Figure 14. Hypothetical role of H^+ in relation to analysis of annual mortality associations 98 U.S. SMAS by Ozkaynak and Thurston, Risk Anal. 7:449-461 (1987).

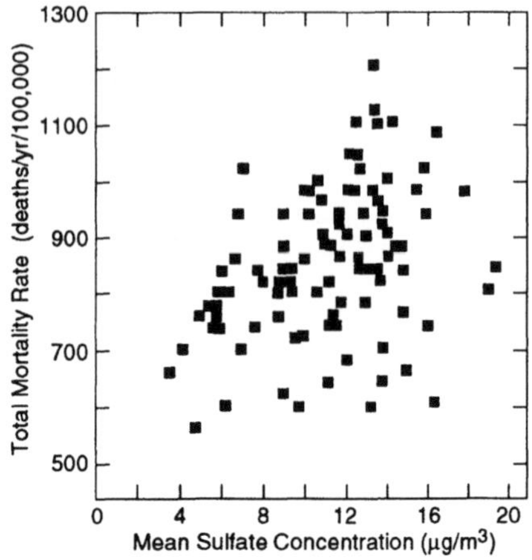

Figure 15. Plot of total mortality rate vs. annual mean SO4= concentra-in tion in 98 U.S. SMSAs in 1980. From: Ozkaynak and Thurston, Risk Anal. 7:449-461 (1987).

Much of the recent literature has been summarized by Pope et al. (1995a). They converted historically measured values for CoH and TSP to estimated levels of PM_{10}, and remarked that very similar coefficients of response were determined in all locations. Table 4 shows Thurston's (1995) independent analysis of acute mortality studies in nine communities with measured PM_{10} concentrations, including four of the ten studies cited by Pope et al. (1995a). As indicated in this table, the coefficients of response tend to be higher when the PM_{10} is expressed as a multiple-day average concentration, and lower when other air pollutants are included in multiple-regression analyses. In any case, the results in each city (except for the very small city of Kingston, TN) indicate a statistically significant association.

While there is mounting evidence that excess daily mortality is associated with short-term peaks in PM_{10} pollution, the public health implications of this evidence are not yet clear. Key questions remain, including:

- which specific components of the PM_{10} are most influential in producing the responses?
- do the effects of the PM_{10} depend on co-exposure to irritant vapors, such as ozone, sulfur dioxide, or nitrogen oxides?
- what influences do multiple day pollution episode exposures have on daily responses and response lags?
- does long-term chronic exposure predispose sensitive individuals being "harvested" on peak pollution days?
- how much of the excess daily mortality is associated with life-shortening measured in days or weeks vs. months, years, or decades?

TABLE 4

COMPARISON OF TIME-SERIES STUDY ESTIMATES

TOTAL MORTALITY RELATIVE RISK (RR) FOR A 100 $\mu g/m^3$ PM_{10} INCREASE

Study Area (Reference)	Measured PM10 Concentrations Mean ($\mu g/m^3$)	Maximum ($\mu g/m^3$)	RR for 100 $\mu g/m^3$	95% CI for 100 $\mu g/m^3$
1. Utah Valley, UT (Pope et al., 1992)	47	297	1.16*††	(1.10-1.22)
2. St. Louis, MO (Dockery et al., 1992)	28	97	1.16*†	(1.01-1.33)
3. Kingston, TN (Dockery et al., 1992)	30	67	1.17*†	(0.88-1.57)
4. Birmingham, AL (Schwartz, 1993)	48	163	1.11*††	(1.02-1.20)
5. Athens, Greece (Touloumi et al, 1994)	78	306	1.07*†	(1.05-1.09)
			1.03**†	(1.00-1.06)
6. Toronto, Can. (Özkaynak et al., 1994)	40	96	1.07*†	(1.05-1.09)
			1.05**†	(1.03-1.07)
7. Los Angeles, CA (Kinney et al., 1995)	58	177	1.05*†	(1.00-1.11)
			1.04**†	(0.98-1.09)
8. Chicago, IL (Ito et al., 1995)	38	128	1.05**†	(1.01-1.10)
9. Santiago, Chile (Ostro et al., 1995)	115	367	1.08*†	(1.06-1.12)
			1.15*††	(1.08-1.22)

* Single pollutant model (i.e. PM_{10}).
** Multiple pollutant model (i.e. PM_{10} and other pollutants simultaneously).
† One-day mean PM_{10} concentration employed.
†† Multiple-day mean PM_{10} concentration employed.

The first four questions above are complex, and difficult to answer at this time on the basis of current knowledge. The Discussion section, which follows this review of the Recent Past, will examine them in greater detail.

The last question above is a critical one in terms of the public health impact of excess daily mortality. If, in fact, the bulk of the excess daily mortality were due to "harvesting" of terminally ill people who would have died within a few days, then the public health impact would be much less than if it led to prompt mortality among acutely ill persons who, if they did not die then, would have recovered and lived productive lives for years or decades longer.

An indirect answer to this question is provided by the results of two recent prospective cohort studies of annual mortality rates in relation to long-term pollutant exposures.

Dockery et al. (1993) reported on a 14-to-16 year mortality follow-up of 8,111 adults in six U.S. cities in relation to average ambient air concentrations of TSP $PM_{2.5}$, fine particle $SO_4^=$, O_3, SO_2 and NO_2. H^+ data were available for only ~ 1 year in each city, while concentration data were available for the other pollutant variables for most or all of the 14-16 years. The mortality rates were adjusted for cigarette smoking, education, body mass index and other influential factors not associated with pollution. The two pollutant variables that best correlated with total mortality (which was mostly attributable to cardiopulmonary mortality) were $PM_{2.5}$ and $SO_4^=$. The overall mortality rate ratios were expressed in terms of the range of air pollutant concentrations in the six cities. The rate-ratios for both $PM_{2.5}$ and $SO_4^=$ were 1.26 (1.08-1.47) overall, and 1.37 (1.11-1.68) for cardiopulmonary. The mean life-shortening was in the range of 2-3 years.

Pope et al. (1995a) linked ambient air pollution data from 151 U.S. metropolitan areas in 1980 with individual risk factor on 552,138 adults who resided in these areas when enrolled in a prospective study in 1982. Deaths were ascertained through December, 1989. Exposure to $SO_4^=$ and $PM_{2.5}$ pollution was estimated from national data bases. The relationships of air pollution to all-cause, lung cancer, and cardiopulmonary mortality was examined using multivariate analysis which controlled for smoking, education, and other risk factors. Particulate air pollution was associated with cardiopulmonary and lung cancer mortality, but not with mortality due to other causes. Adjusted relative risk ratios (and 95% confidence intervals) of all-cause mortality for the most polluted areas compared with the least polluted equaled 1.15 (1.09 to 1.22) and 1.17 (1.09 to 1.26) when using $SO_4^=$ and $PM_{2.5}$ respectively. The mean life-shortening in this study was between 1.5 and 2 years. Figure 16 shows the range of values for the adjusted mortality rates in the various communities versus annual average $SO_4^=$ concentrations. The results appear, both by inspection and analysis, to be quite similar to those found in the previous studies of Ozkaynak and Thurston (1987) (See Figure 15) and Lave and Seskin (1970). Thus, the results of these earlier studies provide some confirmatory support for the findings of Pope et al. (1995b), while the Pope et al. (1995b) results indicate that the concerns about the credibility of the earlier results, due to their inability to control for potentially confounding factors such as smoking and socioeconomic variables, can be eased.

The findings of Dockery et al. (1993) and Pope et al. (1995b), in carefully controlled prospective cohort studies, that mean lifespan shortening is of the order of two years implies that many individuals in the population have lives shortened by many years, and that there is excess mortality associated with fine particle exposure greater than that implied by the cumulative results of the time-series studies of daily mortality.

An important recent study addressed morbidity in a large number of individuals also adjusted for individual risk factors. Ostro (1990) examined lost-time due to respiratory causes vs. ambient particulate matter. It was based on interview data on a random sample of U.S. households in 25 communities in the national Health Interview Survey (HIS) of 1979-1981. The lost-time was most closely related to $SO_4^=$. As shown in Figure 17, the associations, in terms of exposure-response slopes and scatter, were similar in nature to those seen for mortality in Figures 15 and 16.

In terms of hospital admission, Table 5 indicates that the strongest association with a measured component of PM was with the hydrogen content of the aerosol in those studies where H^+ data were available, and that $SO_4^=$ was significantly associated with hospital admissions in all of the studies for which is was available.

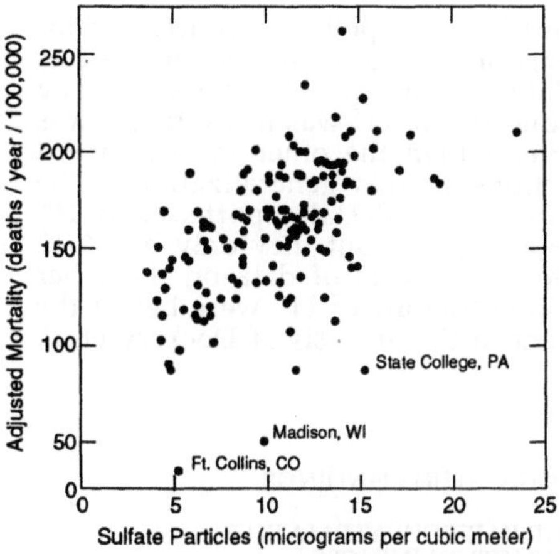

Figure 16. Age-, sex-, and race-adjusted population-based mortality rates for 1980 plotted against mean sulafe air pollution levels for 1980. Data from metropolitan areas that correspond approximately to areas used in prospective cohort analysis. Adapted from: Pope, C.A. et al., Am. J. Respir. Crit. Care Med. 151:669-674 (1995).

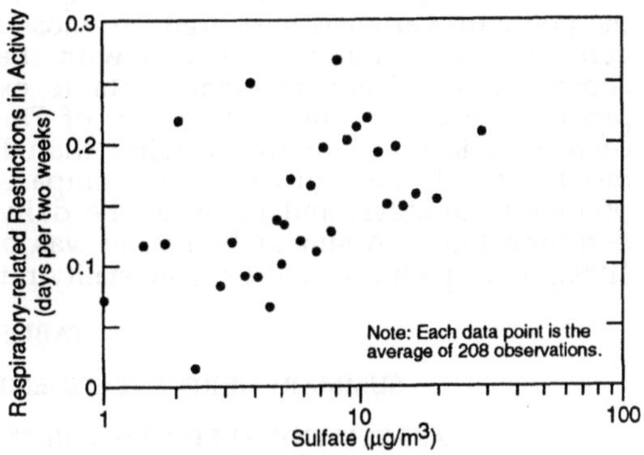

Figure 17. Association between respiratory morbidity and sulfates, controlling for covariates. From: Ostro, B.D., Risk Anal. 10:421-427 (1990).

TABLE 5

SUMMARY OF RECENT EPIDEMIOLOGIC STUDY FINDINGS

ASSOCIATIONS BETWEEN AMBIENT AIR PARTICULATE MATTER AND MORBIDITY INDICES

General Population

Population	Pollutants Monitored	Effects Reported	Associated Pollutant(s)*	Reference
Metropolitan Toronto Summers 1986-1988	TSP, PM_{10}, $PM_{2.5}$, $SO_4^=$, H^+, O_3	Increased hospital admissions for respiratory diseases	**O_3, H^+, $SO_4^=$**	Thurston et al. (1994) Environ. Res. 65:271-290
New York State: Buffalo, Albany Bronx, Westchester Year-round 1988-1990	$SO_4^=$, H^+, O_3	Increased hospital admissions for respiratory diseases in Buffalo and Bronx	**O_3, H^+, $SO_4^=$**	Thurston et al. (1992) IEAEE 2:429-450
Detroit Summers 1986-1989	PM_{10}, O_3	Increased hospital admissions for respiratory diseases	**PM_{10}, O_3**	Schwartz (1994) Am. J. Crit. Care Med. 150:648-655
Ontario Summers 1983-1988	$SO_4^=$, O_3	Increased hospital admissions for respiratory diseases	**O_3**, $SO_4^=$	Burnett et al. (1994) Environ. Res. 65:172-194
Montreal Summers 1984-1988	PM_{10}, $SO_4^=$, O_3	Increased hospital admissions for respiratory diseases	**PM_{10}, $SO_4^=$**	Delfino et al. (1994) Environ. Res. 67:1-19

* Associations with $p \leq 0.05$. Boldface indicates pollutants most highly associated with the effects.

Pope et al. (1995a) also noted associations between pollutant concentrations and respiratory symptoms and respiratory function. Table 6 summarizes results from recent studies, as well as showing the relative strengths of association among the pollutant variables measured. In those studies where H^+ was measured, it was generally more highly correlated with the effects than the other measured PM constituents. There are some notable exceptions to this generalization. The prospective cohort mortality study of Dockery et al. (1993a) reported that H^+ correlated less well with mortality than $PM_{2.5}$ or $SO_4^=$, but they only had 9-12 months of H^+ data in each city, compared to 14-16 years of data on the other pollutant variables, and many of the daily concentrations of H^+ were below the detection limit. A similar limitation was present in the analysis of Dockery et al. (1992) of air pollution and daily mortality rates.

TABLE 6

SUMMARY OF RECENT EPIDEMIOLOGIC STUDY FINDINGS

ASSOCIATIONS BETWEEN AMBIENT AIR PARTICULATE MATTER AND MORBIDITY AND LUNG FUNCTION INDICES

Children

Population	Pollutants Monitored	Effects Reported	Associated Pollutant(s)*	Reference
Asymptomatic and symptomatic children Uniontown, PA Summer 1990	PM_{10}, $PM_{2.5}$, $SO_4^=$, H^+, O_3, SO_2	Reduced PEFR Symptoms	$\mathbf{H^+}$, $SO_4^=$, O_3 $\mathbf{H^+}$, $PM_{2.5}$, $SO_4^=$, SO_2, O_3	Neas et al..(1995) <u>Am. J. Epid. 141</u>:111-122
Asymptomatic and symptomatic children Austrian Alps Summer 1991	PM_{10}, $SO_4^=$, H^+, NH_4^+, O_3, Pollen	Reduced FVC and FEV_1 Increased medication usage	$\mathbf{H^+}$, $\mathbf{PM_{10}}$ (not specified)	Studnicka et al. (1995) <u>Am. J. Respir. Crit. Care Med. 151</u>:423-430
Asthmatic children Conn. River Valley Summers 1991-1993	$SO_4^=$, H^+, O_3, Pollen	Increased medication usage Symptoms Reduced PEFR	$\mathbf{O_3}$, $\mathbf{H^+}$, $SO_4^=$ $\mathbf{O_3}$, $\mathbf{H^+}$, $SO_4^=$ $\mathbf{O_3}$, $\mathbf{H^+}$, $SO_4^=$	Thurston et al. (1994) Presented at 1994 Joint ISEA-ISEE Conf.
Children in 7 U.S. cities Yearlong 1985-1988	PM_{10}, $PM_{2.5}$, $SO_4^=$, H^+, O_3, SO_2, NO_2	Increased symptoms of bronchitis	$\mathbf{H^+}$, $\mathbf{SO_4^=}$, $\mathbf{PM_{2.5}}$, PM_{10}	Damokosh (1993) <u>ARRD 147</u>(4):A632
Children in 24 North American cities Year-round 1988-1990	PM_{10}, $PM_{2.1}$, $SO_4^=$, H^+, O_3, SO_2, NO_2	Increased symptoms of bronchitis Decreased FVC and FEV_1	$\mathbf{H^+}$, $\mathbf{SO_4^=}$, $\mathbf{PM_{2.1}}$, PM_{10} $\mathbf{H^+}$, $\mathbf{SO_4^=}$, $\mathbf{PM_{2.1}}$, PM_{10}	Dockery et al. (1993) <u>ARRD 147</u>(4):A633 Raizenne et al. (1993) <u>ARRD 147</u>(4):A635
Children in 10 rural Canadian Communities Year-round 1985-1986	PM_{10}, $SO_4^=$, O_3, NO_2, SO_2	Decreased FVC and FEV_1	$\mathbf{SO_4^=}$, $\mathbf{O_3}$	Stern et al. (1994) <u>Environ. Res. 66</u>:125-142

* Associations with $p \leq 0.05$. **Boldface indicates pollutants most highly associated with the effects.**

Discussion

The results of studies in recent years, summarized above, have made it possible to frame the remaining issues in a more coherent and focussed manner.

One key issue is the role of $SO_4^=$, and why it consistently correlates with mortality and morbidity as well as, or better than, other metrics of PM pollution. It is extremely unlikely that $SO_4^=$, per se, is a causal factor. If it is not, then it must be acting as a surrogate index for one or more other components in the PM mixture.

One possibility is that the effects are really due to the $PM_{2.5}$ mass, irrespective of particle composition, and that $SO_4^=$ is a more stable measurement of airborne $PM_{2.5}$ than is the reported $PM_{2.5}$ itself. The ambient $PM_{2.5}$ includes nitrates (primarily ammonium nitrate) and organics formed by photochemical reactions in the atmosphere. There can be considerable volatilization of these species on sampling filters, resulting in negative mass artifacts whose magnitude varies with source strengths and ambient temperature.

Another possibility is that $SO_4^=$ is serving as a surrogate for H^+, a more likely active agent on the basis of the results of controlled exposure studies in humans and animals. The support for this hypothesis is summarized in Tables 5, 6, and 7. The utility of $SO_4^=$ as a surrogate for H^+, especially for time-series studies in a given region without complex topography, is illustrated in Figure 18, which demonstrates that both H^+ and $SO_4^=$ concentrations are almost the same at two sites sixty miles apart, and that the concentration of both ions rise and fall together.

TABLE 7

COMPONENTS OF AMBIENT AIR PARTICULATE MATTER (PM) THAT MAY ACCOUNT FOR SOME OR ALL OF THE EFFECTS ASSOCIATED WITH PM EXPOSURES

Component	Evidence for Role in Effects	Doubts
Strong Acid (H^+)	• Statistical associations with health effects in most recent studies for which ambient H^+ concentrations were measured	• Similar PM-associated effects observed in locations with low ambient H^+ levels
	• Coherent responses for some health endpoints in human and animal inhalation and *in vitro* studies at environmentally relevant doses	• Very limited data base on ambient concentrations
Ultrafine Particles ($D \leq 0.2$ μm)	• Much greater potency per unit mass in animal inhalation studies (H^+ and TiO_2 aerosols) than for same materials in larger diameter fine particle aerosols	• Absence of relevant data on responses in humans
	• Concept of "irritation signalling" in terms of number of particles per unit airway surface	• Absence of relevant data base on ambient concentrations

A third possibility is that the causal factor is the number concentration of irritating particles, which would be dominated by the particles in the ultrafine mode (diameters below 50 nm) (Oberdörster et al., 1995).

Perhaps the most likely candidate is a hybrid of H^+ and ultrafines, i.e., acid-coated ultrafine particles. As shown in Figure 19, sulfuric acid coatings on ultrafine zinc oxide particles produce about the same responses as pure sulfuric acid for a given number of equivalent sized particles, yet the coated particles only had one-tenth of the acid content per unit volume of air. Thus, the response may be related to the number of acidic particles that deposit on the lung surfaces rather than the amount of acid deposited. In other words, the total concentration of H^+ may be a better surrogate of the active agent than $SO_4^=$ or $PM_{2.5}$, but it still is a crude index for the number concentration of irritant particles. Amdur and Chen (1989) suggested

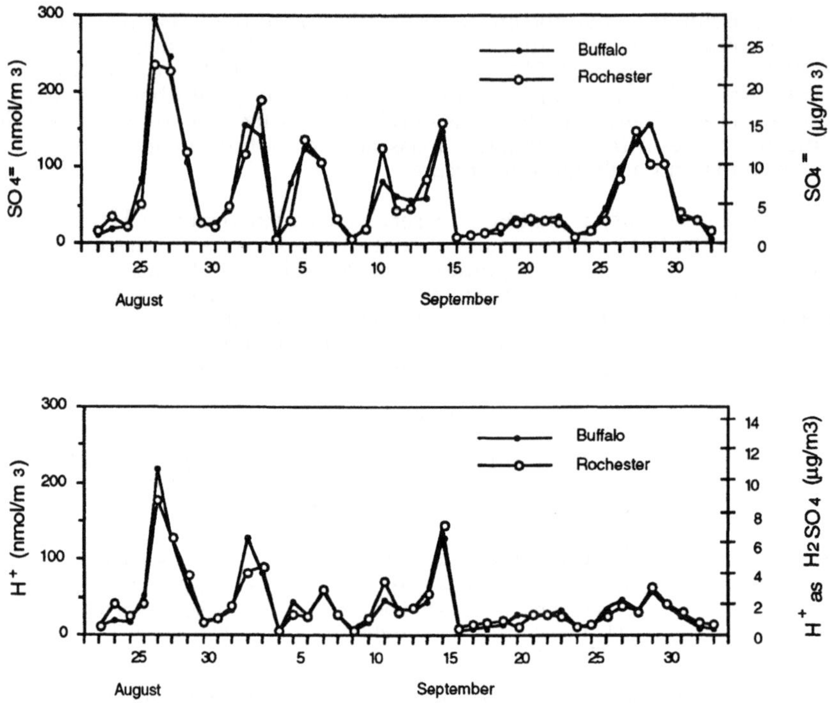

Figure 18. Intercomparison of Rochester, NY and Buffalo, NY sulfate and daily acid aerosol concentrations (August 22 - October 2, 1990).

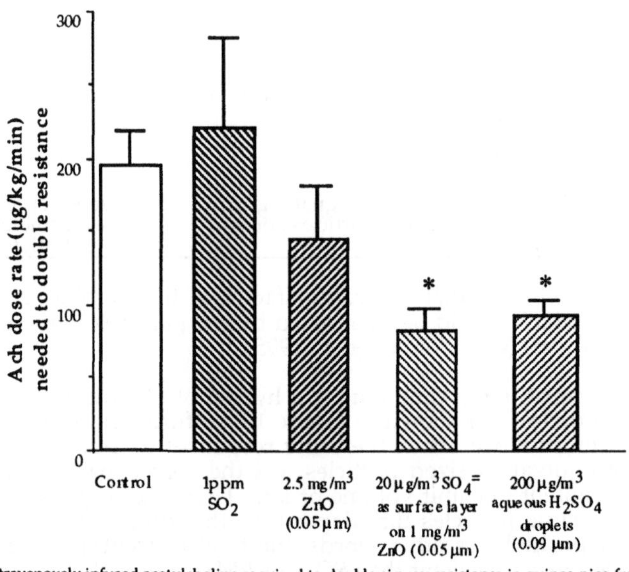

Figure 19. Dose of intravenously infused acetylcholine required to double airway resistance in guinea pigs from baseline levels 2 hours after a 1 hour inhalation exposure. Values are mean ± S.E. The asterisks indicate reductions significant at $p < 0.05$. For the aerosols, the median particle diameters are indicated in parentheses.

that number concentration was important for sulfuric acid aerosol, and Hattis et al. (1987, 1990) gave the concept a name, i.e., "irritation signalling". Research of Chen et al. (1995) indicate that acid-coated particles much smaller than those discussed by Hattis et al. (1987, 1990) were capable of producing lung responses.

If the number concentration of acid-coated particles is the most relevant index of the active agent in ambient PM, then new sampling techniques will be needed to characterize ambient air concentrations and personal exposures.

Other components of the ambient ultrafine aerosol have not been well characterized either, and they may also be important health stressors. One class is the volatile trace metals (such as As, Cd, Cu, Pb, Zn) which condense as ultrafine particles in the effluent airstream of fossil fuel combusters (Amdur et al., 1986) and are inefficiently captured by air cleaners for fly ash collection. Another class is the ultrafine organics from atmospheric photochemical reaction sequences.

Any remaining inconsistency between the epidemiological findings and the results of the controlled exposure studies may be explicable on the basis that the relatively rare individuals who respond in the epidemiological populations are an especially responsive subset of the overall population, and the low probability that such sensitive individuals would be included in the controlled exposure studies in the laboratory. An alternative hypothesis is that the controlled exposure atmospheres do not contain highly toxic components that may be present in ambient atmospheres.

In summary, excess daily mortality and morbidity have been related to ambient pollution at current levels in many communities in the U.S. and around the world using available pollutant concentration data. However, it is not at all clear whether any of the pollutant indices used are causally related to the health effects or, if none of them are, which is the best index or surrogate measure of the causal factor(s). This gap can best be addressed by analyses of pollutant associations with mortality and morbidity in locations where a number of different pollutant metrics are available simultaneously, using analytic methods not dependent on arbitrary model assumptions.

While more research is needed on causal factors for the excess mortality and morbidity associated with PM in ambient air, and on the characterization of susceptibility factors, responsible public health authorities cannot wait for the completion and peer review of this research. It is already clear that the evidence for adverse health effects attributable to PM challenges the conventional paradigm used for setting ambient air standards and guidelines, i.e., that a threshold for adversity can be identified, and a margin of safety can be applied. Excess mortality is clearly an adverse effect, and the epidemiological evidence is consistent with a linear non-threshold response for the population as a whole.

The Air Quality Guidelines of the World Health Organization-Europe (WHO-EURO) are currently undergoing revision. The Working Group of WHO-EURO on PM, meeting in October 1994 in Bilthoven, The Netherlands, determined that it could not recommend a PM Guideline. Instead, it prepared a tabular presentation of the estimated changes in daily average PM concentrations needed to produce specific percentage changes in: 1) daily mortality; 2) hospital admissions for respiratory conditions; 3) bronchodilator use among asthmatics; 4) symptom exacerbation among asthmatics; and 5) peak expiratory flow. The concentrations needed to produce these changes were expressed in PM_{10} for all five response categories. For mortality and hospital admisssions, they were also expressed in terms of $PM_{2.5}$ and $SO_4^=$. Using this guidance, each national or local authority setting air quality standards can decide how much adversity is acceptable for its population. Making such a choice is indeed a challenge.

References

Amdur,M.O. and Chen,L.C., *Environ. Health Perspect.* **79** (1989) 147

Amdur,M.O., Sarofim,A.F., Neville,M., Quann,R.J., McCarthy,J.F., Elliott,J.F., Lam,H.F., Rogers,A.E., and Conner,M.W., *Environ. Sci. Technol.* **20** (1986) 138

Anon, "The effects of the fog on cattle in London", *The Veterinarian* **XLVII** (No. 553), (Fourth Series, No. 229) (Jan. 1874), pp. 1-4, 32, 33

Bates,D.V., personal communication (1987)

Bauer,M.A., Utell,M.J., Speers,D.M., Gibb,F.R., and Morrow,P.E., *Am. Rev. Respir. Dis.* **137** (1988) A167

Burnett,R.T., Dales,R.E., Raizenne,M.E., Krewski,D., Summers,P.W., Roberts,G.R., Raad-Young,M., Dann,T., and Brook,J., *Environ. Res.* **65** (1994) 172

Chen,L.C., Wu,C.Y., Qu,Q.S., and Schlesinger,R.B. *Inhal. Toxicol.* **7** (1995) 577

Chinn,S., Florey,C.duV., Baldwin,I.G., and Gorgol,M., *J. Epidemiol. Commun. Health* **35** (1981) 174

Ciocco,A. and Thompson,D.J., *Am. J. Public Health* **51** (1961) 155

Commins,B.T. and Waller,R.E., *Analyst* **88** (1963) 364

Damokosh,A.I., Spengler,J.D., Dockery,D.W., Ware,J.H., and Speizer,F.E., *Am. Rev. Respir. Dis.* **147** (1993) A632

Delfino,R.J., Becklake,M.R., and Hanley,J.A., *Environ. Res.* **67** (1994) 1

Dockery,D.W., Damokosh,A.I., Neas,L.M., Raizenne,M., Spengler,J.D., Koutrakis,P., Ware,J.H., and Speizer,F.E., *Am. Rev. Respir. Dis.* **147** (1993a) A633

Dockery,D.W., Pope,III,C.A., Xu,X., Spengler,J.D., Ware,J.H., Fay,M.E., Ferris,Jr.,B.G., and Speizer, F.E., *N. Engl. J. Med.* **329** (1993) 1753

Dockery,D.W., Schwartz,J., and Spengler,J.D., *Environ. Res.* **59** (1992) 362

Dohan,F.C., Everts,G.S., and Smith,R., *J. Air Pollut. Control Assoc.* **12** (1962) 418

Fairley,D., *Environ. Health Perspect.* **89** (1990) 159

Firket,J., *Trans. Faraday Soc.* **32** (1936) 1191

Hattis,D., Wasson,J.M., Page,G.S., Stern,B., and Franklin,C.A., *J. Air Pollut. Control Assoc.* **37** (1987) 1060

Hattis,D.S., Abdollahzadeh,S., and Franklin,C.A., *J. Air Waste Manage. Assoc.* **40** (1990) 322

Holland,W.W. and Reid,D.D., *Lancet* **1** (1965) 445

Ito,K., Kinney,P.L., and Thurston,G.D., *Inhal. Toxicol.* **7** (1995) 735

Ito,K., Thurston,G.D., Hayes,C., and Lippmann,M., *Arch. Environ. Health* **48** (1993) 213

Kinney,P.L., Ito,K., and Thurston,G.D., *Inhal. Toxicol.* **7** (1995) 69

Koenig,J.Q., Pierson,W.E., and Horike,M., *Am. Rev. Respir. Dis.* **128** (1983) 221.

Lave,L.B. and Seskin,E.P., *Science* **169** (1970) 723

Lawther,P.J., *J. Inst. Fuel* **36** (1963) 341

Lawther,P.J., Waller,R.E., and Henderson,M., *Thorax* **25** (1970) 525

Leikauf,G., Yeates,D.B., Wales,K.A., Spektor,D., Albert,R.E., and Lippmann,M., *Am. Ind. Hyg. Assoc. J.* **42** (1981) 273

Leikauf,G.D., Spektor,D.M., Albert,R.E., and Lippmann,M., *Am. Ind. Hyg. Assoc. J.* **45** (1984) 285

Lippmann, M., Schlesinger,R.B., Leikauf,G., Spektor,D., and Albert,R.E., *Ann. Occup. Hyg.* **26** (1982) 677

Lippmann,M. and Ito,K., *Inhal. Toxicol.* **7** (1995) 85

Lippmann,M., *Environ. Health Perspect.* **79** (1989) 3

Lippmann,M., Gearhart,J.M., and Schlesinger,R.B., *Appl. Ind. Hyg.* **2** (1987) 188

Lippmann,M., Kleinman,M.T., Bernstein,D.M., Wolff,G.T., and Leaderer,B.P., *Ann. NY Acad. Sci.* **322** (1979) 29

Martin,A.E. and Bradley,W.H., *Mon. Bull. Minist. Health-Public Health Lab. Service* **19** (1960) 57

Mazumdar,S., Schimmel,H., and Higgin,I., *Arch. Environ. Health* **37** (1982) 213

Ministry of Health, "Mortality and Morbidity During the London Fog of December 1952", Her Majesty's Stationary Office, London (1954)

Neas,L.M., Dockery,D.W., Koutrakis,P., Tollerud,D.J., and Speizer,F.E., *Am. J. Epidemiol.* **141** (1995) 111

Oberdörster,G., Gelein,R.M., Ferin,J., and Weiss,B., *Inhal. Toxicol.* **7** (1995) 111

Ostro,B., Sanchez,J.M., Aranda,C., and Eskeland,G.S., *J. Expos. Anal. Environ. Epidemiol.* (in press, 1995)

Ostro,B.D., *Risk Anal.* **10** (1990) 421

Ozkaynak,H. and Thurston,G.D., *Risk Anal.* **7** (1987) 449

Ozkaynak,H., Xue,J., Severance,P., Burnett,R., and Raizenne,M., "Associations between daily mortality, ozone and particulate air pollution in Toronto, Canada", Presented at Colloquium on Particulate Air Pollution and Human Mortality and Morbidity, Irvine, CA (Jan. 1994)

Pope,C.A.,III, Schwartz,J., and Ransom,M.R., *Arch. Environ. Health* **47** (1992) 211

Pope,C.A.,III, Thun,M.J., Namboodiri,M.M., Dockery,D.W., Evans,J.S., Speizer,F.E., and Heath,Jr.,C.W., *Am. J. Respir. Crit. Care Med.* **151** (1995b) 669

Pope,C.A.,III, Dockery,D.W., and Schwartz,J., *Inhal. Toxicol.* **7** (1995a), 1

Raizenne,M., Neas,L., Damokosh,A., Dockberg,D., Spengler,J., Koutrakis,P., Ware,J., and Speizer,F., *Am. Rev. Respir. Dis.* **147** (1993) A635

Roth,D.H., Wyzga,R.E., and Hayter,A.J., "Method and Problems in Estimating Health Risks from Particulates", pp. 1047-1063 in *Aerosols*, Eds. Lee,S.D., Schneider,T., Grant,L.D., and Verkerk,P.J., Lewis, Chelsea, MI (1986)

Schlesinger,R.B., Chen,L.C., Finkelstein,I., and Zelikoff,J.T., *Environ. Res.* **52** (1990) 210

Schlesinger,R.B., Halpern,M., Albert,R.E., and Lippmann,M., *J. Environ. Pathol. Toxicol.* **2** (1979) 1351

Schlesinger,R.B., Naumann,B.D., and Chen,L.C., *J. Toxicol. Environ. Health* **12** (1983) 441

Schrenk,H.H., Heimann,H., Clayton,G.D., and Gafater,W.M., "Air Pollution in Donora, Pennsylvania", Public Health Bull. No. 306, U.S. Government Printing Office, Washington, DC (1949)

Schwartz,J. and Marcus,A.H., *Am. J. Epidemiol.* **131** (1990) 185

Schwartz,J., *Am. J. Epidemiol.* **137** (1993) 1136

Schwartz,J., *Am. J. Respir. Crit. Care Med.* **150** (1994) 648

Shumway,R.H., Tai,R.Y., Tai,L.P., and Pawitan,Y., "Statistical Analysis of Daily London Mortality and Associated Weather and Pollution Effects", Report submitted to California Air Resources Board, Sacramento, CA (1983)

Stern,B.R., Raizenne,M.E., Burnett,R.T., Jones,L., Kearner,J., and Franklin,C.A., *Environ. Res.* **66** (1994) 125

Studnicka,M.J., Frischer,T., Meinert,R., Studnicka-Benke,A., Hajek,K., Spengler,J.D., and Neumann,M.G., *Am. J. Respir. Crit. Care Med.* **151** (1995) 423

Thurston,G.D. and Kinney,P.L., *Inhal. Toxicol.* **7** (1995) 71

Thurston,G.D., Ito,K., Hayes,C.G., Bates,D.V., and Lippmann,M., *Environ. Res.* **65** (1994) 271

Thurston,G.D., Ito,K., Kinney,P.L., and Lippmann,M., *J. Expos. Anal. Environ. Epidemiol.* **2** (1992) 429

Thurston,G.D., Ito,K., Lippmann,M., and Hayes,C., *Environ. Health Perspect.* **79** (1989) 73

Thurston,G.D., Personal communication of table prepared for draft EPA Criteria Document on Particulate Matter (1995)

Touloumi,G., Pocock,S.J., Katsouyanni,K., and Trichopoulos,D., *Int. J. Epidemiol.* **23** (1994) 957

U.S.EPA, "Protecting Visibility", EPA/450/5-79-008, Office of Air Quality Planning and Standards, Research Triangle Park, NC

Ware,J.H., Thibodeau,L.A., Speizer,F.E., Colome,S., and Ferris,B.G.,Jr. *Environ. Health Perspect.* **41** (1981) 255

A Mechanistic Approach to Assessing the Lung Cancer Risk of Diesel Exhaust and Carbon Black

Roger O. McClellan
Chemical Industry Institute of Toxicology
Research Triangle Park, NC, USA
Telephone: (919) 558-1202
Fax: (919) 558-1400

J. C. M. Marijnissen and L. Gradoń (eds.), Aerosol Inhalation: Recent Research Frontiers, 27–77.

Abstract

Diesel exhaust soot particles, consisting principally of a carbonaceous core and absorbed organic compounds, are ubiquitous contaminants of the environment because of the widespread use of diesel engines. Carbon black is produced in large quantities and has many commercial uses around the world. Because the particle size of both materials is small, they are readily inhaled and deposited in the lung, giving rise to concern for induction of lung cancer in humans. Concern that exposure to diesel exhaust or carbon black may cause human lung cancer has been heightened by the finding that long-term exposure of rats to high concentrations of either diesel exhaust or carbon black causes an increased incidence of lung cancer. In this article, the available data are reviewed on diesel exhaust and carbon as agents that cause lung cancer in humans and laboratory animals from the perspective of using the data to assess human lung cancer risk. Current mechanistic data on the pathogenesis of lung cancer induced in rats by diesel soot and carbon black are reviewed. The mechanism of lung tumor induction in rats is apparently unique to high-level exposures, and a threshold relationship exists between exposure and lung tumor induction. Furthermore, mice and Syrian hamsters do not respond with increased lung tumors when exposed to high levels of diesel exhaust, suggesting that the lung tumor response may be unique to the rat, even at high levels of exposure. Thus, in my opinion, the rat lung tumor data from high level exposures to either diesel soot or carbon black should not be used in assessing the human lung cancer risk of exposure to either material. The available human data indicate that the human lung cancer risk will not be increased if exposures to these agents are controlled to low levels.

Introduction

Human health risks from exposure to airborne materials encountered in the workplace or general environment have been of concern for many years. To limit human health risks, responsible companies have voluntarily sought to control air pollutants in the workplace. These efforts have been greatly facilitated by guidance provided by the American Conference of Governmental Industrial Hygienists (ACGIH), which has established threshold limit values (TLV). In the United States, federal government regulations for occupational exposures to airborne materials are established by the Occupational Safety and Health Administration and the Mine Safety and Health Administration. Environmental exposures to airborne materials are regulated in the United States principally by the U.S. Environmental Protection Agency (EPA), primarily based on authority provided by the Clean Air Act.

The scientific basis for developing guidance and regulations for airborne materials includes data from human studies, in-vivo investigations conducted in laboratory animals, and in-vitro experiments with cells and tissues from both humans and laboratory animals. Human data are always used to the maximum extent feasible, with data obtained from other approaches used to complement and extend the human data. If human data are available, they frequently were obtained at high exposure concentrations, necessitating extrapolation to lower exposure concentrations for development of environmental exposure limits. When human data are not available for a material, as will be the case for new materials, then reliance must be placed on the use of data from in-vivo laboratory animal studies and in-vitro studies with cells and tissues. If laboratory animal data are used, then a second extrapolation from laboratory animals to humans is required.

The now-classic 1983 report on risk assessment by the National Research Council (NRC) (1983) provided guidance for the conduct of risk assessments, including the use of inference guidelines, or, as they are more frequently called, default options. The NRC report defined a default option as "the option chosen on the basis of risk assessment policy that appears to be the best choice in the absence of data to the contrary." (NRC, 1983). The default assumptions, which are based on general knowledge, are considered to be conservative (more likely to overestimate rather than underestimate risk). A recent NRC report (NRC, 1994) recommends the continued use of the risk assessment paradigm advanced earlier (NRC, 1983) and suggests that the default options be replaced with specific scientific information when such information is available. Two key default options for cancer risk assessment are (1) that humans are as sensitive as the most sensitive animal species, strain, or sex evaluated; and (2) that cancers produced by chemical agents are produced by the same mechanisms by which radiation causes cancer, and hence the lowest dose of a chemical is assumed to have a cancer risk that can be calculated.

In this article, I will examine the available information relevant to assessing the human cancer risks of exposure to diesel exhaust and carbon black, directing special attention to the use of default assumptions. The discussion will consider qualitative assessments of cancer hazards such as those developed by the International Agency for Research on Cancer (IARC) (1987, 1989) and quantitative cancer risk assessments as developed by the U.S. EPA (1986, 1994a, 1994b). A more detailed treatment of this topic has been published elsewhere (McClellan, 1996).

Diesel Engine Exhaust Studies

Diesel engines have found increasing use over the past century because of their reliability and the efficiency of hydrocarbon fuel combustion. However, there is concern that diesel engine exhaust may pose a carcinogenic risk to humans. This concern relates to the small size of diesel soot particles, which renders them readily inhalable, and their substantial content of polycyclic aromatic hydrocarbons, many of which are mutagenic and in some cases known to be human carcinogens (McClellan, 1987; U.S. EPA, 1994b). The energy crisis of the 1970s served as a stimulus for increased use of diesel engines, including increased use in light-duty vehicles, because of the known fuel efficiency. This, in turn, stimulated critical examination of the potential lung cancer risk of diesel exhaust exposures. The diesel exhaust health effects research program and results have been the subject of a number of conferences (Pepelko et al., 1980; Lewtas, 1982; and Ishinishi et al., 1986a). In addition, a number of groups and individuals have reviewed the resulting data (NRC, 1981; McClellan, 1986a, 1986b, 1986c; McClellan, 1987; McClellan et al., 1989; NIOSH, 1988; IARC, 1989; U.S. EPA, 1990, 1994b; Mauderly, 1992; California Environmental Protection Agency, 1994; and the Health Effects Institute (HEI) (1995).

Epidemiological investigations were a central part of past research on diesel exhaust. Key epidemiological findings have recently been reviewed by Cohen and Higgins (1995) in a report prepared by the Health Effects Institute (HEI) (1995) (Fig. 1). Cohen and Higgins (1995) concluded that "occupational exposure to diesel exhaust from diverse sources increases the rate of lung cancer by 20% to 40% in exposed workers generally and to a greater extent among workers with prolonged or intense exposures, or both." The summary portion of the HEI report (Nauss and The HEI Diesel Working Group, 1995) concludes, "The epidemiologic studies are consistent in showing a weak association between exposure to diesel exhaust and lung cancer, but vary in the strength of the statistical association; only a few studies showed elevated relative risks that were statistically significant." Both Cohen and Higgins (1995) and Nauss and The HEI Diesel Working Group (1995) call attention to the limitations of the epidemiological studies. The three major limitations are (1) the lack of control in many of the studies for confounding factors such as cigarette smoke,

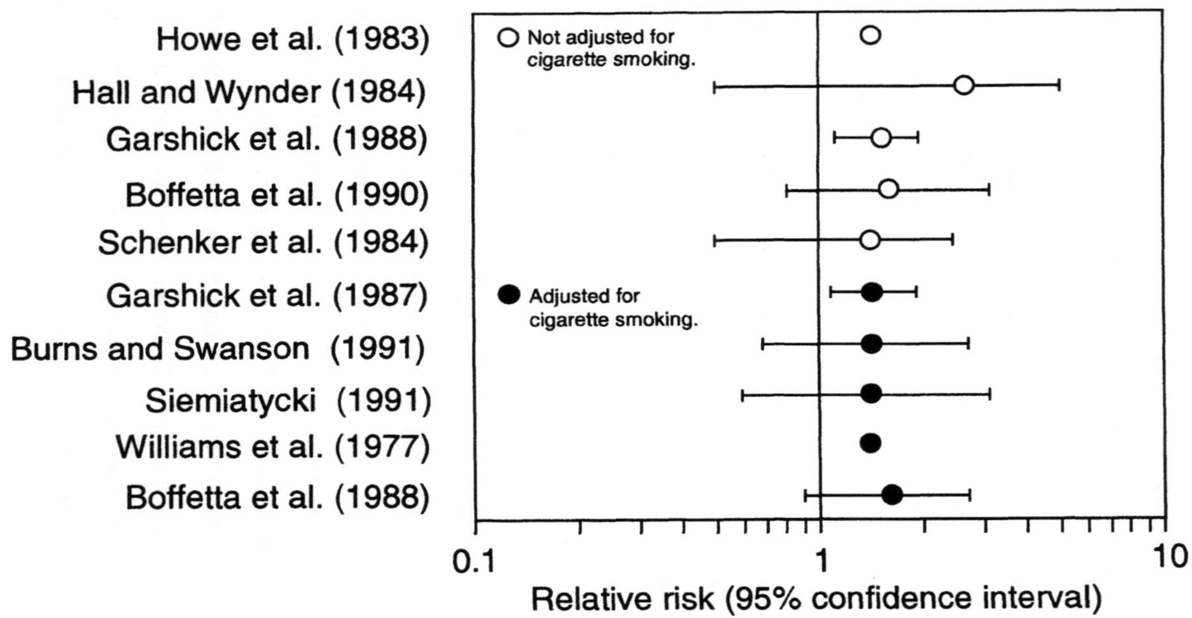

Figure 1. Lung cancer and exposure to diesel exhaust in railroad workers. From Cohen and Higgins (1995).

(2) the lack of measurements of exposure, and (3) the lack of any characterization data on the diesel exhaust particulates for exposures that occurred decades ago. The characteristics of diesel exhaust and especially the content of aromatic hydrocarbons differ among locomotives, trucks, and light-duty vehicles; moreover, emissions have been steadily decreasing as engine performance and fuel quality have improved (Sawyer and Johnson, 1995).

Absent a clear signal for an excess cancer risk associated with occupational exposure, the results of laboratory animal investigations with controlled long-term exposures to graded levels of diesel exhaust are an important consideration. In the 1980s, four laboratories reported almost simultaneously that long-term exposure of rats to high levels of diesel exhaust resulted in an increased risk of lung cancer. These data, as summarized in the HEI (1995) report, are shown in Fig. 2. The initial reaction of many scientists to the findings of an excess of lung tumors in rats exposed to diesel exhaust was that the results matched expectations: the rats inhaled large quantities of particles that contained mutagenic chemicals, and lung tumors would be expected to arise via a classical genotoxic mode of action. However, some individuals raised the possibility of an alternative hypothesis for the pathogenesis of the rat lung tumors that has important implications for assessing the human cancer risks of exposure to diesel exhaust.

At the first major conference (Ishinishi et al., 1986a) where the results of the long-term rat studies were discussed, Vostal (1986) speculated that the tumors arose via another mechanism, calling attention to an increased incidence of lung tumors observed in rats exposed to high levels of other particulate materials generally viewed as nuisance dusts. This included studies with solvent refined coal solids (Drozdowicz et al., 1986), raw spent shale (Holland et al., 1985), titanium dioxide (Lee et al., 1985), titanium tetrachloride hydrolysis products (Lee et al., 1985), and coal dust (Martin et

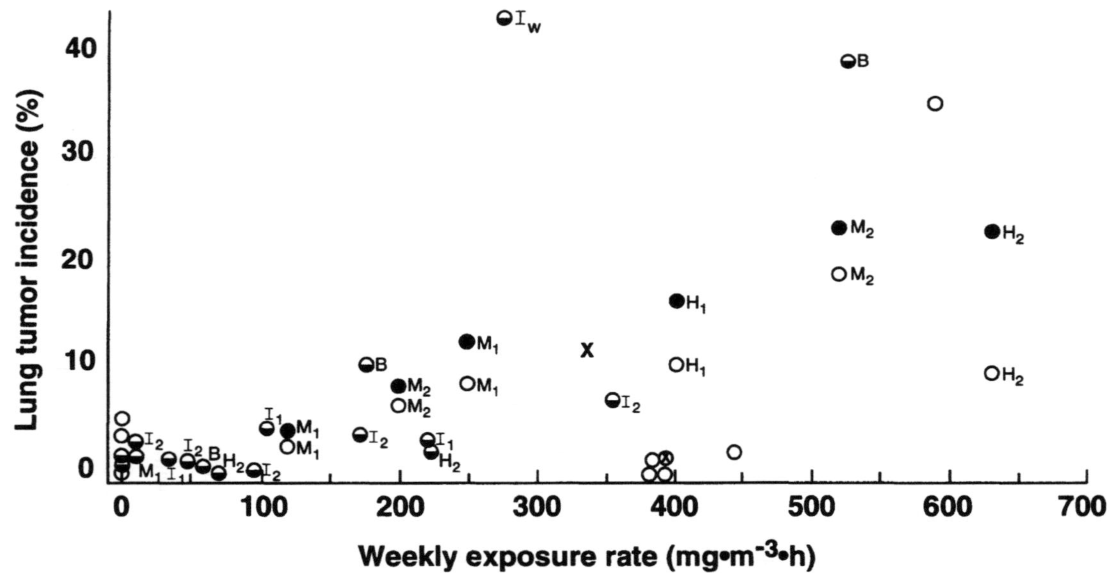

Figure 2. Relation between rat lung tumor incidence and exposure rates for diesel exhaust particulate matter. Data point code: B = Brightwell et al. (1989); H_1 = Heinrich et al. (1986); H_2 = Heinrich et al. (1995); I_1 = Ishinishi et al. (1986b) (exhaust from 1.8-L engine); I_2 = Ishinishi et al. (1986b) (exhaust from 11-L engine); I_w = Iwai et al. (1986); M_1 = Mauderly et al. (1987); M_2 = Mauderly et al. (1994). ●, Includes lesions identified by the investigator as "benign squamous tumors"; ○, excludes these lesions. Reproduced from Nauss and the HEI Diesel Working Group (1995).

al., 1977). Vostal argued that "the action of particles is mediated primarily by the epigenetic mechanism of carcinogenesis, can be characterized more by the promotional than the initiation stages of carcinogenesis, and probably does not involve primary damage to the genome." He concluded that the threshold nature of the exposure-response relationship for lung tumors and the finding of lung tumors in only a single species provided a basis for not using the rat results for estimating human lung cancer risk at low ambient levels of exposure to diesel exhaust. Another report at the same conference provided support for Vostal's thesis. This was the finding by Kawabata et al. (1986) that intratracheal instillation of either carbon black or diesel soot caused a high incidence of lung cancer in rats. As I recall, the findings were at least partially discounted at the time the data were presented because of concern for the large quantities of particles delivered via a nonphysiological mode of administration, intratracheal instillation. Vostal noted that a threshold relationship between exposure and rat lung tumor response was observed for all of these dusts (Fig. 3).

After Vostal (1986) called attention to this nonchemical-specific effect of high-level particulate exposure, others (Morrow, 1988, 1994; McClellan, 1990, 1996; Mauderly et al., 1990; Mauderly, 1996; and McConnell, 1996) have commented on the plausibility of the "lung or clearance overload hypothesis" as a mechanism for induction of lung cancer in rats exposed to high levels of diesel exhaust. Data developed from the Inhalation Toxicology Research Institute studies (McClellan et al., 1986) provided strong support for the overload hypothesis. Rats exposed concurrently with those for which Mauderly et al. (1987) reported an increased incidence of lung cancer were used to evaluate exposure-related changes in soot lung burden, particle clearance, lung inflammation, and lung function. Wolff et al. (1987) reported a distinct difference in the soot burden buildup at the lowest exposure concentration (0.35

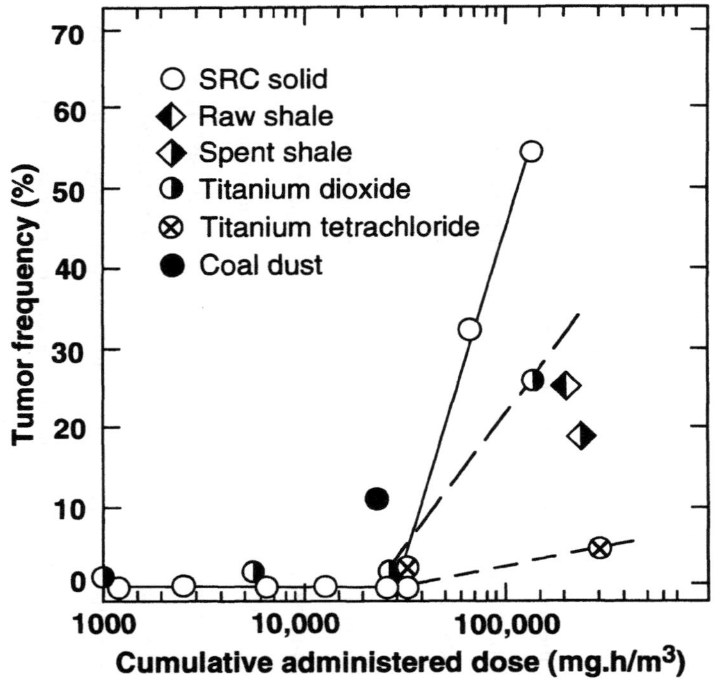

Figure 3. Relationship between the estimated administered dose of airborne particles and tumor frequency. From Vostal (1986).

mg/m^3), as contrasted to the high exposure levels (3.5 and 7.0 mg/m^3) with the lung burdens of soot. At the higher exposure concentrations, the lung burdens were greater than predicted by a proportional extrapolation from the lowest exposure concentration based on the air concentration of soot, i.e., 1 to 10 to 20 based on the exposure concentrations of 0.35, 3.5, and 7.0 mg/m^3 (Fig. 4). To evaluate the kinetics of particle clearance from the lungs, rats were exposed to radiolabeled particles. This revealed a decrease in long-term clearance from the lungs of rats exposed to medium and high concentrations compared with rats exposed to the low concentration and to control rats (Fig. 5). Henderson et al. (1988) reported on biochemical and cellular measurements made on bronchoalveolar lavage fluid and lung tissue. They observed no changes at the lowest exposure concentration (0.35 mg/m^3), while marked changes indicative of inflammation were seen at the two higher exposure concentrations. Mauderly et al. (1988) reported changes in lung structure and function that closely paralleled the changes in particle accumulation and inflammation. Taken collectively, these data all indicate the existence of a threshold in exposure-response relationships for very fine diesel soot particles (~0.25 μm in diameter) at some exposure concentration in excess of 0.35 mg/m^3 with exposures of 7 hours/day, 5 days/week.

In addition to the studies with rats exposed to diesel exhaust, a number of studies have been conducted with Syrian hamsters and several strains of mice. These studies have recently been reviewed (U.S. EPA, 1994b; HEI, 1995) and collectively are considered as being negative for an association between exposure to diesel exhaust and lung cancer induction. In some cases, such as the study recently reported by Mauderly et al. (1996), the mice were exposed concurrently with rats that showed an increase in lung cancer incidence (Mauderly et al., 1987). Since the mice and rats were exposed to the same test atmosphere, the findings emphasize the extent

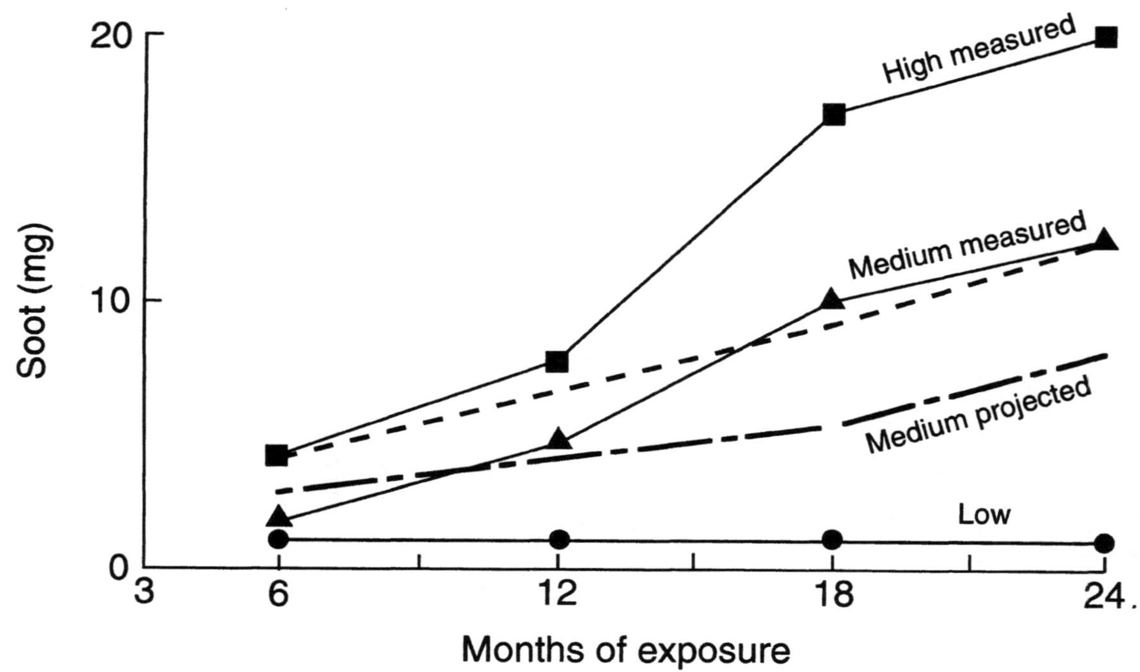

Figure 4. Lung burden of soot in rats chronically exposed to diesel exhaust.

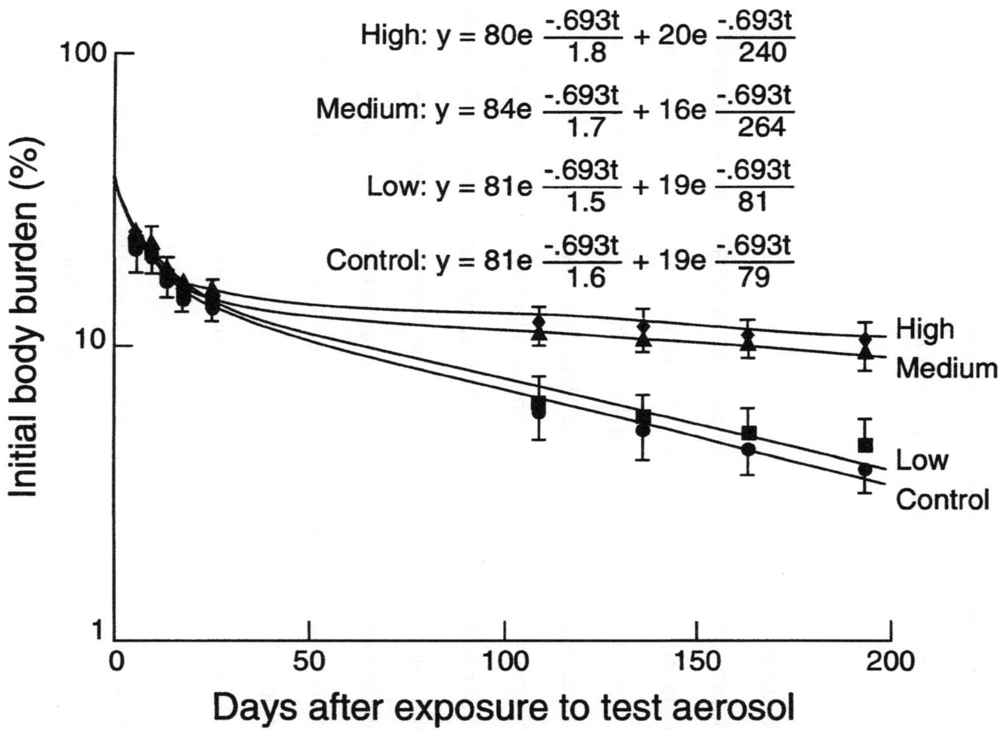

Figure 5. Long-term clearance in rats of ^{134}Cs-FAP particles inhaled after 24 months of diesel exhaust exposure. From Wolff et al. (1987).

to which there is a fundamental difference in the responses of rats and mice to diesel exhaust. Similar differences were reported by Heinrich (1996). Mauderly (1996) reviewed the species differences and called attention to the fact that the extent of inflammation, cytotoxicity, and cell proliferation in terminal bronchiolar and alveolar epithelium was less in mice than in rats, even when lung burdens in the two species were quite similar.

Carbon Black Studies

The carbonaceous core of diesel soot particles is believed to be involved mechanistically in the induction of lung tumors in rats exposed to high concentrations of diesel exhaust. An approach to evaluating the hypothesis is to expose rats via inhalation to carbonaceous particles devoid of absorbed polycyclic aromatic hydrocarbons. This has now been done, and the results have been reported by Heinrich et al. (1995) and Nikula et al. (1995). Both studies used carbon black particle exposure conditions and concentrations similar to the diesel soot exposures that produced an excess of lung tumors. In both studies, an excess of lung tumors was observed that was equal to that produced by diesel exhaust exposures with similar particle concentrations. The results reported by Nikula et al. (1995) are shown in Figure 6. The similarity of the findings for diesel exhaust and carbon black exposures is striking.

Since carbon black is not known to directly react with DNA, a critical question is how the lung cancers arise. Driscoll (1996) and Oberdörster (1996) have addressed this question by exposing rats to graded concentrations of carbon black and evaluating lung inflammation and lung epithelial cell mutagenicity. Their results are summarized in Table 1. They interpret their results as indicating a threshold exposure-response relationship for inflammation and mutagenicity, with the responses

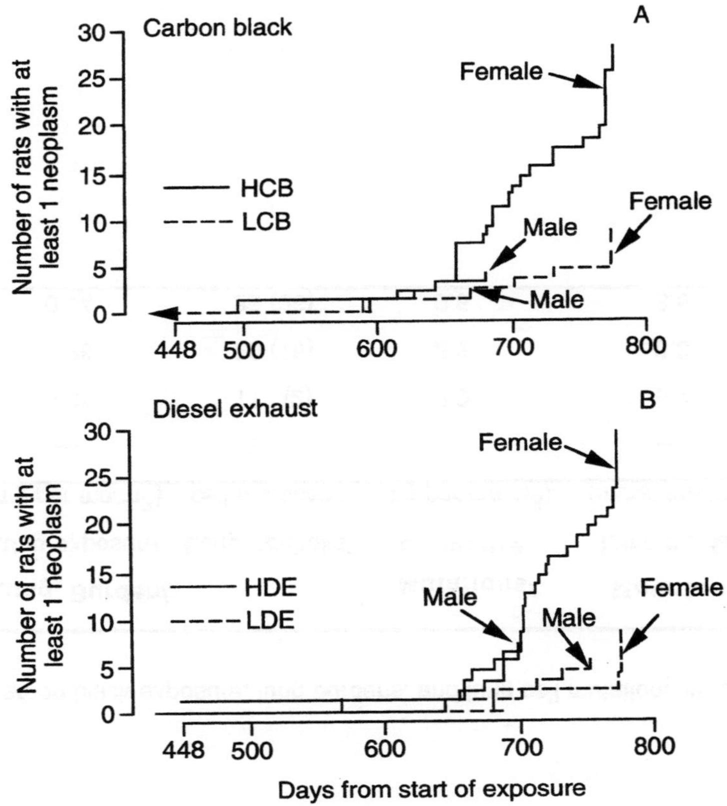

Figure 6. Cumulative number of rats in each exposure group observed to have at least one malignant or benign lung neoplasm with long-term exposure to either carbon black or diesel exhaust. From Nikula et al. (1995).

Table 1. Relationship of carbon black exposure, lung burdens, and lung cell mutations in rats.

Exposure concentration (mg/m^3)	Lung burden (mg)	Lung Burden/ Lung exposure (mg per mg/m^3)	Lung epithelial cell mutations	Mutations/ Exposure (mg per mg/m^3)	Mutations/ Lung burden (mutations/mg)
0	—	—	9	—	—
1	0.35	0.35	11 (2)	2.0	5.7
7	1.92	0.28	25 (16)	2.3	8.3
50	7.11	0.14	35 (26)	0.5	3.6

at the level of 1 mg/m^3 being indistinguishable from those of controls and clear differences being observed at the two high exposure concentrations. Driscoll (1996) and Oberdörster (1996) have pointed out that there is a close correlation between the surface area of the deposited particles and lung inflammation and lung tumorigenicity for a diverse array of particles.

Pathogenesis of Particle-Induced Rat Lung Tumors

The available data related to the mechanisms of lung cancer induction in the rats exposed to high concentrations of either diesel exhaust or carbon black have been integrated in Table 2 and Figure 7. At low exposure concentrations, the rates of disposition and clearance of particles are such that the lung burden of particles increases very slowly and, most importantly, inflammation is minimal. When inflammation is held in check by defense mechanisms, oxidant damage is avoided, or the damage is rapidly repaired. With higher exposure, presumably above 1 mg/m^3 for 6 hr/day, lung burdens reach sufficient levels that the inflammatory process is no longer held in check. Indeed, with sufficiently intense exposure, the lung clearance processes are overwhelmed, the inflammation persists, and the lung burden increases to levels above those predicted from the kinetics at lower exposure concentrations (Wolff et al., 1987; Snipes, 1996; and Yu, 1996). As illustrated in Figure 8, the alterations in clearance serve to drive the development of pulmonary pathology, which, in turn, further affects clearance. Therefore once alterations begin, progression to more serious alterations in both clearance and pulmonary pathology and an associated increased likelihood of lung tumor induction occur.

The schematic representation of the pathogenesis of lung tumors depicted in Figure 7 shows a series of events linked in a linear, unidirectional fashion with two outcomes, a lack of lung tumor induction with low-level exposures and induction of

Table 2. Schematic representation of responses of the lung of laboratory animals and humans to prolonged inhalation of high concentrations of diesel soot or carbon black particles.

Response	Rats	Mice	Hamsters	Humans
Clearance overload	+++	+++	++++	?
Inflammation	+++	+	+	?
Cell proliferation	+++	+	+	?
Fibrosis	+++	+	+	?
Mutations	+++	?	?	?
Lung cancer	+++	+/-	-	?

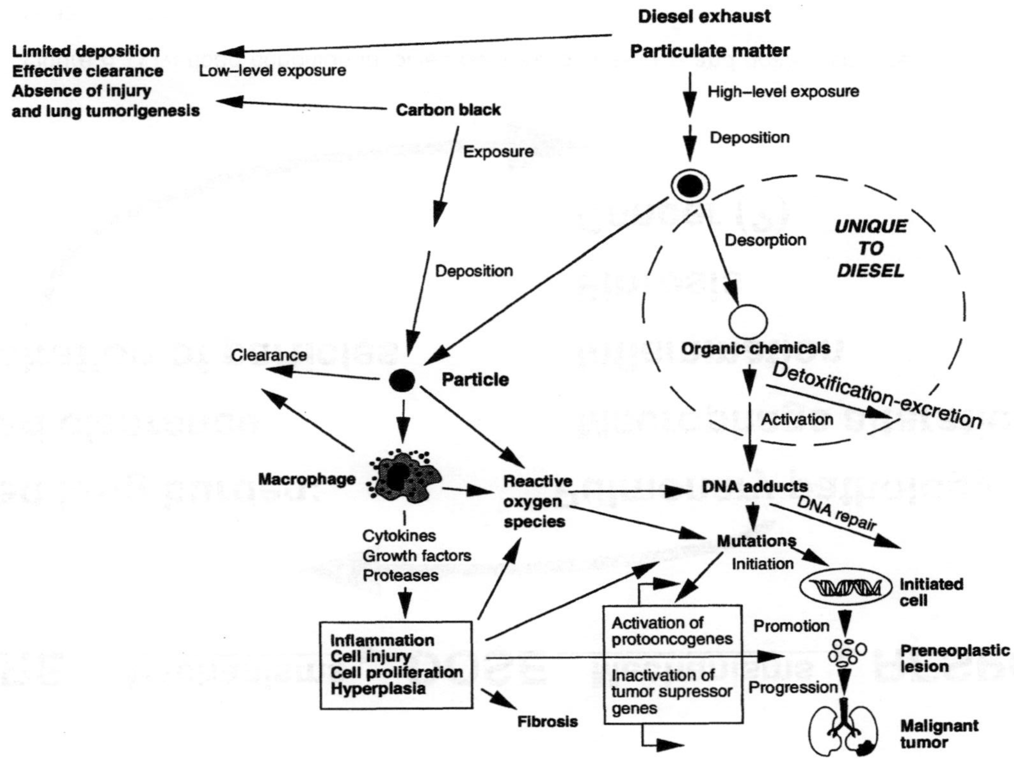

Figure 7. Schematic representation of the pathogenesis of lung cancer in rats with prolonged exposure of high concentrations of diesel exhaust or carbon black particles. From HEI (1995).

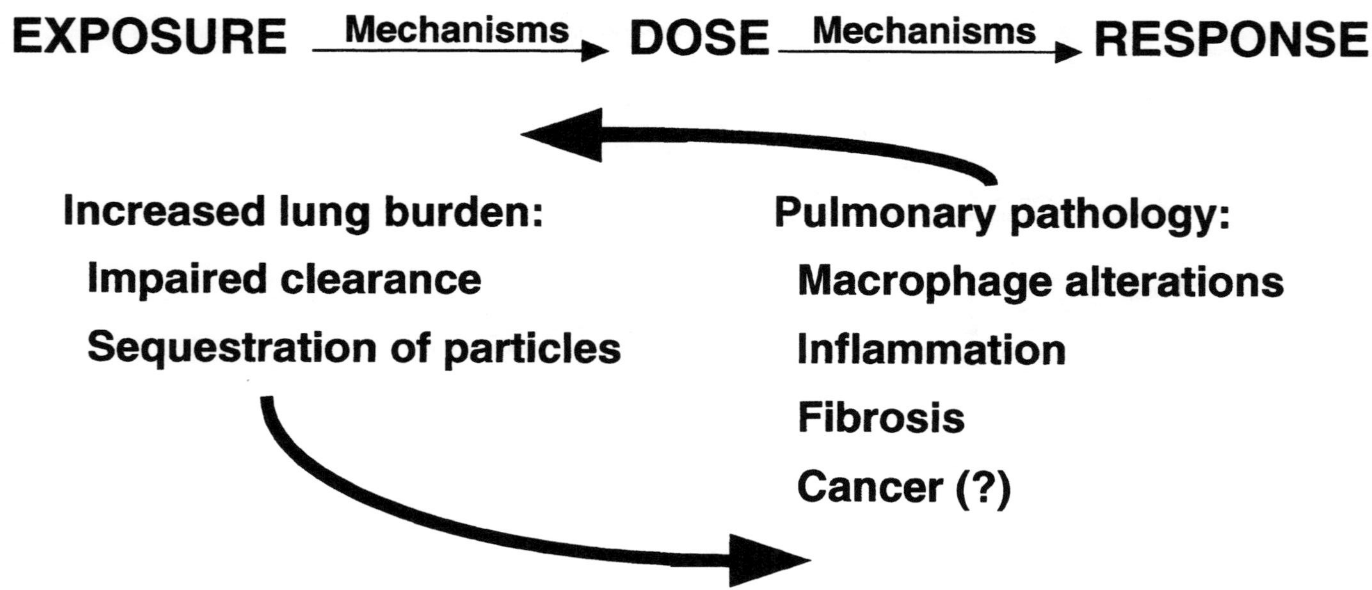

Figure 8. Overload condition involves both exposure-dose and dose-response relationships.

tumors with high-level exposures. This is very likely an oversimplification since an opportunity exists in many of the steps for repair or modulating processes to be operative and counter some of the effects. Indeed, the key difference between the effect of low- versus high-level exposure is related to the effectiveness of the clearance process as a modulating influence on lung burden and the opportunity for interactions between particles and inflammatory cells.

Several key issues in the hypothesized pathogenesis scheme are worthy of note. One issue relates to the modulation of reactive oxygen species, cytokines, growth factors, and proteases. A thesis advanced here is that a threshold exists for as-yet-not-identified cellular, biochemical, and molecular events that must be exceeded before the processes proceed to the induction of lung cancer (Warheit et al., 1996). For both carbon black and diesel soot particles, the process begins when the quantity of particles deposited and retained exceeds the normal protective capacity of the lung. I use the phrase *quantity of particles* in a broad sense, not being specific as to the particle metric that is most appropriate. The role of particle mass has been noted by Morrow (1988, 1994), while Driscoll (1996) and Oberdörster (1996) both note the close correlation between particle surface area and lung tumor induction. The close correlation between particle surface area and particle number is also important to recognize, leaving open the possibility that the number of particles insulting the lung rather than particle surface area is more critical.

The term *protective capacity* is intended to include both clearance mechanisms that are attempting to reduce the particle burden (and insult) and the cellular and molecular events that serve to counteract the effects of proteases, reactive oxygen species, and nitric oxide. All these factors have been postulated to have a role in inflammation and, most significantly for the present discussion, in producing mutations (Cerutti and Trump, 1991; Wink et al., 1991; Nguyen et al., 1992; Maki et al., 1992;

Driscoll et al., 1994; Driscoll, 1996; Oberdörster, 1994, 1996). The role of increased cell proliferation in amplifying mutations that occur spontaneously as well as mutational events initiated by agents of exogenous origin has been noted by a number of investigators (Ames and Gold, 1990; Cohen and Ellwein, 1991; Preston-Martin et al., 1990; and Butterworth et al., 1995). The significance of chemically induced cell proliferation for risk assessment was the topic of a major conference (Butterworth et al., 1991). Other individuals have argued that cell proliferation is not a major risk factor for induction of cancer (Ward et al., 1993; Huff, 1995; and Farber, 1995).

In considering the proposed scheme, it should be noted that not all exposures of rats to high concentrations of particles produce a high incidence of lung tumors. Muhle et al. (1991) and Bellmann et al. (1991) reported on a study in which rats were exposed to high concentrations (up to 16 mg/m^3) of toner (pigmented plastic) particles. A threshold was observed for induction of inflammation, fibrosis, and altered particle clearance, just as others have found for diesel exhaust soot and carbon black. Even at the highest concentration, however, they did not observe an increase in lung tumor incidence. Of significance in considering these findings is that the toner aerosol had a mass median aerodynamic diameter of 4.0 μm with a standard deviation of 1.5.* Thus the number of particles deposited and their surface area would be quite low as compared with similar lung burdens of carbon black or diesel exhaust soot particles, which are much smaller in size and also have a higher specific surface area.

An understanding of the specific mechanisms involved in induction of lung tumors in rats is crucial in deciding how these data can be used for assessing human lung cancer risk for exposure to these agents (McClellan, 1996; Driscoll, 1996; Watson and Valberg, 1996). This is schematically shown in Figure 9 with two hypothetical exposure-response relationships. One curve has a zero slope at low levels of exposure based on the assumption of a threshold for induction, inflammation,

* the geometric standard deviation

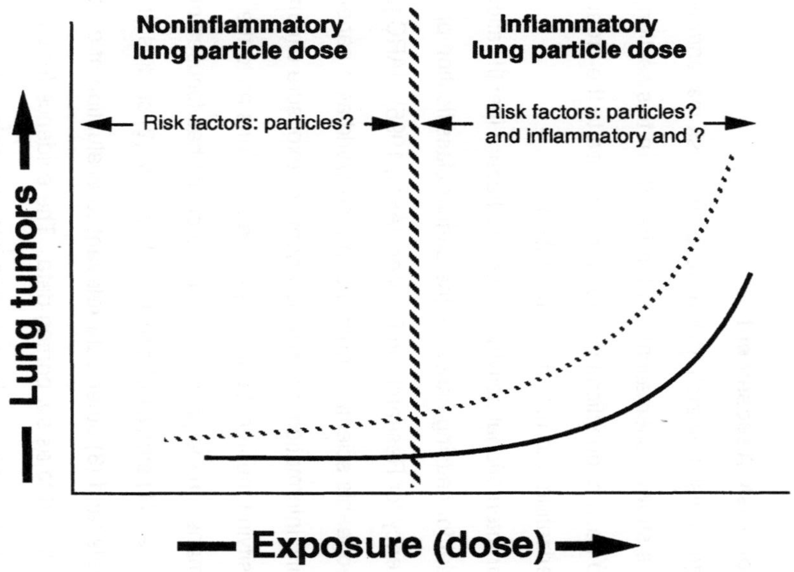

Figure 9. Schematic representation of exposure-response relationship for inhaled particles producing lung cancer via (a) nonspecific inflammation-mediated mechanism or (b) specific genotoxic mechanism.

mutations, and lung tumors. The other curve has a shallow slope at low levels of exposure since the toxicant is assumed to have some inherent genotoxicity that is manifest even in the absence of inflammation. The implications of these two alternative exposure-response relationships will be addressed in the next section of this article.

Implications for Risk Assessment

A paradigm of risk research, risk assessment, risk management, and risk communication is shown schematically in Figure 10. In this section, I will first discuss issues related to hazard identification and then proceed to the broader matters of hazard characterization and risk characterization.

The major international activity relating to hazard identification for chemicals and their potential for causing cancer is the cancer classification program of the International Agency for Research on Cancer (IARC, 1995). IARC periodically brings together knowledgeable scientists from around the world as a panel to review the available scientific information on specific agents or exposure circumstances for the purpose of classifying the agents and exposure situations as to the carcinogenic potential for humans. In the past, the review process has focused on consideration of three types of data: (1) studies of cancer in humans, (2) studies of cancer in laboratory animals, and (3) other data relevant to evaluation of carcinogenicity, which are frequently referred to as supporting data. The evidence for carcinogenicity in humans and laboratory animals is classified into four categories: sufficient evidence of carcinogenicity, limited evidence of carcinogenicity, inadequate evidence of carcinogenicity, or evidence suggesting lack of carcinogenicity. The relevant supporting data include consideration of preneoplastic lesions, tumor pathology, genetic and related effects, structure-activity relationships, metabolism and

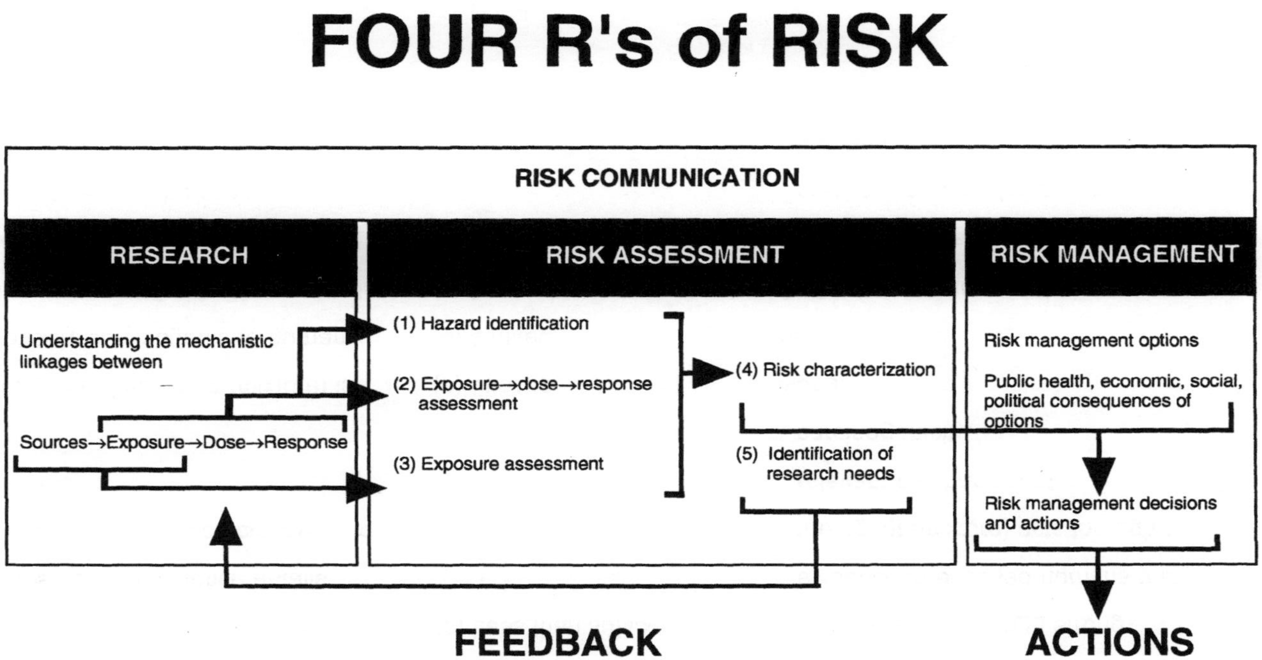

Figure 10. The four R's of risk. The risk research-risk assessment paradigm, as modified from McClellan and North (1994) and the original NRC-NAS paradigm (NRC, 1983).

Table 3. IARC carcinogen evaluation scheme[a].

Category	Human evidence	Experimental animal evidence
Group 1. The agent (mixture) is carcinogenic to humans. The exposure circumstance entails exposures that are carcinogenic to humans.	(a) Sufficient.	No animal evidence required.
	(b) Less than sufficient.	Sufficient evidence and strong evidence in exposed humans that the agent (mixture) acts through a relevant mechanism of carcinogenicity.
Group 2a. The agent (mixture) is probably carcinogenic to humans. The exposure circumstance entails exposures that are probably carcinogenic to humans.	(a) Limited.	None.
	(b) Limited.	Sufficient.
	(c) Inadequate.	Sufficient and strong evidence that the carcinogenesis is mediated by a mechanism that also operates in humans.

Table 3. (continued)

Group 2b. The agent (mixture) is possibly carcinogenic to humans. The exposure circumstance entails exposures that are possibly carcinogenic to humans.	(a) Limited.	Less than sufficient.
	(b) Inadequate.	Sufficient.
	(c) Inadequate.	Limited together with supporting evidence from other relevant data.
Group 3. The agent (mixture or exposure circumstance) is not classifiable as to its carcinogenicity.	(a) Inadequate.	Inadequate or limited.
	(b) Inadequate.	Sufficient and strong evidence that the mechanism of carcinogenicity in experimental animals does not operate in humans.
Group 4. The agent (mixture) is probably not carcinogenic to humans.	(a) Lack of carcinogenicity.	Lack of carcinogenicity.
	(b) Inadequate.	Lack of carcinogenicity consistently and strongly supported by a broad range of other relevant data.

[a]Abstracted from the Preamble of Volume 62 (IARC, 1995).

pharmacokinetics, physicochemical parameters, and analogous biological agents. The several types of data are then considered as a whole in arriving at an overall evaluation (Table 3). The IARC process is a qualitative assessment oriented to classifying an agent or exposure circumstance as to its human carcinogenicity. It explicitly does not seek to establish the carcinogenic potency of agents.

Since 1991, IARC has explicitly made provision for the use of mechanistic data in the classification process (IARC, 1992). Mechanistic data can be used to upgrade the classification of a chemical. For example, ethylene oxide was reclassified to Group 1 (a human carcinogen) from Group 2a (a probable human carcinogen) based on mechanistic considerations (IARC, 1994). Presumably a chemical might be placed in a lower category than would otherwise be the case based on mechanistic data. An example of such a situation, which draws on experience in the United States, is the review of chemicals shown to produce kidney tumors in male rats via a mechanism involving a male rat-specific protein, α_{2u}-globulin. The U.S. EPA (1991) reviewed the evidence for tumor induction by chemicals acting via interactions with α_{2u}-globulin, accumulation of the protein in the male rat kidney, toxicity, and, ultimately, kidney tumor formation and determined that this pathway is unlikely to be operative in humans. Consequently, the rat data need not be considered in assessing human cancer risks from chemicals shown to act via this mechanism. The basis for this decision has been strengthened by the finding of Borghoff and Lagarde (1993) that neither α_{2u}-globulin nor analogous proteins are present in humans.

In a previous review, diesel exhaust was classified as a probable human carcinogen (Group 2a) based on limited human evidence and sufficient laboratory animal evidence of carcinogenicity (IARC, 1989). Most of the evidence on the mode of induction of rat lung tumors by diesel exhaust soot has been obtained since that review. In addition, the recent HEI (1995) review of the epidemiological evidence raises questions as to the strength of the association between diesel engine exhaust exposure and human lung cancer. If diesel engine exhaust were being reviewed today by the IARC panel, several outcomes are possible. It could reaffirm a Group 2a categorization, or an argument could conceivably be made for downgrading to Group 2b based on a weaker epidemiological data set and lack of relevance of the rat data based on mechanistic considerations. Alternatively, some individuals might argue for classification in Group 1 based on the strength of the human evidence, even absent convincing data from the animal studies.

Carbon black was reviewed in the 1980s by IARC (1984, 1987) as to its carcinogenicity and was reviewed again in October 1995. The reviews in the 1980s placed it in Group 3, not classifiable as to carcinogenicity, based on inadequate evidence of both human and animal carcinogenicity.

The key epidemiological data considered in the 1987 review was a study by Hodgson and Jones (1985) of European carbon black workers. It did not show an association between carbon black exposure and lung cancer. More recently, Robertson (1996) summarized the data on North American carbon black workers and also noted no association between carbon black exposure and lung cancer. However, he noted the small population sizes and the low power of the studies to detect an increased risk attributable to carbon black exposures. An additional piece of evidence comes from the community-based epidemiological studies of Siemiatycki (1991). He noted a small increase in the relative risk of lung cancer for individuals who reported working in occupations with carbon black exposures. In my view, the Siemiatycki (1991) study should be viewed as being of an exploratory nature. It points to the need to conduct definitive studies in the industries suspected of yielding the excess lung cancer cases. Such studies could more adequately evaluate exposure to carbon black, possible confounders such as cigarette smoking, and exposure to other agents suspected of causing lung cancer.

Let us turn now to the animal data. In its most recent review of carbon black, IARC (1996) considered the body of epidemiological evidence for carbon black as a carcinogen in humans to be weak and the results conflicting. The review panel concluded that "there is *inadequate evidence* in humans for carcinogenicity of carbon black." The positive findings of excess lung cancer in rats in the well-conducted studies of Nikula et al. (1995) and Heinrich et al. (1995) were considered by the recent IARC (1996) "as *sufficient evidence* in experimental animals for the carcinogenicity of carbon black." This, coupled with insufficient evidence in humans, led to an overall evaluation "that carbon black is possibly carcinogenic to humans (Group 2b). The IARC (1996) report contains an excellent discussion of the current mechanistic data on carbon black induction of tumors in rats and the issue of their relevance to human risk assessment. After considering all the data, the IARC Working Group concluded, "Thus, whether humans respond to chronic inhalation of particles, including carbon black, more like a rat or more like a mouse cannot be decided at present." This conclusion strongly influenced the Group 2b (possibly carcinogenic to humans) classification. If the group had felt the mechanistic data were compelling in indicating that the mechanisms occurring in rats do not occur in humans, they could have classified carbon black as Group 3 (not classifiable as to carcinogenicity in humans). This is a position that Levi (1996) has advanced. Conversely, if they had concluded that the mechanisms operative in rats are operative in humans, they could have moved carbon black to a Group 2a classification (possibly carcinogenic to humans).

In another article (McClellan, 1994), I discussed the evolving cancer risk assessment practices of the U.S. EPA and recommendations of a recent committee of the National Research Council, National Academy of Science (NRC, 1994) concerning cancer classification schemes. The NRC has recommended and the EPA appears to be implementing a broader hazard characterization, an approach that goes beyond the relatively simple classification scheme of IARC, with its clear focus on hazard identification. A key aspect of this approach is the use of a narrative statement to complement any categorization scheme. Specifically, the NRC (1994) recommended that the statement contain at least the following: (1) "An evaluation of the strength of the available human and animal evidence"; (2) "A weight-of-evidence evaluation of any available information on the relevance to humans of the animal models used and the results obtained from them and on the conditions of exposure (route, dose, duration, and timing) under which carcinogenic responses to other conditions of exposure (usually conditions that could exist in human populations exposed environmentally) have been measured (either in humans or in laboratory animals)."

Based on the foregoing, I propose the narrative statements shown below for both agents.

Diesel Exhaust: Example Narrative Statement

Numerous epidemiological studies have been conducted on populations exposed for years to high concentrations of diesel exhaust during the course of employment in the railroad, trucking, and mining industries. Taken in aggregate, these studies suggest a weak association between exposure to diesel exhaust and an increased risk of lung cancer of 20% to 40%. The results of long-term studies with laboratory animals exposed to diesel exhaust have been variable. Studies with mice and hamsters have been negative or equivocal. Rats have developed an increased incidence of lung cancer; however, experimental data suggest that the lung cancers arise by mechanisms that are not likely to be relevant to humans and especially to those exposed for short periods of time or at low concentrations. Taken in aggregate, the data indicate that prolonged exposure to high levels of diesel exhaust (in excess of 100-1000 μg/m³) may be carcinogenic to humans. Such exposures should be avoided. Low levels of exposure (less than 100 μg/m³) typical of well-controlled occupational and environmental situations are unlikely to pose a carcinogenic risk.

Carbon Black: Example Narrative Statement

A limited number of epidemiologic studies of workers exposed to high levels of carbon black have been conducted and have not indicated any cancer risk attributable to the exposures; however, the small size of the populations under study limits their sensitivity for detecting a small increase in cancer risk. Rats exposed for their lifetimes to very high concentrations of carbon black develop an increased incidence of lung cancer. Experimental data suggest that the lung cancers arise by mechanisms not likely to be relevant to human populations and especially those exposed for short periods of time or at low concentrations. Taken in aggregate, the data indicate that carbon black exposures are not likely to pose a cancer risk even to occupationally exposed populations when exposures are controlled to less than 1000 μg/m³.

In my view, these narrative statements offer a better opportunity than the sole use of a classification category for communicating to the public and workers the potential lung cancer risk of exposure to either diesel exhaust soot or carbon black. The guidance level for diesel exhaust (100 μg/m³) has purposefully been set lower than for carbon black (1000 μg/m³) in recognition of the potential for some unquantifiable hazard related to the organic compounds associated with the diesel soot.

The foregoing discussion has focused on a qualitative hazard identification or characterization. An additional consideration is the potential for conducting a quantitative risk assessment. Based on knowledge of the induction mechanism of lung tumors in rats at high levels of exposure, I believe that quantitative extrapolation of these results to humans exposed to low levels of either diesel exhaust soot or carbon black would be inappropriate. If the rat data were to be extrapolated to humans, recognition that the rat tumors arise via a nongenotoxic mechanism argues for using a threshold exposure-response model rather than the linearized multistage model traditionally used for genotoxic carcinogens. Chen and Oberdörster (1996) have recently proposed such a model. The proposed revisions to the U.S. EPA (1994a) cancer risk assessment guidelines would appear to accommodate both the use of narrative hazard characterization statements and the use of models other than the linearized multistage extrapolation model.

Research Needs

Although much research has been conducted on both diesel exhaust and carbon black, a number of issues still exist that, if addressed, could serve to substantially reduce uncertainties in our current human health risks of these two airborne particulate materials. In establishing the research agenda, I advocate the use of Figure 10 as a framework. The risk assessment process should be used to identify the uncertainties, which would then serve as a basis for prioritizing the research to be conducted so that it has the greatest potential for impact. This approach requires that one consider whether the research addresses significant uncertainties in regard to human risk assessment as well as the potential for the research to be conducted and successfully carried to completion. In developing the research agenda, there is a need to recognize that the findings may affect how lung cancer risks for a broad range of particles are assessed (Vu, 1996). There is a critical need for such information, recalling that an increased incidence of lung tumors was observed in rats exposed to high concentrations of several different kinds of particles. These findings must be placed in perspective with regard to their relevance for assessing human cancer risks. In addition, recent reports suggest a range of human health effects associated with exposure to quite low concentrations of particles in the ambient air (Pope et al., 1995; Schwartz, 1994; Dockery et al., 1993; Dockery and Pope, 1994), findings that are not currently supported by biologically plausible mechanisms. Moolgavkar et al. (1995, 1996) reanalyzed a portion of the same data sets and come to different conclusions, namely, that the health effects are related to air pollution in general rather than specifically to the particulate fraction of the air pollution. Additional laboratory animal studies may help provide insight into the pathobiological basis for the human effects and thus contribute to the scientific basis for setting ambient air quality standards for particulate materials. The animal studies provide the opportunity for experimental testing of alternative hypotheses linking exposure to dose, to response.

The following list of research needs is reproduced from a recent review published elsewhere (McClellan, 1996, pp. 219-220).

1. Continued clinical surveillance and epidemiological follow-up of workers exposed to high levels of either diesel exhaust or carbon black and appropriate control populations should be carried out. Particular attention should be given to estimating exposures and to long-term follow-up of individuals who have left the industry.
2. Additional mechanistic studies of the pathogenesis of particle-induced lung cancer in rats using multiple exposure concentrations and long-term exposure would be valuable. Their value could be enhanced by the parallel conduct of studies of one or more species such as mice or Syrian hamsters that do not develop lung cancer with prolonged exposure to high levels of particulate

material. Such an approach should aid in identifying critical, obligatory steps in the pathogenesis of particle-induced lung cancer in rats.

3. Additional long-term, in-vivo mechanistic studies in laboratory animals should be conducted with aerosols that contain particles of varied size to elaborate the importance of particle size and other parameters.

4. In-vitro studies should be conducted with cells and lung tissue from mice, rats, and humans to elucidate possible species differences in various components of the hypothesized mechanisms by which prolonged, high-level exposure to particles causes cancer in rats and the extent to which these mechanisms occur in other species. Attention should be given to mechanisms of both injury and repair at exposure levels likely to be relevant to humans.

5. Investigations with contemporary molecular biology techniques should be conducted using specimens obtained from in-vivo, in-vitro, and ex-vivo studies with different kinds of particles to assess the specificity or lack of specificity of molecular markers of injury among the various kinds of particles. Designing and conducting studies using multiple exposure levels will be especially important to determine the influence of particle exposure rate and particle burden on phagocyte activation, release of reactive oxygen species, and other downstream events.

6. Controlled-exposure, human studies with radioactively tagged plastic particles and a subsequent long-term study of particle clearance should be conducted to provide improved input data for construction and validation of particle clearance models in humans.

7. Substantial attention should be given to synthesizing the data obtained from the complementary studies (epidemiological, controlled-exposure human, in-vivo animal, and in-vitro studies using both animal and human materials) into

models designed for estimating human cancer risk at relevant levels of human exposure.

The reader interested in considering research on these issues is also referred to the paper by Mauderly (1996), who identifies opportunities for resolution of the current uncertainties in using the rat data to assess human risk.

Summary

A remarkably large body of information is available today for use in evaluating the potential human cancer risk from exposure to diesel exhaust particles and carbon black. A substantial number of epidemiological studies with individuals occupationally exposed indicate a weak association between diesel exhaust exposure and increased incidence of lung cancer. A less extensive epidemiological data set for carbon black is negative with the exception of the results of an exploratory, community-based epidemiological study, which showed a weak association between carbon black exposure and increased risk of lung cancer.

Numerous studies have been conducted with mice, Syrian hamsters, and rats exposed to diesel exhaust. The studies with mice and Syrian hamsters have generally yielded negative findings, whereas an increased incidence of lung tumors has been observed in all the rat studies. Rats have also been observed to develop an increased incidence of lung tumors when exposed to carbon black and, most significantly, to other nuisance dusts.

A growing body of information is accumulating on the pathogenesis of rat lung tumors produced by long-term, high-concentration exposure to particles that do not contain chemicals directly interacting with DNA. The evidence indicates that high-level exposures to particles leads to alterations in clearance of particles from the lung, increased lung burdens of particles, persistent inflammation, damage to DNA by oxidants, increased lung cell mutations, increased cell proliferation, and, finally, lung tumors. These alterations observed with high-level exposures are not observed at low exposure concentrations, suggesting a threshold for injury. This leads to the conclusion that the exposure-response relationship observed at high exposure levels should not be extrapolated to low-level exposures typical of well-controlled work places or the general environment. Moreover, the likelihood that the effects are also specific to the rat raises a broader question as to the appropriateness of extrapolating the rat findings to other species, including humans.

References

Ames, B. N., and Gold, L. A. 1990. Chemical carcinogenesis: Too many rodent carcinogens. *Proc. Natl. Acad. Sci.* 87:7772-7776.

Bellmann, B., Muhle, H., Creutzenberg, O., Dasenbrock, C., Kilpper, R., MacKenzie, J.C., Morrow, P., and Mermelstein, R. 1991. Lung clearance and retention of toner, utilizing a tracer technique, during chronic inhalation exposure in rats. *Fundam. Appl. Toxicol.* 17:300-313.

Boffetta, P., Stellman, S. D., and Garfinkel, L. 1988. Diesel exhaust exposure and mortality among males in the American Cancer Society Prospective Study. *Am. J. Ind. Med.* 14:403-415.

Boffetta, P., Harris, R. E., and Wynder, E. L. 1990. Case-control study of occupational exposure to diesel exhaust and lung cancer risk. *Am. J. Ind. Med.* 17:577-591.

Borghoff, S. J., and Lagarde, W. H. 1993. Assessment of binding of 2,4,4-trimethyl-2-pentanol to low-molecular-weight proteins isolated from kidneys of male rats and humans. *Toxicol. Appl. Pharmacol.* 119:228-235.

Brightwell, J., Fouillet, X., Cassano-Zoppi, A. L., Bernstein, D., Crawley, F., Duchosal, F., Gatz, R., Percel, S., and Pfeifer, H. 1989. Tumours of the respiratory tract in rats and hamsters following chronic inhalation of engine exhaust emissions. *J. Appl. Toxicol.* 9:23-31.

Butterworth, B. E., Conolly, R. B., and Morgan, K. T. 1995. A strategy for establishing mode of action of chemical carcinogens as a guide for approaches to risk assessments. *Cancer Lett.* 93:129-146.

Butterworth, B. E., Slaga, T. J., Farland, W., and McClain, M. (eds.). 1991. *Chemically induced cell proliferation: implications for risk assessment.* New York: Wiley-Liss.

Burns, P. B., and Swanson, G. M. 1991. The Occupational Cancer Incidence Surveillance Study (OCISS): Risk of lung cancer by usual occupation and industry in the Detroit metropolitan area. *Am. J. Ind. Med.* 19:655-671.

California Environmental Protection Agency (CEPA). 1994. Health Risk Assessment for Diesel Exhaust (Preliminary Draft). Office of Environmental Health Hazard Assessment, Sacramento, CA.

Cerutti, P. A., and Trump, B. F. 1991. Inflammation and oxidative stress in carcinogenesis. *Cancer Cells* 3:1-7.

Chen, C.W., and Oberdörster, G. 1996. Selection of models for assessing dose-response relationships for particle-induced lung cancer. *Inhal. Toxicol.*, in press.

Cohen, S. M., and Ellwein, L. B. 1991. Genetic error, cell proliferation, and carcinogenesis. *Cancer Res.* 51:6493-6505.

Cohen, A. J., and Higgins, M. W. P. 1995. Health effects and diesel exhaust: Epidemiology. In *Diesel exhaust: A critical analysis of emissions, exposure, and health effects.* A Special Report of the Institute's Diesel Working Group, Health Effects Institute, Cambridge, MA, April 1995, pp. 251-292.

Dockery, D. W. and Pope, C. A., III. 1994. Acute respiratory effects of particulate air pollution. Annu. Rev. Public Health 15:107-132.

Dockery, D. W., Pope, C. A., III, Xu, X., Spengler, J. D., Ware, J. H., Fay, M. E., Ferris, B. G., Jr., and Speizer, F. E. 1993. An association between air pollution and mortality in six U.S. cities. N. Engl. J. Med. 329:1753-1759.

Driscoll, K. E. 1996. The role of inflammation in the development of rat lung tumors in response to chronic particle exposure. *Inhal. Toxicol.*, in press.

Driscoll, K. E., Maurer, J. K., Hassenbein, D., Carter, J., Janssen, Y. M. W., Mossman, B. T., Osier, M., and Oberdörster, G. 1994. Contribution of macrophage-derived cytokines and cytokine networks to mineral dust-induced lung inflammation. In

Toxic and carcinogenic effects of solid particles in the respiratory tract, eds. U. Mohr, D. L. Dungworth, J. L. Mauderly, and G. Oberdörster, pp. 177-189. Washington, DC: International Life Sciences Institute Press.

Drozdowicz, B. Z., Ulrich, C. E., Boothe, A. D., Procter, J. E., and Golberg, L. 1986. Chronic inhalation toxicity and carcinogenicity study of SRC-1 (solvent refined coal) solid in rats — Results from the 6- and 12-month interim sacrifices. *The Toxicologist* 6(1):132.

Farber, E. 1995. Cell proliferation as a major risk factor for cancer: a concept of doubtful validity. *Cancer Res.* 55:3759-3762.

Garshick, E., Schenker, M. B., Muñoz, A., Segal, M., Smith, T. J., Woskie, S. R., Hammond, S. K., and Speizer, F. E. 1987. A case-control study of lung cancer and diesel exhaust exposure in railroad workers. *Am. Rev. Respir. Dis.* 135:1242-1248.

Garshick, E., Schenker, M. B., Muñoz, A., Segal, M., Smith, T. J., Woskie, S. R., Hammond, S. K., and Speizer, F. E. 1988. A retrospective cohort study of lung cancer and diesel exhaust exposure in railroad workers. *Am. Rev. Respir. Dis.* 137:820-825.

Hall, N. E. L., and Wynder, E. L. 1984. Diesel exhaust exposure and lung cancer: A case-control study. *Environ. Res.* 34:77-86.

Health Effects Institute (HEI). 1995. *Diesel exhaust: A critical analysis of emissions, exposure, and health effects.* A Special Report of the Institute's Diesel Working Group, Health Effects Institute, Cambridge, MA, April 1995.

Heinrich, U. 1996. Comparative response to long-term particle exposure among rats, mice and hamsters. *Inhal. Toxicol.* 8 (Suppl.):51-72.

Heinrich, U., Fuhst, R., Rittinghausen, S., Creutzenberg, O., Bellmann, B., Koch, W., and Levsen, K. 1995. Chronic inhalation exposure of Wistar rats and two different strains of mice to diesel engine exhaust, carbon black, and titanium dioxide. *Inhalation Toxicol.* 7:533-556.

Heinrich, U., Muhle, H., Takenaka, S., Ernst, E., Fuhst, R., Mohr, U., Pott, F., and Stöber, W. 1986. Chronic effects on the respiratory tract of hamsters, mice and rats after long-term inhalation of high concentrations of filtered and unfiltered diesel engine emissions. *J. Appl. Toxicol.* 6:383-395.

Henderson, R. F., Pickrell, J. A., Jones, R. K., Sun, J. D., Benson, J. M., Mauderly, J. L., and McClellan, R. O. 1988. Response of rodents to inhaled diluted diesel exhaust: Biochemical and cytological changes in bronchoalveolar lavage fluid and in lung tissue. *Fundam. Appl. Toxicol.* 11:546-567.

Hodgson, J. T., and Jones, R. D. 1985. A mortality study of carbon black workers employed at five United Kingdom factors between 1917 and 1989. *Arch. Environ. Health* 40:261-266.

Holland, L. M., Wilson, J. S., Tillery, M. I., and Smith, D. M. 1985. Lung cancer in rats exposed to fibrogenic dusts. In *Silica, silicosis and cancer*, eds. D. Goldsmith, D. Winn, and C. Shy, pp. 267-279. New York: Praeger.

Howe, G. R., Fraser, D., Lindsay, J., Presnal, B., and Yu, S. Z. 1983. Cancer mortality (1965-77) in relation to diesel fume and coal exposure in a cohort of retired railway workers. *JNCI* 70:1015-1019.

Huff, J. 1995. Mechanisms, chemical carcinogenesis and risk assessment: cell proliferation and cancer. *Am. J. Ind. Med.* 27:293-300.

International Agency for Research on Cancer (IARC). 1984. *IARC Monographs on the Evaluation of the Carcinogenic Risk of Chemicals to Humans, Volume 33, Polynuclear Aromatic Hydrocarbons, Part 2, Carbon Blacks, Mineral Oils (Lubricant Base Oils and Derived Products) and Some Nitroarenes*, IARC

Working Group on the Evaluation of Carcinogenic Risk of Chemicals to Humans, Lyon, France, June 7-14, 1983, 245 pp.

International Agency for Research on Cancer (IARC). 1987. Supplement No. 7, *Overall Evaluations of Carcinogenicity: An Updating of IARC Monographs Volumes 1-42*, 440 pp.

International Agency for Research on Cancer (IARC). 1989. *IARC Monographs on the Evaluation of Carcinogenic Risks to Humans, Volume 46, Diesel and Gasoline Engine Exhausts and Some Nitroarenes*, Lyon, France, 458 pp.

International Agency for Research on Cancer (IARC). 1992. No. 116, *Mechanisms of Carcinogenesis in Risk Identification* (H. Vainio, P.N. Magee, D.B. McGregor, and A.J. McMichael, Eds.), 616 pp.

International Agency for Research on Cancer (IARC). 1994. *IARC Monographs on the Evaluation of Carcinogenic Risks to Humans, Volume 60, Some Industrial Chemicals*, Lyon, France.

International Agency for Research on Cancer (IARC). 1995. *IARC Monographs on the Evaluation of Carcinogenic Risks to Humans, Volume 62, Wood Dust and Formaldehyde*. IARC Working Group on the Evaluation of Carcinogenic Risks to Humans, Lyon, France, October 11-18, 1994, 406 pp.

International Agency for Research on Cancer (IARC). 1996. *IARC Monographs on the Evaluation of Carcinogenic Risks to Humans, Volume 65, Printing Processes, Printing Inks, Carbon Black, and Some Nitro Compounds,* Lyon, France.

Ishinishi, N., Koizumi, A., McClellan, R. O., and Stöber, W., Eds. 1986a. Carcinogenic and mutagenic effects of diesel engine exhaust. In *Proceedings of the International Satellite Symposium on Toxicological Effects of Emissions from Diesel Engines* held in Tsukuba Science City, Japan, July 26-28, 1986. Amsterdam: Elsevier.

Ishinishi, N., Kuwabara, N., Nigase, S., Suzuki, T., Ishiwata, S., and Kohno, T. 1986b. Long-term inhalation studies on effects of exhaust from heavy and light duty diesel engines on F344 rats. In *Carcinogenic and mutagenic effects of diesel engine exhaust*, eds. N. Ishinishi, A. Kozumi, R. O. McClellan, and W. Stöber, pp. 329-348. New York: Elsevier Science Publishers.

Iwai, K., Udagawa, T., Yamagishi, M., and Yamada, H. 1986. Long-term inhalation studies of diesel exhaust on F344 SPF rats. Incidence of lung cancer and lymphoma. In *Carcinogenic and mutagenic effects of diesel engine exhaust*, eds. N. Ishinishi, A. Koizumi, R., McClellan, and W. Stöber, pp. 349-360. Amsterdam: Elsevier.

Kawabata, Y., Iwai, K., Udagawa, T., Tukagoshi, K., and Kazue, H. 1986. Effects of diesel soot on unscheduled DNA synthesis of tracheal epithelium and lung tumor formation. In *Carcinogenic and mutagenic effects of diesel engine exhaust*, eds. N. Ishinishi, A. Koizumi, R., McClellan, and W. Stöber, pp. 213-222. Amsterdam: Elsevier.

Lee, K. P., Trochimowicz, H. J., and Reinhardt, C. F. 1985. Pulmonary response of rats exposed to titanium dioxide (TiO_2) by inhalation for two years. *Toxicol. Appl. Pharmacol.* 79:179-192.

Levi, L. S. 1996. Differences between rodents and humans in lung tumour response — lessons from recent studies with carbon black. *Inhal. Toxicol.*, in press.

Lewtas, J., Ed. 1982. Toxicological effects of emissions from diesel engines. In *Proceedings of the Environmental Protection Agency 1981 Diesel Emissions Symposium,* held in Raleigh, NC, USA, October 5-7, 1981, 380 pp. New York: Elsevier Biomedical.

Maki, A., Berezesky, I. K., Iargnoli, J., Holbrook, N. J., and Trump, B. F. 1992. Role of [Ca^{++}] in induction of c-fos, c-jun and c-myc mRNA in rat PTE after oxidative stress. *FASEB J.* 6:919-924.

Martin, J.C., Daniel, H., and Le Bouffant, L. 1977. Short- and long-term experimental study of the toxicity of coal-mine dust and some of its constituents. In *Inhaled particles IV*, Vol. 1, ed. W. H. Walton, pp. 361-371. Oxford: Pergamon.

Mauderly, J. L. 1992. Diesel exhaust. In *Environmental toxicants — Human exposures and their health effects,* ed. M. Lippmann, Chapter 5, pp. 119-162. New York: Van Nostrand Reinhold.

Mauderly, J. L. 1996. Lung overload: the dilemma and opportunities for resolution. *Inhal. Toxicol.* 8(Suppl.): in press.

Mauderly, J. L., Banas, D. A., Griffith, W. C., Hahn, F. F., Henderson, R. F., and McClellan, R. O. 1996. Diesel exhaust is not a pulmonary carcinogen in CD-1 mice exposed under conditions carcinogenic to F344 rats. *Fundam. Appl. Toxicol.* (in press).

Mauderly, J. L., Cheng, Y. S., and Snipes, M. B. 1990. Particle overload in toxicological studies: friend or foe? *J. Aerosol Med.* 3:5169-5187.

Mauderly, J. L., Gillett, N. A., and Henderson, R. F. 1988. Relationship of lung structural and functional changes to accumulation of diesel exhaust particles. *Ann. Occup. Hyg.* 32:659-668.

Mauderly, J. L., Jones, R. K., Griffith, W. C., Henderson, R. F., and McClellan, R. O. 1987. Diesel exhaust is a pulmonary carcinogen in rats exposed chronically by inhalation. *Fundam. Appl. Toxicol.* 9:208-221.

Mauderly, J. L., Snipes, M. B., Barr, E. B., Belinsky, S. A., Bond, J. A., Brooks, A. L., Chang, L.-Y., Cheng, Y. S., Gillett, N. A., Griffith, W. C., Henderson, R. F., Mitchell, C. E., Nikula, K. J., and Thomassen, D. G. 1994. Pulmonary Toxicity of

Inhaled Diesel Exhaust and Carbon Black in Chronically Exposed Rats, Part 1, Neoplastic and Nonneoplastic Lung Lesions. Research Report Number 68, Health Effects Institute, Cambridge, MA.

McClellan, R. O. 1986a. Conclusions and recommendations: Toxicological effects of emissions from diesel engines. In *Carcinogenic and mutagenic effects of diesel engine exhaust*, eds. N. Ishinishi, A. Koizumi, R.O. McClellan, W. Stöber, pp. 529-534. Proceedings of the International Satellite Symposium on Toxicological Effects of Emissions from Diesel Engines held in Tsukuba Science City, Japan, July 26-28, 1986. Amsterdam: Elsevier.

McClellan, R. O. 1986b. Health effects of diesel exhaust: A case study in risk assessment. *Amer. Ind. Hyg. Assoc. J.* 47:1-13.

McClellan, R. O. 1986c. Opening remarks: Toxicological effects of emissions from diesel engines. In *Carcinogenic and mutagenic effects of diesel engine exhaust*, eds. N. Ishinishi, A. Koizumi, R. O. McClellan, W. Stöber, pp. 3-8. Proceedings of the International Satellite Symposium on Toxicological Effects of Emissions from Diesel Engines held in Tsukuba Science City, Japan, July 26-28, 1986. Amsterdam: Elsevier.

McClellan, R. O. 1987. Health effects of exposure to diesel exhaust particles. *Ann. Rev. Pharmacol. Toxicol.* 27:279-300.

McClellan, R.O. 1990. Particle overload in the lung: Approaches to improving our knowledge. *J. Aerosol Med.* (Suppl.) 3:197-207.

McClellan, R. O. 1994. A Commentary on the NRC Report "Science and Judgment in Risk Assessment". *Reg. Toxicol. Pharmacol.* 20:S142-S168.

McClellan, R. O. 1996. Lung cancer in rats from prolonged exposure to high concentrations of carbonaceous particles: implications for human risk assessment. *Inhal. Toxicol.* 8 (Suppl.):193-226.

McClellan, R. O., Bice, D. E., Cuddihy, R. G., Gillett, N. A., Henderson, R. F., Jones, R. K., Mauderly, J. L., Pickrell, J. A., Shami, S. G., and Wolff, R. K. 1986. Health effects of diesel exhaust. In: *Aerosols* (S. D. Lee, T. Schneider, L. D. Grant, and P. J. Verkerk, eds.), pp. 597-615, Lewis Publishers, Chelsea, Michigan.

McClellan, R. O., Cuddihy, R. G., Griffith, W. C., and Mauderly, J. L. 1989. Integrating diverse data sets to assess the risk of airborne pollutants. In *Assessment of inhalation hazards: Integration and extrapolation using diverse data*, eds. U. Mohr, D. V. Bates, D. L. Dungworth, P. N. Lee, R. O. McClellan, and F. J. C. Roe, pp. 3-22, ILSI Monographs. Heidelberg: Springer-Verlag.

McClellan, R. O., and North, D. W. 1994. Making full use of scientific information in risk assessment. In *Science and judgment in risk assessment*, Appendix N-2, pp. 629-640. National Research Council, Committee on Risk Assessment of Hazardous Air Pollutants, Washington, DC: National Academy Press.

McConnell, E. E. 1996. Maximum tolerated dose in particulate inhalation studies. A pathologist's point of view. *Inhal. Toxicol.* 8 (Suppl.): in press.

Moolgavkar, S. H., Luebeck, E. G., Hall, T. A., and Anderson, E. L. 1995. Particulate air pollution, sulfur dioxide, and daily mortality: a reanalysis of the Steubenville data. *Inhalation Toxicol.* 7:35-44.

Moolgavkar, S. H., Luebeck, E. G., Hall, T. A., and Anderson, E. L. 1996. Air pollution and daily mortality in Philadelphia. *J. Exposure Anal. Environ. Epidemiol.*, submitted.

Morrow, P. E. 1988. Possible mechanisms to explain dust overloading of the lungs. *Fundam. Appl. Toxicol.* 10:369-384.

Morrow, P. E. 1994. Mechanisms and significance of "particle overload." In *Toxic and carcinogenic effects of solid particles in the respiratory tract*, eds. Mohr, U.,

Dungworth, D. L., Mauderly, J. L., Oberdörster, G., pp. 17-25. Washington, DC: International Life Sciences Institute Press.

Muhle, H., Bellmann, B., Creutzenberg, O., Dasenbrock, C., Ernst, H., Kilpper, R., MacKenzie, J. C., Morrow, P., Mohr, U., Takenaka, S., and Mermelstein, R. 1991. Pulmonary response to toner upon chronic inhalation exposure in rats. *Fundam. Appl. Toxicol.* 17:280-299.

National Institute of Occupational Safety and Health (NIOSH). 1988. *Carcinogenic effects of exposure to diesel exhaust.* NIOSH Current Intelligence Bulletin 50. DHHS (NIOSH) Publication No. 88-116, Center for Disease Control, Atlanta, GA.

National Research Council (NRC). 1981. *Health effects of exposure to diesel exhaust: Impacts of diesel-powered light-duty vehicles.* Washington, DC: National Academy Press.

National Research Council (NRC). 1983. Committee on the Institutional Means for Assessment of Risks to Public Health: *Risk assessment in the federal government: Managing the process.* Washington, DC: National Academy Press.

National Research Council (NRC). 1994. Committee on Risk Assessment of Hazardous Air Pollutants: *Science and judgment in risk assessment.* Washington, DC: National Academy Press.

Nauss, K. M., and The HEI Diesel Working Group. 1995. Critical issues in assessing the carcinogenicity of diesel exhaust: A synthesis of current knowledge. In *Diesel exhaust: A critical analysis of emissions, exposure, and health effects.* Health Effects Institute, Cambridge, MA, pp. 11-61.

Nguyen, T., Brunson, D., Crespi, C. L., Penman, B. W., Wishnok, J. S., and Tannenbaum, S. R. 1992. DNA damage and mutation in human cells exposed to nitric oxide *in vitro. Biochemistry* 285:1173-1180.

Nikula, K. J., Snipes, M. B., Barr, E. B., Griffith, W. C., Henderson, R. F., and Mauderly, J. L. 1995. Comparative pulmonary toxicities and carcinogenicities of chronically inhaled diesel exhaust and carbon black in F344 rats. *Fundam. Appl. Toxicol.* 25:80-94.

Oberdörster, G. 1994. Extrapolation of results from animal inhalation studies with particles to humans? In *Toxic and carcinogenic effects of solid particles in the respiratory tract*, eds. U. Mohr, D. L. Dungworth, J. L. Mauderly, and G. Oberdörster, pp. 335-353. Washington, DC: International Life Sciences Institute Press.

Oberdörster, G. 1996. Significance of particle parameters in the evaluation of exposure-dose-response relationships of inhaled particles. *Inhal. Toxicol.* 8 (Suppl.): in press.

Pepelko, W. E. 1996. Evaluation of the carcinogenic risk of biochemically inert, insoluble particles by the EPA using rat inhalation data. *Inhal. Toxicol.*, in press.

Pepelko, W. E., Danner, R. M., and Clarke, N. A., eds. 1980. *Health effects of diesel engine emissions*, Vol. 1 and 2, U.S. Environmental Protection Agency.

Pope, C. A., III, Dockery, D. W., and Schwartz, J. 1995. Review of epidemiological evidence of health effects of particulate air pollution. *Inhalation Toxicol.* 7:1-18.

Preston-Martin, S., Pike, M. C., Ross, R. K., Jones, P. A., and Henderson, B. E. 1990. Increased cell division as a cause of human cancer. *Cancer Res.* 50:7415-7421.

Robertson, J. McD. 1996. Epidemiologic studies in North American carbon black workers. *Inhal. Toxicol.*, in press.

Sawyer, R. F., and Johnson, J. H. 1995. Diesel emissions and control technology. In *Diesel exhaust: A critical analysis of emissions, exposure, and health effects.* A Special Report of the Institute's Diesel Working Group, April 1995, pp. 66-81. Cambridge, MA: Health Effects Institute.

Schenker, M. B., Smith, T., Muñoz, A., Woskie, S., and Speizer, F. E. 1984. Diesel exposure and mortality among railway workers: result of a pilot study. *Br. J. Ind. Med.* 41:320-327.

Schwartz, J. 1994. Air pollution and daily mortality: a review and meta analysis. *Environ. Res.* 64:36-52.

Siemiatycki, J., ed. 1991. *Risk factors for cancer in the workplace.* Boca Raton, FL: CRC Press.

Snipes, M. B. 1996. Current information on lung overload in non-rodent mammals: contrast with rats. *Inhal. Toxicol.* 8 (Suppl.):91-140.

U.S. Environmental Protection Agency (U.S. EPA). 1986. *Guidelines for carcinogen risk assessment. Fed. Reg.* 51:33992-34003.

U.S. Environmental Protection Agency (U.S. EPA). 1990. *Health assessment document for diesel emissions*, Workshop Review Draft, EPA-600/8-90/057A. Office of Research and Development, Office of Health and Environmental Assessment, Washington, DC.

U.S. Environmental Protection Agency (U.S. EPA). 1991. *Alpha-2u-globulin: association with chemically-induced renal toxicity and neoplasia in the male rat*, EPA-625/3-91/019F. U.S. Environmental Protection Agency, Washington, DC.

U.S. Environmental Protection Agency (U.S. EPA). 1994a. *Draft revisions to the guidelines for carcinogen risk assessment:* External Review draft, August 1994. EPA/600/BP-92/003. Office of Research and Development, Washington, DC.

U.S. Environmental Protection Agency (U.S. EPA). 1994b. *Health assessment document for diesel emissions*: External Review Draft. EPA 600/8-90/057. Office of Research and Development, Washington, DC.

Vostal, J. J. 1986. Factors limiting the evidence for chemical carcinogenicity of diesel emissions in long-term inhalation experiments. In *Carcinogenic and mutagenic effects of diesel engine exhaust*, eds. N. Ishinishi, A. Koizumi, R. O. McClellan, W. Stöber, pp. 381-396, Proceedings of the International Satellite Symposium on Toxicological Effects of Emissions from Diesel Engines held in Tsukuba Science City, Japan, July 26-28, 1986. Amsterdam: Elsevier.

Vu, V. T. 1996. Use of hazard and risk information in risk management decisions: solid particles and fibers under EPA's TSCA and EPCRA. *Inhal. Toxicol.* 8 (Suppl.):181-192.

Ward, J. M., Uno, H., Kurata, Y., Weghost, C. M., and Jang, J.-J. 1993. Cell proliferation not associated with carcinogenesis in rodents and humans. *Environ. Health Perspect.* 101 (Suppl. 5):125-136.

Warheit, D. B., Hansen, J. F., Yuen, I. S., Snajdr, S. I., and Hartsky, M. A. 1996. Prolonged cellular proliferation and pulmonary inflammation is produced in rats inhaling high concentrations of low toxicity, insoluble particulates: comparisons with cytotoxic dusts. *Inhal. Toxicol.* 8 (Suppl.):155-168.

Watson, A. Y. and Valberg, P. A. 1996. Particle-induced lung tumors in rats: evidence for species-specificity in mechanisms. *Inhal. Toxicol.* 8 (Suppl.):227-258.

Williams, R. R., Stegens, N. L., and Goldsmith, J. R. 1977. Associations of cancer site and type with occupation and industry from the Third National Cancer Survey Interview. *JNCI* 59:1147-1185.

Wink, D. A., Kasprzak, K. S., Maragos, C. M., Elespuru, R. K., Misra, M., Dunams, T. M., Cebula, T. A., Koch, W. H., Andrews, A. W., Allen, J. S., et al. 1991. DNA deaminating ability and genotoxicity of nitric oxide and its progenitors. *Science* 254:1001-1003.

Wolff, R. K., Henderson, R. F., Snipes, M. B., Griffith, W. C., Mauderly, J. L., Cuddihy, R. G., and McClellan, R. O. 1987. Alterations in particle accumulation and clearance in lungs of rats chronically exposed to diesel exhaust. *Fundam. Appl. Toxicol.* 9:154-166.

Yu, C. P. 1996. Extrapolation modeling of particle deposition and retention from rats to humans. *Inhal. Toxicol.* 8 (Suppl.):279.

Measurement of Highly Concentrated Fine Aerosols

Wladyslaw Witold Szymanski

Institute of Experimental Physics, University of Vienna
Boltzmanngasse 5, A-1090 Vienna, Austria

Keywords

Fine Aerosols, Nanometer-Sized Particles, Airborne Particulate Matter, Optical Measurement, Particle Sizing, Absolute Concentration Determination, In-Line Monitoring

Introduction

The presence of airborne particulate contaminants and exposure of humans to the elevated levels of such pollutants has been linked to a wide spectrum of diseases. The knowledge of airborne particle sizes and concentrations is substantial for the assessment of a possible health hazard, as well as for the biomedical research linked to inhalation and regional deposition of those contaminants within the human respiratory system.

The size range of common airborne particulate pollutants - aerosols - extends from about 0.001 um in particle radius, for e.g. combustion nuclei or metallurgical fumes, over more than 5 orders of magnitude. Observing characteristic ambient particle size distributions, frequently more than 50% of the total ambient mass can be related to fine aerosols - particles smaller than about 1 um radius. Furthermore, typically over 80% of the total number of atmospheric particles is smaller than about 0.1 um (Whitby and Cantrell, 1976, Sioutas et al. 1994). Moreover, there is a reason for the assumption that fine aerosols are responsible for excess mortality in urban regions. Recently established correlation between an increased mortality and the presence of remarkable concentrations of fine particles in a six city study in the US (Dockery et al.,1993) motivated an increased interest in the health impact of fine aerosols. However, concentration determination of fine aerosols based on current mass concentration standards is facing a severe measurement problem, since number concentration of 10^6 cm^{-3} of 0.01 um particles in diameter yields a mass concentration of about only 500 ng / m^3.

Because health effects and toxicology of aerosols seem to depend on the total number of particles deposited in the respiratory system on a unit airway surface per breath (Hofmann, 1992), there is an increasing concern regarding the particle residence time in the pulmonary system and the catalytic activity of fine aerosols down to nanometer-sized particles (Friedlander, 1995). At least for those reasons it appears advantageous to utilize techniques allowing real-time precise measurement of aerosols over an extensive range of concentrations and capable of detecting and sizing of airborne particles including nanometer-sized aerosols.

Among various methods for measurement of aerosol sizes and concentrations, the intensity based optical techniques provide basis to perhaps most widely used real-time measuring instrumentation. The rationale for that is the possibility of an in situ and real-time determination of aerosol size distributions and concentrations, whereby the accessible

J. C. M. Marijnissen and L. Gradoń (eds.), Aerosol Inhalation: Recent Research Frontiers, 79–90.

size range starts with about 0.001um in particle radius. Optically determined particle size corresponds well to the geometrical particle size for not too irregular particles (Gebhart, 1991; Liu et al. 1986). Since recent research indicate that lung retention is rather a function of the geometrical and not the aerodynamic particle size (Stahlhofen, 1994), it seems useful to explore the opportunities provided by the optical particle measurement.

At comparatively low concentrations, single optical particle measurement can be applied (Gebhart, 1986; Szymanski, 1986; Wen and Kasper, 1986). However, for concentrations beyond several 10^3 cm^{-3} optical integral measuring methods have to be used. Those techniques require usually an empirical calibration procedure. Furthermore, with increasing aerosol concentrations the influence of multiple scattering (Szymanski, 1992) may affect the measurement to the extend that data interpretation becomes rather difficult. The uppermost directly measurable particle number concentrations are currently of the order of 10^8 cm^{-3}. The purpose of this article is to discuss some direct-reading optical methods yielding the absolute determination of particle number concentration, the assessment of particle sizes, as well as the possible influence of multiple scattering on optical particle measurement.

Light intensity based real-time measurement techniques

The instrumentation for characterization of aerosol sizes and concentrations by means of optical, intensity based methods can be divided into two classes: single particle measuring devices and collective particle measuring devices. Assuming that particle sampling and transport is not biasing the measurement, the quality of information from an optical measurement depends firstly on the ability of the instrument to detect all particles in question, particularly at the lower sensitivity limit of an instrument. Furthermore, the response of an optical measuring system may depend greatly on the optical properties of the investigated aerosol and finely the range of concentrations in which a reliable measurement can be performed must be established. The determination of these conditions is crucial, because many instruments would „merrily report erroneous data and not notifying the user“ (Hirleman, 1988). In following, based on the above named criteria, some selected optical measuring systems will be presented and analyzed for their capability to measure high concentrations and sizes of fine aerosols.

a. Single optical particle measurement

Single optical particle counters detect the scattered light intensity from one particle at a time, while the particle is passing through a well defined sensing volume illuminated by a light source such as laser. There are generally two optical designs for measurement of the scattered light intensity - the forward angle scattering and the wide angle scattering. The forward scattering range is usually confined to narrow angles, the wide angle scattering range allows a variety of either on- or off-axis designs with respect to the optical axis - the direction of the incoming radiation (Gebhart, 1986; Szymanski and Liu, 1986). In an ideal case a response of an instrument would be a monotonic function of the particle size independent on the optical property - refractive index - of particles. In reality, depending on the optical configuration, different optical behavior can be expected from a measuring system when the refractive index of particles varies.

In following the sensitivity of two optical designs to the absorbing and non-absorbing particles will be shown. Although this discussion does not refer to any particulate instrument the results will be representative for any instrument with a similar optical design.

The response R is defined here by the scattered light flux normalized to the incident radiation. For an unpolarized laser beam irradiating a particle the response R is given by:

$$R = \frac{\lambda^2}{8\pi^2} \cdot \int_{\theta_1}^{\theta_2} \left(I_1(x,m,\theta) + I_2(x,m,\theta) \right) \sin\theta \, d\theta \qquad (1)$$

where I_1 and I_2 are scattered light intensities with parallel and perpendicular polarization, respectively. m is the refractive index and x is the size parameter defined by $\pi Dp/\lambda$, with Dp being the particle diameter and λ the wavelength if the irradiation. θ_1,θ_2 are scattering angles limiting the light collecting solid angle. The calculations here are based on the Mie theory of light scattering (van d. Hulst, 1957). The response for non-absorbing particles with the complex refractive index of m = 1.50 - 0.0i and for absorbing particles with m = 1.5 - 1.0i shown below emphasizes the influence of optical properties on particle sizing.

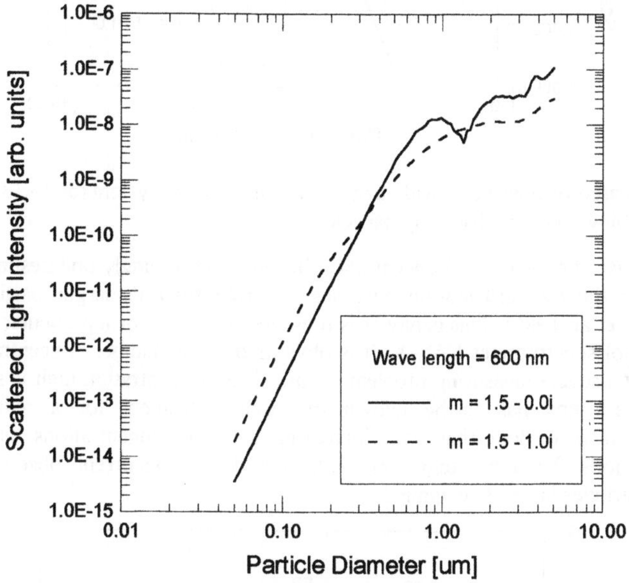

Fig. 1. Calculated response of a forward scattering optical system (10° - 30°) for absorbing and non-absorbing particles.

It can be seen in Fig.1, that the forward scattering design (10° - 30°) is more suitable for measurement of absorbing particles, whereby for non-absorbing particles a multi-valued response for particle sizes of the order of 1 um results in a substantial sizing error. Analyzing the performance of a wide angle scattering system (50° - 130) it is apparent that such a design provides an accurate sizing possibility for non-absorbing fine aerosols and rather limited size resolution for absorbing particles in the size range about 0.2 - 0.7 um, as shown in the Fig.2. It is also worthwhile to note that the lower sensitivity of both optical systems discussed here improves for absorbing particles compared to non-absorbing particles. This has to be taken into account particularly for concentration measurement at a lower sizing limit of any single optical particle counter. When operated under proper conditions optical particle counters provide a direct measurement of number concentration. However, with increasing concentration those instruments are prone to coincidence errors, which occur when more than one particle is present in the sensing volume at the same time.

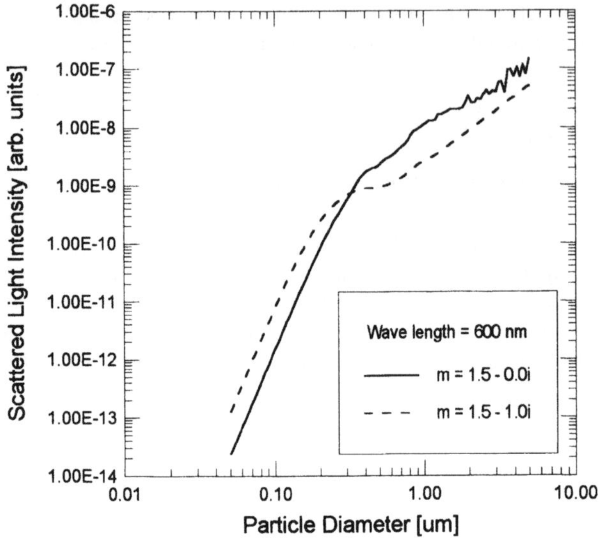

Fig. 2. Calculated response of a wide angle scattering optical system (50° - 130°) to absorbing and non-absorbing particles.

In such a case the measured concentration becomes increasingly underestimated and at the same time the size distribution becomes distorted towards larger particles (Liu at al.,1986, Janka et al.,1986). This behavior is presented in Fig.3. for a particular instrument - the laser aerosol spectrometer LAS-X. It is obvious that for number concentrations above about 10^4 cm^{-3} a severe measuring problem occurs. This concentration limit is specific for a given instrument and has to be known for each instrument for a reliable aerosol measurement. It is evident that with increasing aerosol concentrations single particle detection methods fail and integral methods, based on concurrent observation of an ensemble of particles have to be applied.

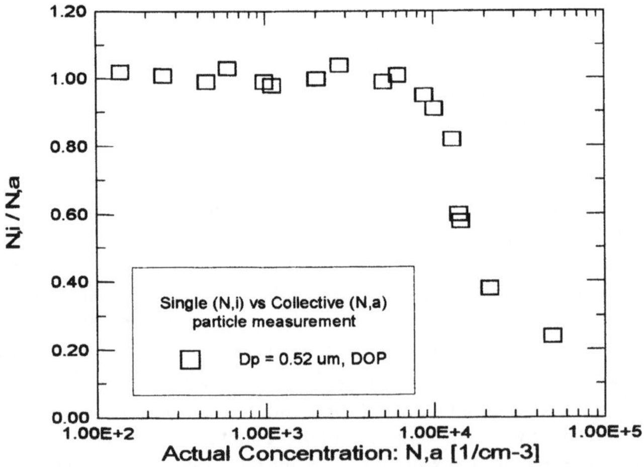

Fig. 3. Experimentally determined influence of coincidence on the concentration measurement by means of LAS-X particle counter.

b. Collective particle measurement

There is a wide variety of instrumental possibilities to assess the concentration of aerosols based on the measurement performed on an ensemble of particles and a number of extensive publications concerning these techniques (ACGIH, 1994; Gressel et al. 1988; Gebhart et al. 1988; Hirleman and Bohren, 1991). The common disadvantage of those so-called photometric or nephelometric methods is the necessity of an empirical calibration procedure. This task is becoming particularly difficult when the observed aerosol concentrations reach such high values that the total scattered light flux can not be expressed as the sum of fluxes scattered by individual particles. In such a case multiple scattering becomes appreciable (Szymanski, 1992). The effect of multiple scattering in most cases results in a non-linear change of the measured integral signal along with the linear increase of the actual particle concentration. In the following experimental procedures will be presented which allow an absolute, that means calibration-free determination of number concentrations up to the order of 10^8 cm^{-3}. The measurements are based on a simultaneous detection of scattered and transmitted light fluxes from droplets formed on nuclei particles and growing in supersaturated vapors due to heterogeneous nucleation. They are extensively described elsewhere (Szymanski and Wagner, 1990) that is why here merely the key aspects of the method are given.

The optical arrangement for such a measurement is shown in Fig. 4. The optical, cylindrical measuring chamber acts also as an expansion cloud chamber in which the heterogeneous nucleation and a subsequent growth of droplet aerosol are achieved by means of fast adiabatic expansion. The investigated aerosol is illuminated by a gas laser along the axis of the cylinder and by a laser diode in the perpendicular direction. Both laser systems operate at different wave lengths and are optically de-coupled.

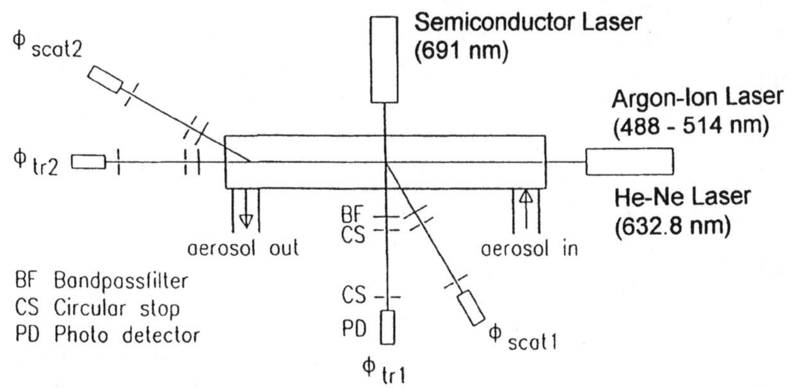

Fig. 4. Schematic diagram of the dual wave length laser optical system for the collective particle measurement.

Experimental transmitted and scattered light flux curves, registered in the optical chamber together with the corresponding theoretical curves calculated by means of the Mie theory (v. d. Hulst, 1981) are shown in Fig. 5. and display a unique matching morphology. Upper curves represent the (inverted) transmitted light (Φ_{tr}) flux attenuated during the droplet growth. Lower, simultaneously measured scattered light flux curves (Φ_{scat}) show a specific resonance sub-structure. They represent scattered light fluxes monitoring the size changes of the growing droplets. It is quite evident that even by visual comparison of the scattered theoretical and experimental light curves a unique identification of droplet sizes at given times is possible. The quantitative evaluation of the transmitted light yields the absolute particle number concentration.

The transmitted light flux through an ensemble of particles can be expressed as:

$$T = \exp(-N \cdot z \cdot R \cdot \int p(D_p) \cdot C_{ext}(D_p, \lambda)\, dD_p)(1 + X(z, C_{ext})) \qquad (2)$$

where $C_{ext}(D_p,\lambda) = (\pi/4).D_p^2.Q_{ext}(D_p,\lambda)$ is the extinction efficiency and $R = R(D_p,\varepsilon)$ is the forward scattering correction factor calculable from the Mie theory for a given optical system with a field of view ε (Szymanski, 1992a). N represents the particle number concentration and z is the path length of the laser beam. X (z, Cext) is the multiple scattering correction term. Physically it represents the amount of multiply scattered light being re-directed back into the transmitted light flux sensor. In the absence of multiple scattering the correction term X vanishes and the transmitted light flux can be described by the Beer-Lambert law. Under such conditions all quantities in the Eq.2. but N and D_p, are known. The independently, but simultaneously measured scattered light flux provides the momentary particle size information D_p needed in the Eq.2 (Wagner, 1985). Together with the size information the Eq.2. yields then the unknown particle concentration N without a requirement of any calibration, hence is can be viewed as an absolute number concentration standard (EXT-Method).

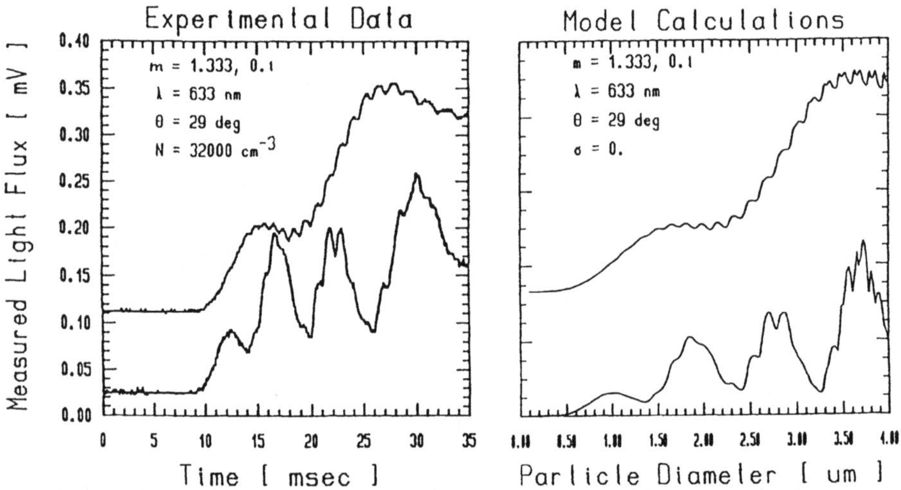

Fig. 5. Experimental and theoretical scattered and transmitted light fluxes. The indicated concentration ($N = 32000\ cm^{-3}$) was obtained from the experimental data set.

Under measuring conditions where multiple scattering in the optical chamber can be neglected, both optical systems must yield the same concentration value. With increasing aerosol concentrations this will no longer be case and a determination of the multiple scattering correction term X is necessary to obtain the actual particle concentration. This evaluation procedure is described elsewhere (Szymanski, 1992).

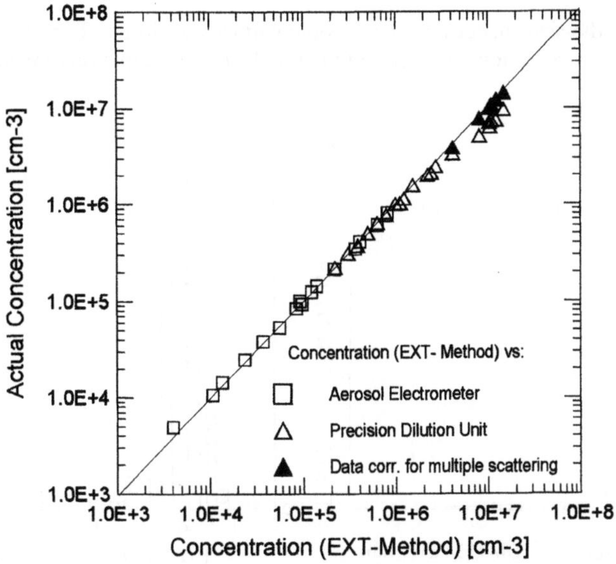

Fig. 6. Optical concentration measurement based on Eq.2 with multiple scattering corrections.

The concentration measuring performance of the dual laser optical system described above is shown in Fig. 6. The actual particle number concentration was determined here experimentally using a precision dilution system (Filipovicova, 1995). Measured data shows a linear relationship for concentrations below about $2 \cdot 10^6$ cm^{-3}. Above this value the influence of multiple scattering was determined and applied to obtain the actual aerosol number concentration resulting in a linear behavior of the measuring system over the investigated range of concentrations.

Combining the optical particle concentration detection with the controlled heterogeneous nucleation, as is the case in the described method, allows to perform concentration measurement down to the nanometer-sized particle range. The smallest measurable particle size is determined by the choice of the supersaturation. Gradually changing of the supersaturation level results in activation to growth of increasingly finer particles, hence also allowing to determine a Kelvin-equivalent particle diameter. The Kelvin-equivalent diameter (with respect to water vapor) represents the diameter of a water drop that would become activated to grow at the same supersaturation as the investigated particles. Such measurements were performed earlier in the size analyzing nuclei counter (Liu et al.,1984) and provide concentration measurement of growing droplets as a function of the supersaturation based on the Kelvin-equivalent particle size distribution. Reasonable measurements can be performed with this technique for supersaturations below about 450%, since above this value homogeneous nucleation takes place. However, the supersaturation level of 400% corresponds already to particle radius of about 0.001 um.

Cumulative and differential size distribution of DOP particles with nominal diameter of 18 nm are shown in Fig.7. The good agreement of the Kelvin particle size with the actual size shown here depends however on the nuclei particle material and surface structure and composition (Porstendörfer et al, 1985). Nevertheless, this behavior is also of substantial interest for e.g. the size changes of hygroscopic particles or droplet formation mechanisms in varying humidity conditions, as may be found in the human respiratory system.

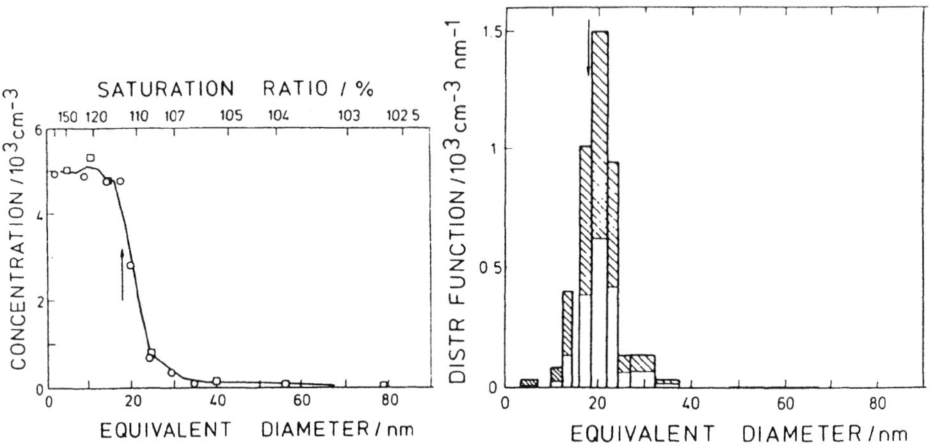

Fig. 7. Kelvin-equivalent particle size distributions

For a number of applications the condensational measurement technique presented above may not be suitable, however the dual wave length laser arrangement can also be applied for monitoring and measurement of aerosols without utilizing the heterogeneous nucleation. The procedure is an extension and modification of an experimental technique presented earlier (Teorell, 1931).

The method discussed here is based on a simultaneous measurement of transmitted light fluxes through an arbitrary aerosol at two, different wave lengths. It can be easily derived from the Eq. 2 that a ratio of such two signals, hereafter called the transmission ratio, depends mainly on the particle size and is virtually independent on the particle concentration, at least as long one can suppress the multiple scattering effects.

For a reliable optical aerosol measurement optical properties of particles should be known, or a particular method has to be insensitive to variation of those properties. Fig. 8 shows clearly a considerable independence of the transmission ratio from the changes of the real part of the complex refractive index and a monotonic dependence on the imaginary part for a given particle size. This specific behavior allows to determine the refractive index of the aerosol in question and to utilize the transmission ratio for aerosol measurement. The functional dependence of the transmission ratio on the particle size is shown in Figs. 9 and 10 for absorbing and non-absorbing aerosols, respectively. The assessment of the particle size from the transmission ratio is obvious. Preliminary experiments indicate the applicability of the method for in-line particle sizing in the range from about 0.01 - 0.5 um in diameter for unimodal aerosols. With the knowledge of the mean particle size also the particle concentration can be continuously monitored and evaluated from the transmitted light flux signal using the methodology presented above. Furthermore, the evaluation of the transmission ratio provides also an information about the extend of multiple scattering affecting a particular optical measuring system and allow to extend the concentration measurement into highly concentrated aerosols.

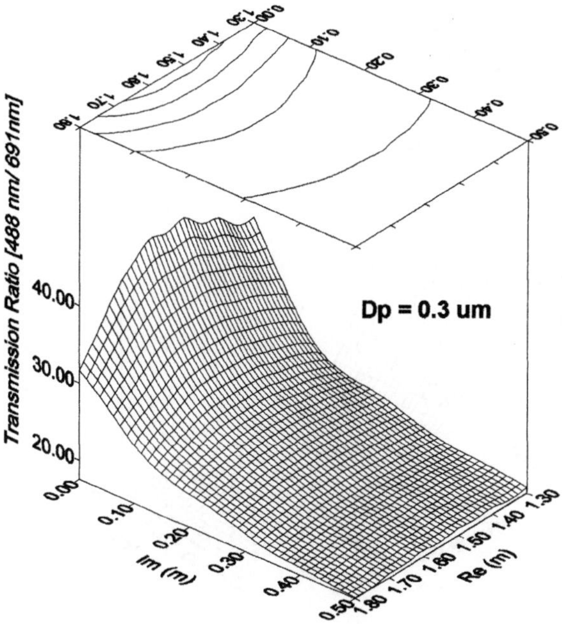

Fig. 8. Variability of the transmission ratio with the complex refractive index.

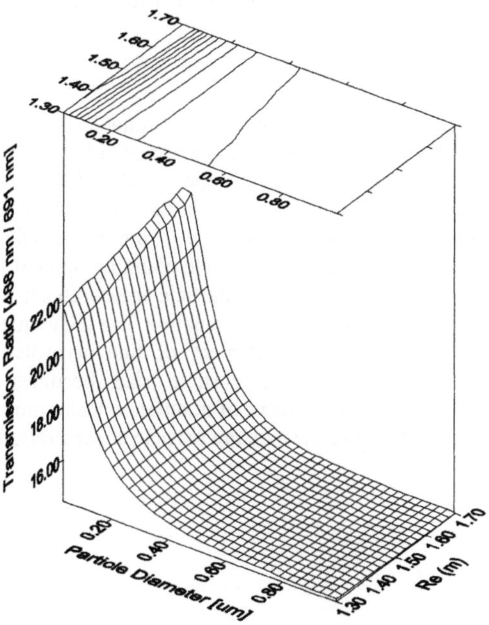

Fig. 9. Changes of the transmission ratio as a function of the particle diameter for absorbing aerosols with m = 1.5 - 1.0i and geometric standard deviation of 1.5.

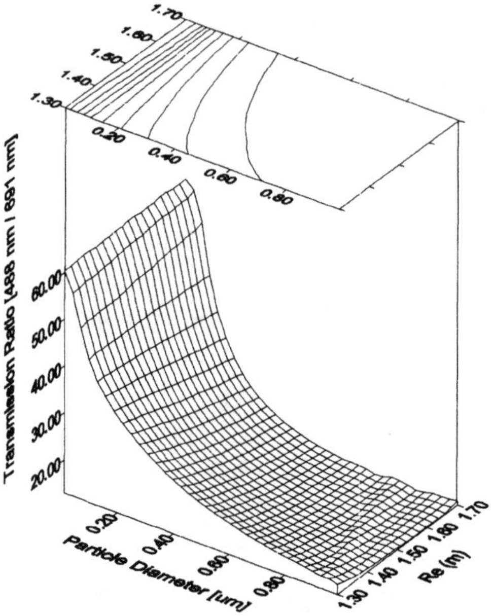

Fig. 10. Changes of the transmission ratio as a function of the particle diameter for non-absorbing aerosols with m = 1.5 - 1.0i and geometric standard deviation of 1.5.

Summary

Optical, intensity based techniques described here are suitable for in situ and real-time measurement of aerosols. They indicate a considerable potential for applications in a health hazard assessment field, or in a biomedical research connected with to aerosols. They allow a reliable measurement of airborne particulate matter in the size range spanning over more than four orders of magnitude starting at about 0.001 um in particle radius. Number concentrations can be determined from a single particle counting upward to concentrations of the order of 10^8 cm^{-3}, whereby the collective particle measurement method presented here yields an absolute number concentration determination The techniques are suitable for both, an extractive, as well as in-line measurement of fine aerosols. Since the mass concentration determination with decreasing particle size becomes an increasingly problematic issue, the size selective measurement of number concentration for fine aerosols appears also as a valuable complement to the existing particle concentration standards.

Acknowledgments

The support of the Jubiläumsfond der Österreichischen Nationalbank, Proj. Nr. 4363 and 5561 for this research is here gratefully acknowledged.

References

ACGIH,"Air Sampling Instruments", 8th Edition, Cincinnati, Ohio, USA (1994)

Dockery, D.W., Pope III, C.A., Xu, X., Spengler, J.D., Ware, J.H., Fay, M.E., Ferris, B.G. and Speizer, F.E. ,"Six Cities Chrinic Mortality Study", *New England J. of Medicine*, 329:1753-1759 (1993)

Filipovicova, D.,"Determination of High Particle Concentrations", *MS Thesis* (in preparation), University of Vienna (1995)

Friedlander, S.K. ,"Nanometer Particles and Their Agglomerates", *Proceedings of International Conference on Aerosols Science and Technology*, Taipei, Taiwan (1995)

Gebhart, J. ,"Background Noise and Counting Efficiency of Single Particle Counters", in *Aerosols: Formation and Reactivity*, p.632-635, Pergamon Press, Oxford. (1986)

Gebhart, J.,"Response of Single Optical Particle Counters to Particles of Irregular Shape", *Part. Syst. Charact.* 8:40-47 (1991)

Gebhart, J., Heigwer, G., Roth, C. and Stahlhofen, W. ,"The Use of Light Scattering Photometry in Aerosol Medicine", *J. Aerosol Med.* 1: 89-112 (1988)

Gressel, M.G., Heitbrink, W.A. and McGlothing, J.D. ,"Advantages of Real-Time Data Acquisition for Exposure Assessment", *Appl. Ind. Hyg.* 3:316-320 (1988)

Hirleman, E.D.," Optical Techniques for Particle Size Analysis", *Laser Topics* 10:7-10 (1988)

Hirleman, E.D. and Bohren, C.F. (Feature Eds.),"Optical Particle Sizing" *Applied Optics* 30:4688-4986 (1991)

Hofmann, W.,"Aerosol Deposition in the Human Respiratory Tract: Theory and Experiment" in *Aerosols -Science, Industry, Health and Environment*, Vol 1, p. 99-103, Pergamon (1990)

Janka, K., Keskinen, J., Lehtimäki, M.,"Size Distribution Dependent Concentration Limitation Caused by Electrical Pulse Processing: PMS LAS-X", in *Aerosols: Formation and Reactivity*, p.632-635, Pergamon Press, Oxford (1986)

Liu, B. Y. H., Pui, D. Y. H., McKenzie, R. L., Agarwal, J. K., Jaenicke, R., Pohl, F. G., Preining, O., Reischl, G. P., Szymanski, W. W. and Wagner, P. E.,"Measurement of Kelvin-Equivalent Size Distributions of Well-Defined Aerosols with Particle Diameters Above 13 nm" *Aerosol Sci. Technol.* 3:107-115 (1984)

Liu, B.Y.H., Szymanski, W.W. and Pui, D.Y.H,"Response of Laser Optical Particle Counter to Transparent and Light Absorbing Particles", *ASHRAE Transactions* V.92, Pt.1 (1986)

Porstendörfer, J., Scheibel, H.G., Pohl, F.G., Preining, O., Reischl, G., Wagner, P.E., "Heterogeneous Nucleation of Water Vapor on Monodispersed Ag and NaCl Particles with Diameters Between 6 and 18 nm", *Aerosol Sci. Technol.* 4:65-79 (1985)

Pui, D.Y.H. and Liu, B.Y.H.,"Advances in Instrumentation for Atmospheric Aerosol Measurement", *Physica Scripta* 37: 252-169 (1988)

Sioutas, C., Koutrakis, P. and Burton, R.M.,"A High-volume Small Cutpoint Impactor for Separation of Atmospheric Particulate from Gaseous Pollutants", *Particulate Sci. and Technol.* 12: 207-221 (1994)

Stahlhofen, W.,"Neuere Untersuchungen und Erkentnisse über das Verhalten von Aerosolteilchen im Atemtrakt des Menschen", *Lecture at the Austrian Academy of Sciences*, Vienna, Austria (1994)

Szymanski, W.W. and Liu, B.Y.H.,"On the Sizing Accuracy of Laser Optical Particle Counters", *Part. Charact.* 3:1-7 (1986)

Szymanski, W.W.,"Multiple Scattering Enhancement of Transmitted Radiation in Water Droplet Aerosols", *J. Aerosol Sci.* 23:425-435 (1992)

Szymanski, W.W. and Wagner, P.E.,"Absolute Aerosol Number Concentration Measurement by Simultaneous Observation of Extinction and Scattered Light", *J. Aerosol Science.* 21:441- 451 (1990)

Szymanski, W.W.,"Investigation of Multiple Scattering Contribution to Tramsmission of Radiation in Scale Model Clouds", in *Atmospheric Propagation and Remote Sensing*, SPIE, Vol. 1688, p.62-73 (1992a)

Teorell, T.,"Photometrische Messung der Konzentration und Dispersität bei kolloiden Lösungen", *Koll.-Ztschr.* 54: 58-66 (1931)

Van de Hulst, H.C.,"Light Scattering by Small Particles", Dover Publications, Inc. New York (1981)

Wagner, P.E.,"A Constant Angle Mie Scattering Method (CAMS) for Investigation of Particle Formation Processes", *J. Coll. Interface Sci.* 105:456-467 (1985).

Wen, H.Y. and Kasper, G.,"Counting Efficiencies of Six Commercial Particle Counters", *J.Aerosol Sci.* 17:947-1961 (1986)

Whitby, K.T. and Cantrell, B.,"Fine Particles" in Proc. of International Conference on Environmental Sensing and Assessment, IEEE, Las Vegas, USA(1976)

Lung Morphometry and Particle Transport and Deposition: Overview of Existing Models

Werner Hofmann

Institute of Physics and Biophysics, University of Salzburg, Hellbrunner Str. 34, A-5020 Salzburg, Austria

Keywords

Computer Modeling; Lung Morphometry; Particle Deposition Mechanisms; Fluid Dynamics; Particle Deposition Patterns.

Introduction

Health effects of inhaled particulate matter in the human respiratory tract depend on the total number (or total mass) of particles deposited on human airway surfaces in a single breath or in a given period of time ("total deposition"). A characteristic feature of these effects, however, is their apparent site selectivity. For example, lung cancers in smokers have been detected primarily in bronchial airways. Consequently, the lung cancer incidence in smokers is more aptly related to bronchial deposition ("regional deposition") rather than to total deposition. Furthermore, the majority of bronchial carcinomas has been found in the segmental and subsegmental bronchi ("differential deposition"), possibly originating at bronchial airway bifurcations ("localized deposition"). While total and regional deposition in human test subjects can be studied by various experimental methods, it is presently not possible to determine experimentally particle deposition in individual airways and within airway bifurcations *in situ*. Such information can only be provided by mathematical modeling.

Modeling particle deposition patterns in the human lungs requires a set of idealized assumptions regarding lung morphometry, gas and particle transport, breathing patterns and the action of physical deposition mechanisms. These assumptions must satisfy two conditions: (i) they must be anatomically and physiologically realistic, and, (ii) they must permit solutions of the equations governing particle transport and deposition. These two intimately connected conditions imply that more realistic morphometric and physiologic models require more sophisticated mathematical models.

Lung morphometry

The theoretical prediction of total, regional and local particle deposition requires the mathematical description of the human airway structure. While conductive airways are represented by straight cylindrical tubes of given diameter and length, alveoli attached to respiratory airways can be approximated by spherical volumes with different levels of truncation. The location of an individual airway within the lungs may be characterized either by an assigned airway generation number (or bifurcation number) or by its diameter. The structural parameters of the branching airway system may be described by average values ("deterministic model") or by probability density distributions and their correlations ("stochastic model").

J. C. M. Marijnissen and L. Gradoń (eds.), Aerosol Inhalation: Recent Research Frontiers, 91–102.

Presently available morphometric models of the human lung have been derived from morphometric measurements of a few individual lung casts. Because these lung casts were prepared from different individuals, Yu and Diu (1982) attributed the reported differences in linear airway dimensions primarily to inter-subject variability in lung volume and structure. The application of different casting and measurement techniques in these studies may add an additional systematic uncertainty to the observed inter-subject variations. An even more fundamental problem, however, is the question whether these few lungs used in the morphometric analyses are indeed representative of the whole population. In other words, can the bronchial doses based on a given morphometric model be correlated with the occurrence of a bronchial tumor in a specific lung?

To illustrate the effect of different lung morphometries on regional particle deposition, different morphometric lung models have been derived from the few morphometric measurements of bronchial and alveolated airway dimensions. The four lung models employed in the present analysis are: (i) the diameters and lengths of the tracheobronchial airways (generations 0-13) are provided by James (1988), while the structure of the pulmonary airways (generations 14-23) is taken from Haefeli-Bleuer and Weibel (1988); (ii) here, the tracheobronchial tree (generations 0-15) as described by James (1988) is connected to the pulmonary airway structure (generations 16-26) proposed by Hansen et al. (1975); (iii) the bronchial airways are characterized by Weibel's Model A (1963) (generations 0-13), supplemented by the acinar structure of Haefeli-Bleuer and Weibel (1988); and, (iv) a slightly modified version of model (ii), where diameters and lengths of bronchial and pulmonary airways refer to the dimensions listed in NRC (1991). The originally published linear dimensions in those models are scaled down to a normal functional residual capacity of 3000 ml by the commonly applied linear scaling procedure (Yu and Diu, 1982).

The effect of different lung morphometries on the deposition of 10 nm particles is illustrated in Fig. 1 for resting breathing conditions with a tidal volume, V_T, of 500 ml and a cycle time, t, of 4 s (symmetric inspiration and expiration without pause). While deposition fractions in individual airway generations are comparable in the four selected morphometric models (Hofmann and Mainelis, 1993), bronchial deposition fractions divided by their respective surface areas show distinct differences between the various bronchial airway models. Assuming that these morphometric differences reflect typical inter-subject variations, then the experimentally observed fluctuations of bronchial deposition in human test subjects (Stahlhofen et al., 1989) may be attributed to differences in airway morphometry rather than to differences in individual breathing patterns.

In the above employed symmetric, deterministic lung models, each airway in a given airway generation has identical linear dimensions, branches symmetrically into two identical daughter airways, and, consequently, all pathways leading to this airway have the same lengths (i.e., the same distance from the trachea). Statistical analyses of morphometric data of two human lungs (Raabe et al., 1976) have revealed, however, that there is a statistical relationship between the geometrical parameters of a given parent airway and those of the two asymmetrically dividing daughter branches. In other words, there is a certain degree of freedom in the continuation of a pathway at a given bifurcation (generally called intra-subject variability), which can be described mathematically by lognormal probability density distributions (Koblinger and Hofmann, 1985). It should be noted, however, that the statistical selection is constrained by correlations among the various geometrical parameters, i.e., these parameters are not statistically independent from each other. In this respect, the human airway network is not a fully stochastic system. Such correlations allow us to compute deposition for the five individual lung lobes, which differ in the linear dimensions of the lobar bronchi leading to the individual lobes and, consequently, in their respective volumes (Hofmann and Koblinger, 1990). Another consequence of the intra-subject variability of airway lengths and diameters and the asymmetric branching pattern is that the number of bronchial airway generations along randomly selected pathways varies from 11 to 21 (Koblinger and Hofmann, 1985), in contrast to deterministic lung models with a specified number of airway generations. For example, the tracheobronchial region in the deterministic lung model of Yeh and Schum (1980), which is based on the same morphometric data as the stochastic model (Koblinger and Hofmann, 1985), consists of 15 bronchial airway generations.

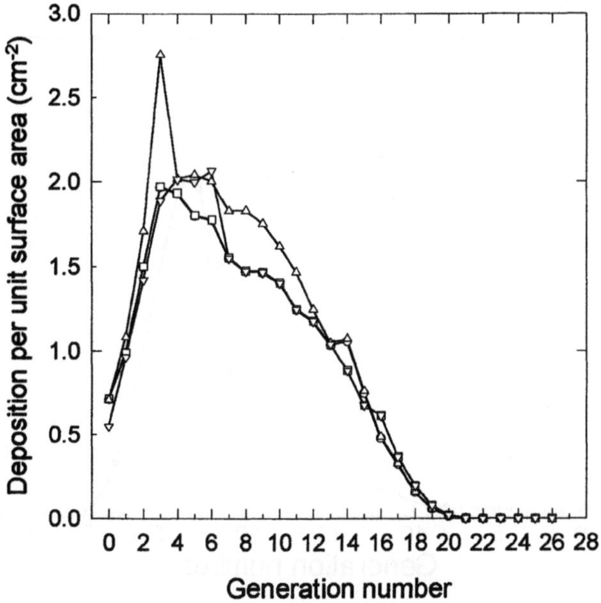

Fig. 1 Deposition per unit surface area for 10 nm particles under sedentary breathing conditions (V_T = 500 ml, t = 4 s) as a function of the airway generation number for four morphometric lung models: (i) generations 0-13 (James, 1988), generations 14-23 (Haefeli-Bleuer and Weibel, 1988) (○); (ii) generations 0-15 (James, 1988), generations 16-26 (Hansen et al., 1975) (□); (iii) generations 0-13 (Weibel, 1963), generations 14-23 (Haefeli-Bleuer and Weibel, 1988) (∇); and (iv) generations 0-26 (NRC, 1991) (Δ) (for more details, see related text).

Stochastic particle deposition calculations have been performed for a wide variety of particle sizes and flow rates (Koblinger and Hofmann, 1990; Hofmann and Koblinger, 1990) and validated by the comparison with published experimental data on total and regional deposition in human test subjects (Hofmann and Koblinger, 1992). Here, the term „stochastic particle deposition" refers to the transport of inspired particles through the above described stochastic lung structure by selecting randomly the sequence of airways for each individual particle. Within a given airway, however, the average deposition behavior of an ensemble of particles is considered, i.e., deposition is computed by deterministic equations (note: at present, the random transport of individual particles within a given bifurcation geometry and the resulting deposition patterns have been studied only for a few selected symmetric and asymmetric bronchial airway bifurcations (Balásházy and Hofmann, 1993a; 1993b; 1994)).

The results of stochastic particle deposition calculations for inhaled 10 nm and 1 µm unit density particles under resting breathing conditions (V_T = 1000 ml, t = 4 s) are displayed in Fig. 2. Though Monte Carlo results are always associated with a statistical uncertainty, they are of the order of some percent and are therefore not included in the figures presented here. A comparison of these stochastic simulations with corresponding calculations based on the deterministic lung model of Yeh and Schum (1982) (note: both models are based on the same morphometric data) reveals that total and regional deposition are similar in both models (Hofmann and Koblinger, 1990). However, in contrast to the deterministic lung model with a fixed number of bronchial and alveolar airway generations, the structural variability in the real lung causes a variable pathlength in both bronchial and respiratory airways; because of this variability in total pathlength, deposition is shifted to more distal airways in the stochastic lung model.

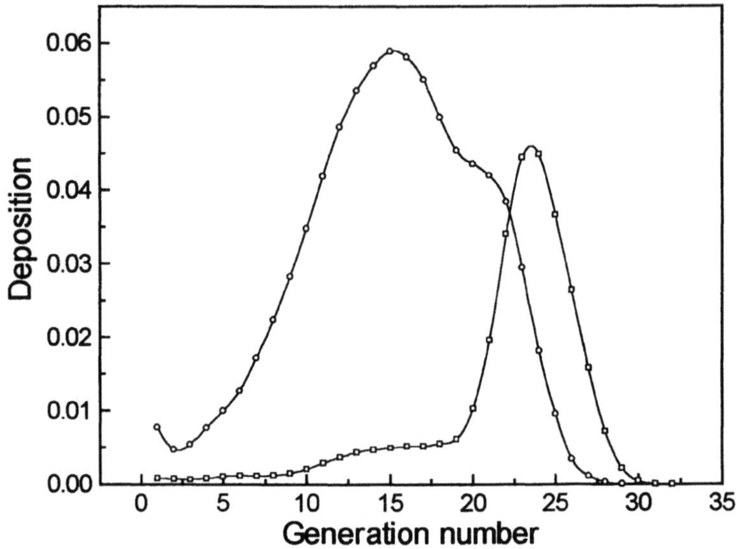

Fig. 2 Deposition patterns of 10 nm (O) and 1 μm (□) particles under resting breathing conditions (V_T = 1000 ml, t = 4 s) in a stochastic lung model applying a linear scaling procedure.

The deposition fractions for the stochastic lung model plotted in Fig. 2 represent statistical averages of widely varying contributions in individual airways belonging to the same airway generation. This variability in deposition can reasonably be approximated by lognormal distributions (Hofmann and Koblinger, 1990) and reflects the variability in geometrical parameters and airflow patterns among individual airway generations.

While the above considerations refer to intra-subject variability, Stahlhofen et al. (1989) have demonstrated that total and regional deposition in different human test subjects can vary substantially over a wide range, even under identical breathing conditions. The effect of inter-subject variability in particle deposition may be modeled by scaling the linear airway dimensions in proportion to the individual total lung volume, while preserving their intra-lobar structural relationships (Hofmann and Koblinger, 1992).

The linear dimensions of present lung models are based on morphometric measurements on fully inflated lungs (e.g. Raabe et al., 1976). To simulate deposition of inhaled particles in human lungs under normal breathing conditions, airway diameters and lengths have to be scaled down to the smaller dimensions at functional residual capacity, which correspond to a total lung volume of about 3000 ml (Yu and Diu, 1982). The commonly applied linear scaling procedure is based on the assumption that all linear dimensions vary with the cube root of the total lung volume (Yu and Diu, 1982). However, recent measurements of the biomechanical properties of bronchial and alveolar tissue have suggested a different scaling procedure (Habib et al., 1994): While airway lengths are scaled from the Horsfield et al. (1971) airway geometry at total lung capacity by a constant scaling factor (i.e., all airway lengths are multiplied by the same linear factor), airway diameters are scaled by both a constant scaling factor and a generation-dependent sigmoidal curve (i.e., the relative changes of airway diameters increase with penetration deeper into the lung). Since the morphometry of the stochastic lung model (Koblinger and Hofmann, 1985) employed in the present study is based on the Raabe et al. (1976) data obtained at total lung capacity, we have defined an additional scaling procedure, which is based on the Habib et al. (1994) model, but normalized to the original Raabe et al. (1976) geometry.

The effects of these three scaling procedures on particle deposition in individual airway generations have been investigated for the inhalation of 10 nm particles under resting breathing conditions with a tidal volume of 1000 ml and a breathing cycle time of 4 s (symmetric inspiration and expiration without pause). These scaled down deposition patterns are compared to that obtained with the originally reported lung morphometry at total lung capacity. The results plotted in Fig. 3 indicate: (i) the various scaling procedures increase bronchial deposition relative to the corresponding simulations based on the original morphometric data, and (ii) the novel scaling procedure suggested by Habib et al. (1994) yields higher bronchial deposition fractions than the commonly adopted linear scaling procedure.

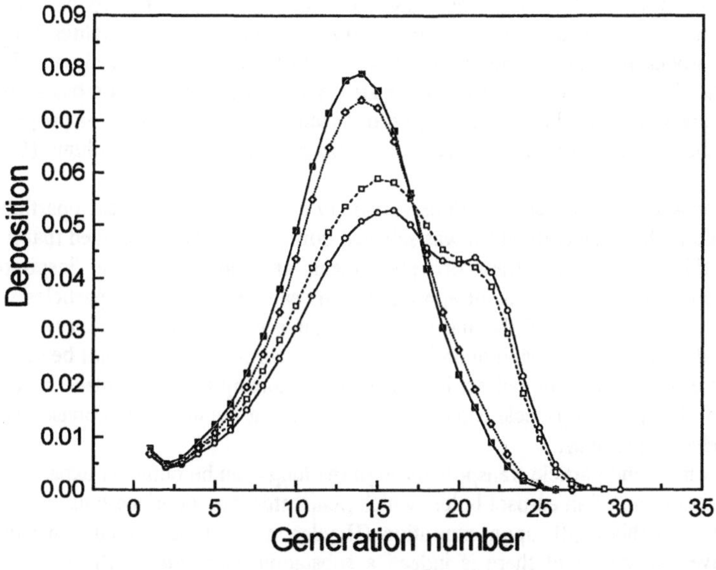

Fig. 3 Deposition patterns of 10 nm particles under resting breathing conditions (V_T = 1000 ml, t = 4 s) for the original Raabe et al. morphometry at total lung capacity (○) and three scaled down morphometries assuming a constant scaling factor (□) and the Habib et al. scaling procedures normalized to the Horsfield et al. model (◊) and to the Raabe et al. data (■) (for more details, see related text).

Particle transport and deposition

For regional deposition in the human respiratory tract, i.e., for the nasal, extrathoracic, fast-cleared thoracic (commonly assumed to be equivalent to bronchial) and slow-cleared thoracic (commonly assumed to be equivalent to alveolar) regions, deposition can be described by semi-empirical equations as functions of particle size and flow rate (Stahlhofen et al., 1989) through best mathematical fits to the available experimental data. In this model, the human respiratory tract is treated as a series of filters through which the aerosol particle passes during inspiration and expiration. The major advantages of such semi-empirical models are that they are based on existing experimental data in human test subjects and no sophisticated computer programs are necessary. By definition, however, they cannot provide information about particle deposition in individual airway generations.

Thus an alternative approach, which is actually used in most particle deposition modeling efforts, is to trace an inspired air volume through the whole airway system upon inspiration and expiration, considering in each airway generation the deposition probability of aerosol particles transported by this air volume. Regional, i.e., bronchial and alveolar, and total deposition can then be obtained by integrating over all individual deposition fractions in the defined range of airway generations during a complete breathing cycle. While this approach requires a rather detailed knowledge of morphometric, physiologic and physical mechanisms and parameters, it permits, on the other hand, a much more detailed study of the various factors involved in particle deposition in the human lungs.

During the transport of an inspired air volume through the branching network of the human lung, this air volume is divided into two smaller volumes at each bifurcation. While these two subvolumes are identical in the case of a symmetric morphometric lung model, e.g. Weibel's (1963) Model A, flow splitting in an asymmetrically dividing model has to be related to a structural parameter of the lung, such as the airway diameter of the daughter branch. For example, in the stochastic transport and deposition model of Koblinger and Hofmann (1990) it is assumed that the airflow rate in a given daughter airway is proportional to its cross section. This assumption is consistent with the view that the cross section of a given airway is proportional to the number of distal ventilatory units which have to be supplied by that air volume (Overton et al., 1989).

Even in a stochastic lung structure with its variability in pathway lengths, each particle returns to the trachea along the same path which was selected during inspiration, provided that it was not deposited. In this case, where transport phenomena are identical during inspiration and expiration, the so-called "first-in/last-out approximation" can be applied, i.e., particles which are inhaled at the very beginning of the inspiration period are exhaled at the very end of the expiration period. Thus the dispersion of an inhaled narrow aerosol bolus will not be altered upon exhalation. There is, however, an indirect influence via the mixing that takes place in the alveoli, because the probability that a particle enters an alveolus during the inspiration phase depends on the randomly selected pathway.

Information on air and particle transport through the lungs can be obtained experimentally by measuring the dispersion of an aerosol bolus, which penetrates into a defined lung volume and is then recovered from this depth upon expiration (Heyder et al., 1988). Such bolus dispersion experiments have indicated that there is indeed a substantial widening of the initially inhaled narrow bolus. Several factors may contribute to the dispersion of an inhaled bolus: asymmetry of the branching network, differences in gas velocity profiles during inspiration and expiration, asymmetry between expansion and contraction of alveoli, local ventilation inhomogeneities, asymmetric and asynchronous ventilation, secondary flows generated at bifurcations, and cardiac oscillations (Scheuch and Stahlhofen, 1991; 1992). At present, the relative contributions of these mechanisms to the overall phenomenon are still not fully understood.

The effects of two of the above factors, namely axial diffusion and alveolar mixing, on bolus dispersion has recently been examined by Hofmann et al. (1994). First of all, the effect of axial diffusion has been considered by introducing an effective axial diffusion coefficient, which is different for inspiration and expiration, thereby reflecting differences in flow velocity profiles (Scherer et al., 1975). Though this effective axial diffusivity has been derived specifically for laminar flow of gases in upper airway generations, it was assumed (in the absence of any other pertinent information) that these parameters can be applied also to small aerosol particles and to distal airway generations (Egan and Nixon, 1985). Secondly, alveolar mixing has been modeled by assuming that a certain fraction of the inhaled tidal air mixes completely with the residual air in the alveoli. Thus, at the onset of expiration, this fraction of the air volume inspired in the preceding inspiration phase is uniformly distributed over the whole alveolar volume.

The dispersion of the expired bolus, i.e., the time distribution of the exhaled bolus, can be characterized by the standard deviation, σ, of the distribution. Figure 4 represents the calculated standard deviations for 0.85 μm aerosol boli with a small initial dispersion, inhaled at a flow rate of 250 $cm^3\ s^{-1}$ during an inspiration time of 4 s, at different times during the inspiration period and therefore leading to different depths of penetration. Consistent with the experimental protocol, expiration was prolonged up to the end of the experimentally observed exhalation phase. These theoretical predictions are compared in this figure with experimental data on human test subjects. In these experiments, the penetration of the inhaled air into the lungs is characterized by its volumetric depth; similarly, dispersions of both inhaled and exhaled boli are described in terms of distribution in volume. In the computer simulations, however, time is used to describe particle penetration, i.e., dispersions of inhaled and exhaled boli are characterized by their distributions in time. For this comparison between experimental observations and theoretical predictions, we therefore converted the experimental volume distributions into time distributions. The time between the inhalation of the peak of the aerosol bolus and the end of the inspiration phase, which is directly proportional to volumetric depth for a constant flow rate, is called here "time left for penetration". Consistent with the experimental observation, the standard deviation of the expired bolus increases with depth of penetration.

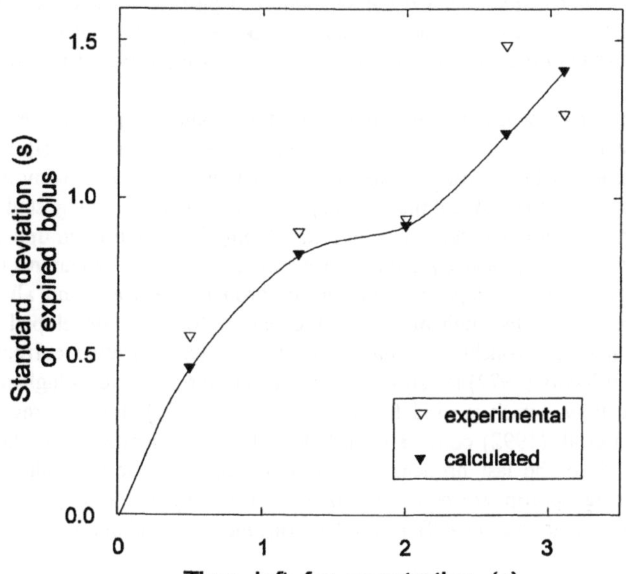

Fig. 4 Comparison of experimental data with theoretical predictions: standard deviations of the distributions characterizing the dispersion of an inspired aerosol bolus upon expiration, plotted as a function of time left for penetration (particle diameter = 0.85 μm, flow rate = 250 $cm^3\ s^{-1}$, inspiration time = 4 s, expiration prolonged up to the end of expiration).

The satisfactory agreement between experiment and theoretical prediction suggests that axial diffusion and alveolar mixing may account for most of the observed width of the expired bolus.

Recently, Edwards (1995) has successfully applied macrotransport theory to the same experimental data analyzed above (Heyder et al., 1988). His macrotransport model of lung dispersion, based on the symmetrical Weibel (1963) Model A, provides a complete theoretical formalism for determining each of the first three moments of an exhaled aerosol bolus, thereby relating aerosol deposition to the mean velocity and dispersivity of the exhaled aerosol bolus for varying depths of bolus penetration into the lung.

Different mathematical formulations of particle deposition efficiencies in human airways as functions of particle size and flow rate have been proposed by various authors for the different physical mechanisms acting upon an inhaled particle. The effect of different equations on particle deposition will be illustrated here for submicron particles, where Brownian motion is the dominant physical deposition mechanism. Analytical solutions of particle deposition by diffusion can be obtained in straight or bent cylindrical tubes for specified flow profiles, such as fully developed laminar flow in cylindrical airways (Ingham, 1975). Depending on the assumed flow profile, however, such as plug flow, turbulent flow, developing laminar flow or fully developed laminar flow, different deposition equations have been derived (Martonen, 1982). Thus the main source of uncertainty in submicron particle deposition calculations is the choice of the relevant flow profile in a given airway generation, which is even further complicated by the modification of the primary flow profile by the complex geometry of the central bifurcation zone in a branching airway system and the formation of secondary flows (Heistracher and Hofmann, 1994). One possibility to consider realistic flow fields in upper bronchial airway bifurcations is to derive an empirical equation for diffusion deposition from measurements of submicron particle deposition in an upper airway cast (Cohen and Asgharian, 1990). An alternative approach is the solution of the Navier-Stokes equations and the simulation of particle trajectories and localized deposition patterns within defined bronchial bifurcation models by numerical methods, varying particle size, flow rate and bifurcation geometry in a systematic fashion (Balásházy and Hofmann, 1993a; 1993b; 1995).

For the comparison of presently available equations for submicron particle deposition, the following diffusion deposition equations are used in the subsequent analysis: (i) first, a fully developed laminar parabolic flow profile is assumed in all bronchial and pulmonary airway generations and deposition in a cylindrical tube is computed by the equation suggested by Ingham (1975); (ii) while enhanced diffusion deposition in a developing laminar flow in upper bronchial airways is described by the empirical equation derived by Cohen and Asgharian (1990) from upper airway cast measurements, deposition in smaller bronchial airways and all pulmonary generations is modeled again by Ingham's (1975) equation for fully developed flow; (iii) enhanced deposition in large bronchi is characterized by the multiplicative correction factor proposed by Martin and Jacobi (1972) for airway generations 0 to 6, relative to Inghams's (1975) equation for parabolic flow, which is used for all distal airways; (iv) the flow regime pattern proposed by Martonen et al. (1992) consists of turbulent flow in generations 0-3, laminar plug flow from generation 4 to the last bronchial generations, and laminar parabolic flow in all pulmonary airways, using appropriate equations for each flow regime; and, (v) the empirical equation of Cohen and Asgharian (1990) is used consistently throughout the human airway system.

Resulting regional deposition patterns of 10 nm particles for these five different sets of deposition equations are plotted in Fig. 5 for sedentary breathing conditions with a tidal volume of 500 ml and a cycle period of 4 s (symmetric inspiration and expiration without pause), based on the James (1988) (for tracheobronchial airways) and Haefeli-Bleuer and Weibel (1988) (for pulmonary airways) models. First of all, a distinct difference between the computations based on the flow regime distribution proposed by Martonen et al. (1992) and the four other simulations assuming essentially laminar parabolic flow profiles can be observed. Secondly, the application of the deposition enhancement factors published by Martin and Jacobi (1972) significantly increases deposition in the large bronchi. While these different sets of deposition equations can produce different regional deposition patterns, total deposition fractions are rather similar for all modeling conditions.

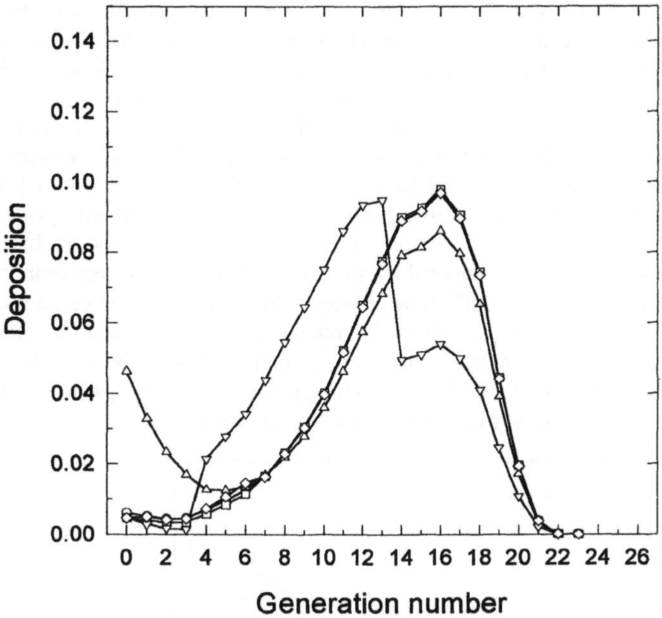

Fig. 5 Deposition patterns of 10 nm particles under sedentary breathing conditions (V_T = 500 ml, t = 4 s) for five sets of diffusion deposition equations: (i) Ingham (1975) (□); (ii) Cohen and Asgharian (1990) and Ingham (1975) (○); (iii) Martin and Jacobi (1972) and Ingham (1975) (Δ); (iv) Martonen et al. (1992) (∇); and (v) Cohen and Asgharian (1990) (◊) (for more details, see related text).

Besides *in vivo* regional deposition data, a new line of experimental evidence has become available in recent years: two-dimensional imaging of the lung with gamma camera (e.g. Newman et al., 1991) and three-dimensional imaging with single photon emission computed tomography (SPECT) (e.g. Chan, 1993). While the general application of gamma camera and SPECT in lung studies has been extensive, their specific use in aerosol research is still uncommon (Martonen et al., 1994).

In current lung deposition models, computed deposition fractions are converted to surface doses (i.e., deposition per unit surface area) by implicitly assuming that particles are uniformly distributed among human airway surfaces. For large particles, it is not surprising that the efficient action of inertial impaction forces produces deposition "hot spots" at the carinal ridge in tracheobronchial airway casts (Schlesinger and Lippmann, 1978) and single bifurcation models (Kim and Iglesias, 1989). But also for diffusion dominated submicron particles, local accumulations at the carinal ridge have been observed experimentally in tracheobronchial airway casts (Cohen et al., 1988), single pathway tracheobronchial models (Martonen et al., 1987), and single bronchial bifurcation models (Kinsara, 1995). These experimental findings are supported by recent theoretical predictions based on a specifically developed numerical model for the computation of air velocity fields and particle trajectories in three-dimensional airway bifurcation models (Balásházy and Hofmann, 1993a; 1993b; 1995).

Analytical solutions of the equations of particle motion and deposition can only be obtained under the assumptions of an idealized airway geometry and idealized flow profiles. As a consequence, such analytical models cannot predict local inhomogeneities of particle deposition within airway bifurcations, such as the formation of hot spots. Thus the analysis of particle deposition patterns in a realistic airway geometry under realistic flow conditions and the simultaneous action of different physical deposition mechanisms require the application of numerical modeling techniques (Gradon and Orlicki, 1990; Balásházy and Hofmann, 1993a).

In order to study particle deposition within bronchial airway bifurcations, we developed a numerical model for the computation of air velocity fields and aerosol particle trajectories in a symmetric three-dimensional bronchial bifurcation model (Balásházy and Hofmann, 1993a; 1993b), which was subsequently extended to incorporate the effects of asymmetry in airway branching and flow division (Balásházy and Hofmann, 1995) and of physiologically realistic shapes of the bifurcation transition zone and the carinal ridge (Heistracher and Hofmann, 1995). In this numerical model, airflow is computed by solving the Navier-Stokes equation in a three-dimensional computer mesh by finite difference or finite volume methods. Trajectories of aerosol particles entrained in the airstream through the bifurcation are then simulated by Monte Carlo techniques, considering the simultaneous effects of gravitational settling, inertial impaction, Brownian motion, and interception. Finally, the resulting particle deposition pattern is determined by the intersection of the particle trajectories with the surrounding surfaces.

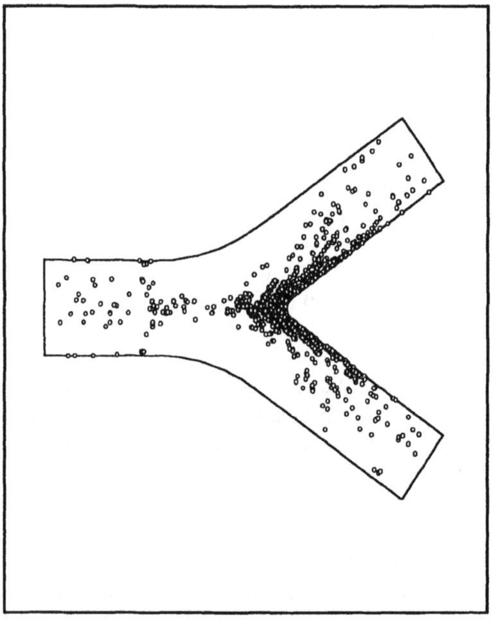

Fig. 6 Inspiratory spatial deposition patterns of 10 μm particles in a symmetric physiologically realistic airway bifurcation (generations 3-4) for over 4,000 randomly selected particle trajectories. The inspiratory flow rate of 60 l min^{-1} corresponds to a respiratory minute volume of 30 l min^{-1}.

As an example, the spatial inspiratory deposition pattern of 10 μm unit density particles in a symmetric, physiologically realistic bifurcation model, representing the airway generation 3-4 juncture in Weibel's (1963) Model A, with an inspiratory flow rate of 60 l min^{-1} is displayed in Fig. 6 for over 4,000 randomly selected particle trajectories. In this simulation, 2,015 out of 4,527 particles entering the bifurcation model will be deposited, resulting in a deposition efficiency of 44.5%. The projection of the three-dimensional deposition pattern onto the y-x plane reveals that deposition is highly inhomogeneous and sites of enhanced deposition can be observed directly at the airstream divider.

Nonuniform deposition patterns were also observed in alveolar ducts and within alveoli (Tsuda et al., 1994). For submicron particle sizes, the deposition pattern was markedly inhomogeneous, with higher deposition near the alveolar entrance ring. Recent numerical fluid dynamics and particle trajectory simulations have further demonstrated that the moving geometry of an alveolus during expansion and contraction produces even more nonuniform deposition patterns of submicron and large particles than for a fixed alveolar geometry.

Conclusions

Current models of particle transport and deposition in the human respiratory tract, ranging from simple semi-empirical to rather sophisticated stochastic and numerical models, permit the prediction of particle deposition patterns at different levels of complexity, ranging from total to differential and localized deposition. All these models have been validated by comparisons with the available experimental data in human test subjects, bronchial airway casts, and bifurcation models, thereby lending further credence to the various model predictions.

Despite the apparent success of particle deposition modeling, three major issues have to be addressed in future efforts: First of all, only very limited information is presently available on the intra- and inter-subject variabilities of the human lung morphometry and their effects on the resulting particle deposition patterns; this requires the statistical analysis of many more lung casts. Secondly, local deposition patterns caused by localized fluid dynamics patterns at airway bifurcations and in alveoli have been studied so far only for very idealized inhalation conditions and thus have to be extended to physiologically realistic *in vivo* breathing conditions. Thirdly, the validation of existing models of particle deposition in individual airway generations has to be improved by comparison with experimental data on bolus dispersion (transport of particles) and computerized tomography (transport and deposition).

Acknowledgements

This research is funded in part by the Austrian Fonds zur Förderung der wissenschaftlichen Forschung, Project P10426-ÖME.

References

Balásházy, I., and Hofmann, W., *J. Aerosol Sci.* **24** (1993) 745-772
Balásházy, I., and Hofmann, W., *J. Aerosol Sci.* **24** (1993) 773-786
Balásházy, I., and Hofmann, W., *J. Aerosol Sci.* **26** (1995) 273-292
Chan, H.K., *J. Aerosol Med.* **6** (1993) 23-36
Cohen, B.S., Harley, N.H., Schlesinger, R.B., and Lippmann, M., *Ann. occup. Hyg.* **32, Suppl. 1** (1988) 1045-1053
Cohen, B.S., and Asgharian, B., *J. Aerosol Sci.* **21** (1990) 789-797
Edwards, D.A., *J. Aerosol Sci.* **26** (1995) 293-317
Egan, M.J., and Nixon, W., *Radiat. Prot. Dosim.* **11** (1985) 5-17
Gradon, L., and Orlicki, D., *J. Aerosol Sci.* **21** (1990) 3-19

Habib, R.H., Chalker, R.B., Suki, B., and Jackson, A.C., *J. Appl. Physiol.* **77** (1994) 441-451
Haefeli-Bleuer, B., and Weibel, E.R., *Anat. Rec.* **220** (1988) 401-414
Hansen, J.E., Ampaya, E.P., Bryant, G.H., and Navin, J.J., *J. Appl. Physiol.* **38** (1975) 983-986
Heistracher, T., and Hofmann, W., *J. Aerosol Sci.* **25, Suppl. 1** (1994) S549-S550
Heistracher, T., and Hofmann, W., *J. Aerosol Sci.* **26** (1995) 497-504
Heyder, J., Blanchard, J.D., Feldman, H.A., and Brain, J.D., *J. Appl. Physiol.* **64** (1988) 321-327
Hofmann, W., and Koblinger, L., *J. Aerosol Sci.* **21** (1990) 675-688
Hofmann, W., and Koblinger, L., *J. Aerosol Sci.* **23** (1992) 51-63
Hofmann, W., and Mainelis, G., *J. Aerosol Sci.* **24, Suppl. 1** (1993) S359-S360
Hofmann, W., Koblinger, L., Brand, P., Ferron, G., and Heyder, J., *Ann. occup. Hyg.* **38, Suppl. 1** (1994) 167-174
Horsfield, K., Dart, G., Olson, D.E., and Cumming, G., *J. Appl. Physiol.* **31** (1971) 207-217
Ingham, D.B., *J. Aerosol Sci.* **6** (1975) 125-132
James, A.C., "Lung Dosimetry", pp. 259-309 in *Radon and Its Decay Products in Indoor Air*, Eds. Nazaroff, W.W., and Nero, E.V., Wiley Interscience, New York (1988)
Kim, C.S., and Iglesias, A.J., *J. Aerosol Med.* **2** (1989) 1-14
Kinsara, A.A., Loyolka, S.K., Tompson, R.V., Miller, W.H., and Holub, R.F., *Health Phys.* **68** (1995) 371-382
Koblinger, L., and Hofmann, W., *Phys. Med. Biol.* **30** (1985) 541-556
Koblinger, L., and Hofmann, W., *J. Aerosol Sci.* **21** (1990) 661-674
Martin, D., and Jacobi, W., *Health Phys.* **23** (1972) 23-29
Martonen, T.B., *Bull. Math. Biol.* **44** (1982) 425-442
Martonen, T.B., Hofmann, W., and Lowe, J.E., *Health Phys.* **52** (1987) 213-217
Martonen, T.B., Zhang, Z., and Yang, Y., *J. Aerosol Sci.* **23** (1992) 389-406
Martonen, T.B., Yang, Y., and Dolovich, M., *J. Thoracic Imaging* **9** (1994) 188-197
National Research Council (NRC), "Comparative Dosimetry of Radon in Mines and Homes", National Academy Press, Washington, DC (1991)
Newman, S.P., Moren, F., Trofast, E., Talee, N., and Clarke, S.W., *Int. J. Pharm.* **74** (1991) 209-213
Overton, J.H., Barnett, A.E., and Graham, R.C., "Significances of the Variability of Tracheobronchial Airway Paths and Their Air Flow Rates to Dosimetry Model Predictions of the Absorption of Gases", pp. 273-291 in *Extrapolation of Dosimetric Relationships for Inhaled Particles and Gases*, Ed. J.D. Crapo, Academic Press, San Diego, CA, USA (1989)
Raabe, O.G., Yeh, H.C., Schum, G.M., and Phalen, R.F., "Tracheobronchial Geometry: Human, Dog, Rat, Hamster", Report LF-53, Lovelace Foundation, Albuquerque, NM, USA (1976)
Scherer, P.W., Shendalman, L.H., Greene, N.M., and Bouhuys, A., *J. Appl. Physiol.* **38** (1975) 719-723
Scheuch, G., and Stahlhofen, W., *Exp. Lung Res.* **17** (1991) 763-787
Scheuch, G., and Stahlhofen, W., *Exp. Lung Res.* **18** (1992) 343-358
Schlesinger, R.B., and Lippmann, M., *Environ. Res.* **15** (1978) 424-431
Stahlhofen, W., Rudolf, G., and James, A.C., *J. Aerosol Med.* **2** (1989) 285-308
Weibel, E.R., "Morphometry of the Human Lung", Springer, Berlin (1963)
Tsuda, A., Butler, J.P., and Fredberg, J.J., *J. Appl. Physiol.* **76** (1994) 2497-2509
Yeh, H.C., and Schum, G.M., *Bull. Math. Biol.* **42** (1980) 461-480
Yu, C.P., and Diu, C.K., *Am. Ind. Hyg. Assoc. J.* **43** (1982) 54-65

The Behavior of Aerosols in the Human Nasal and Oral Airways, Including the Ultrafine Size Region

David L. Swift

Aerosol Research Laboratory, Division of Environmental Health Engineering, Johns Hopkins University School of Hygiene and Public Health, 615 N. Wolfe St., Baltimore, MD 21205, USA

Keywords

Aerosols; Respiratory Tract; Lung Deposition; Ultrafine Particles; Nasal Airway; Oral Airway; Inertia; Diffusion.

Introduction

The role of the upper human airways in modifying incoming air and removing large particles has been well known for many years. Recently, it has become apparent that the upper airways, which include the nasal, oral, pharyngeal, laryngeal and upper tracheal passages, must be viewed as both a filter to protect the more distal airways and a possible site of toxicity. It is also possible that specific deposition sites of these airways may act as portals of entry for molecules which readily pass through the epithelium.

For these reasons, it is important to view these airways not only from a "black box" viewpoint, but as a region of the respiratory tract which requires detailed description of the deposition of particles of various sizes. It is also important to know how the dimensions of the airways influence the flow and deposition of particles and what factors are responsible for the inter-subject variability which has been observed (ICRP, 1995).

In this paper, we shall discuss air flow in the upper airways, the dimensions and shape of the airways, measurement of micrometer size (fine) particles in replicate casts and *in vivo*, measurement of 1-200 nm diameter (ultrafine) particles in replicate airway models, and inter-subject variability.

Air Flow in the Upper Airways

Because of the complex shape of the airways and their inaccessibility without invasive techniques, measurements of air flow must be made in *post mortem* airways (Proetz, 1953) or in replicate airways determined from *post mortem* casting or from *in vivo* imaging such as CT or MRI. Early studies of flow streamlines in nasal casts by Swift and Proctor (1977) showed that flow changed directions abruptly at the nasal valve and became turbulent just distal to this location, about 2 cm behind

J. C. M. Marijnissen and L. Gradoń (eds.), Aerosol Inhalation: Recent Research Frontiers, 103–125.

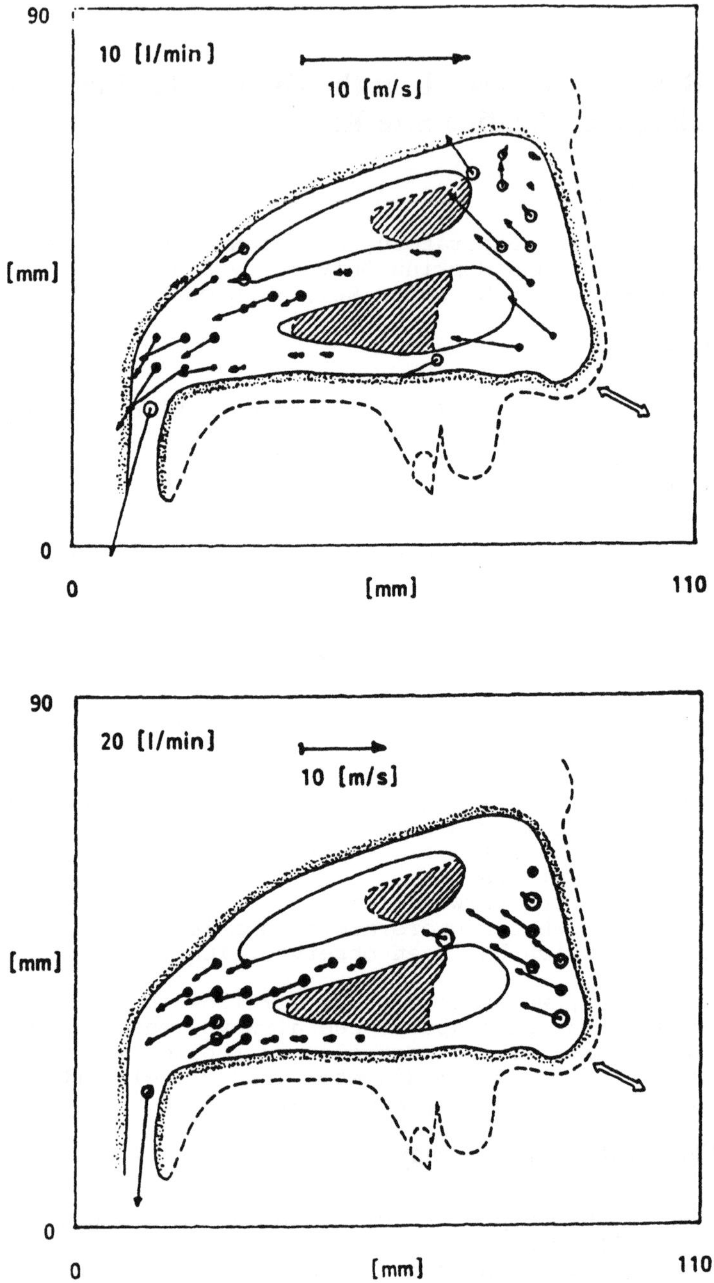

Fig. 1. LDV Measured Velocities in Nasal Replicate Model

the nostril. This is consistent with the fact that heat and moisture transport occurs efficiently in the nasal airway, as described by Proctor and Swift (1977).

Air velocity was studied in a replicate nasal passage by Izundu (1982) using a pitot-static tube, from nostril to mid-trachea. The velocity was found to be highly variable, dependent upon the cross sectional area. The velocity distribution in a nasal-pharyngeal-tracheal replicate model was measured using laser doppler velocimetry (LDV) by Christensen (1988) and found to be turbulent beyond the valve for both inspiration and expiration at volumetric flow rates from 10-30 l min^{-1}. A plot of the velocity distribution for a steady flow of 10 and 20 l min^{-1} in the model is shown in Fig. 1. The hydraulic diameter and Reynolds number throughout the nasal passage was calculated by Cheng *et al* (1988) and is shown in Table 1.

No studies of the flow in the oral airway have been reported in the literature, but one may surmise the nature of flow from the mass and heat transfer characteristics with respect to water vapor and inspired air warming. It is known that the oral passage is not as efficient as the nasal passage in this respect, but the flow may still be assumed to be turbulent.

Dimensions and Shape of the Upper Airways

Several methods have been employed to measure the shape and dimensions of the human upper airways. The nasal airway cross sectional area from the nostril to the nasopharynx can be measured rapidly by acoustic rhinometry (Hilberg *et al*, 1989). However, no information about the shape of the airway can be obtained from this measurement. Nasal airway shape can be determined *in vivo* by computerized tomography (CT) or magnetic resonance imaging (MRI). Because the latter involves no ionizing radiation exposure, it is preferable for normal, healthy subjects.

An MRI scan from nostril to the posterior nasopharynx at 3 mm intervals requires about 40 minutes. Ten normal, adult subjects had their nasal dimensions characterized by this method, as reported by Cheng (1995). A set of nasal passage images for one of the subjects studied by Cheng (1995) is shown in Fig. 2. Such images were used in an *in vivo* deposition study of ultrafine aerosols, in an attempt to find what dimension(s) may account for inter-subject variation. Individual values of the cross sectional area derived from the MRI's are shown in Fig. 3, including the average area-distance for the ten subjects (Cheng, 1995).

Deposition of Fine Particles in a Nasal Passage Replicate

A nasal passage replicate (extending from nostril to mid-trachea)

Table 1. Dimension and Reynolds number of nasal airway cast

Compartments	Cross-section area (cm^2)	Perimeter* (cm)	Hydraulic diameter (cm)	Re† 4 l min^{-1}	Re 20 l min^{-1}	Re 50 l min^{-1}
Vestibule	0.62	4.74	0.52	302	1508	3770
Nasal valve	0.78	5.21	0.60	274	1372	3429
Nasal valve (end)	0.94	8.52	0.44	168	839	2097
Turbinate (beginning)	3.27	16.97	0.77	84	421	1053
Turbinate (middle)	3.46	26.54	0.52	54	269	673
Turbinate (end)	2.69	14.53	0.74	98	492	1230
Nasopharynx	2.12	8.06	1.05	177	887	2217
Mid nasopharynx	0.46	5.02	0.36	285	1424	3559
Pharynx	3.30	9.17	1.44	156	779	1948
Larynx	3.82	10.77	1.42	133	664	1659
Trachea	2.27	5.94	1.53	241	1203	3008

* Data obtained by Izundu (1982).

† $Re = \rho d_h Q/\mu A_c$, where $\rho = 9.82 \times 10^{-4}\,g\,cm^{-3}$ at 23°C and 626 mmHg and $\mu = 1.832 \times 10^{-4}$ poise at 626 mmHg.

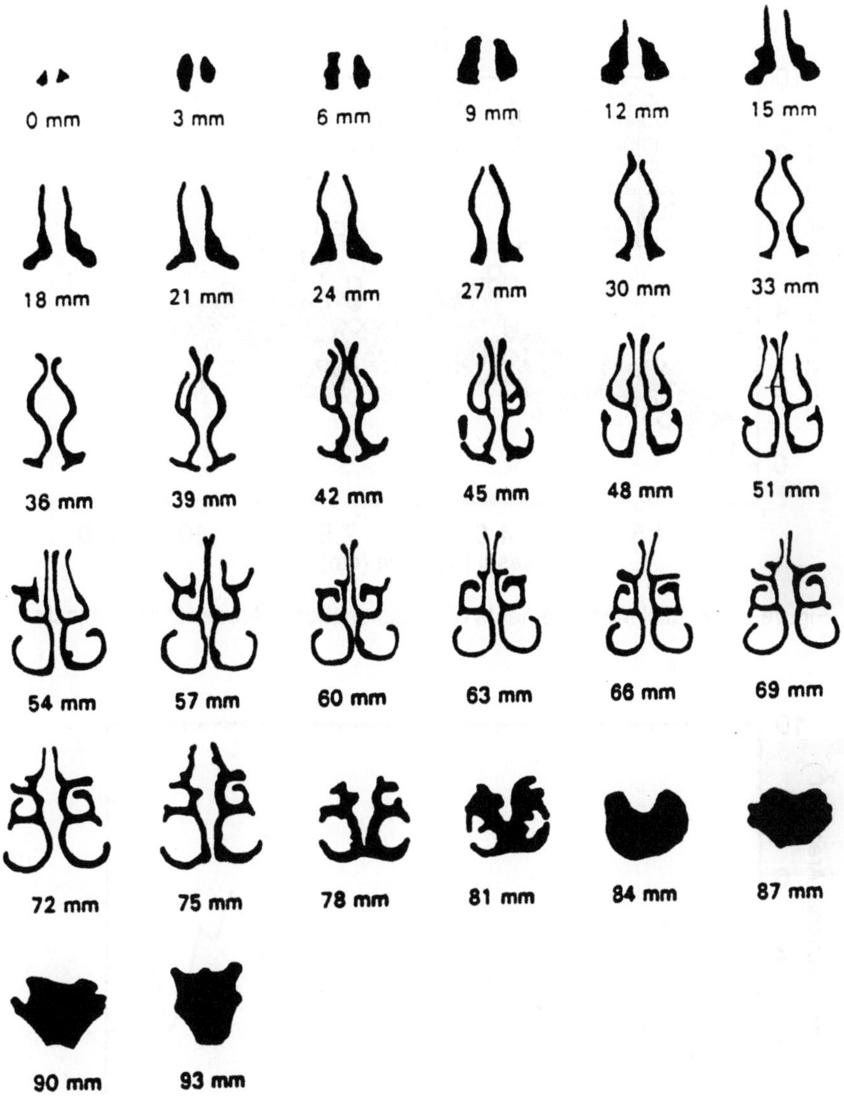

Fig. 2. Coronal Scans of Nasal Passage by MRI

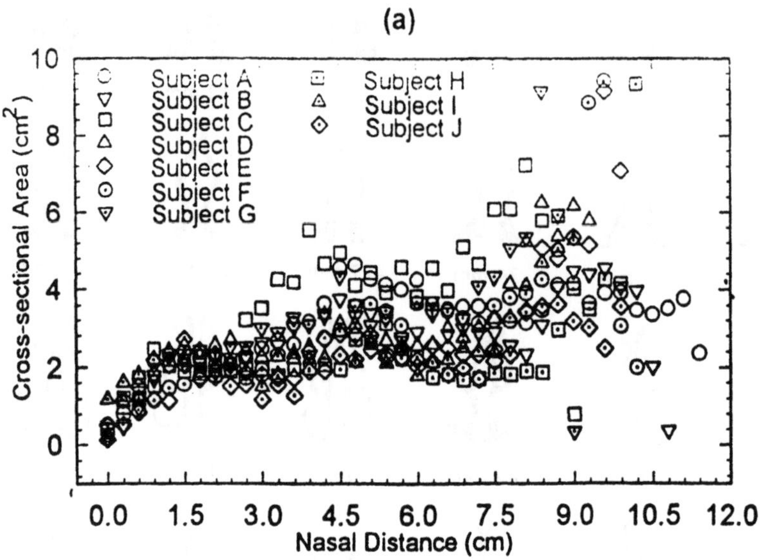

Figure 3 a MRI Measurements of Cross-sectional Areas for 10 Subjects

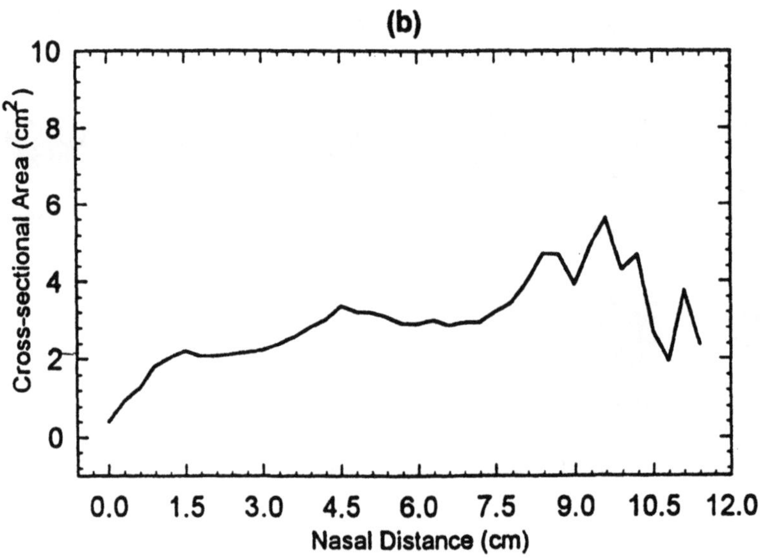

Figure 3 b Mean Values of Cross-sectional Areas for 10 Subjects

Fig. 3. Computed Nasal Passage Cross Sectional Areas for Ten Subjects

of a normal adult, denoted ANT1, was connected to an aerosol storage box into which monodisperse NaCl aerosols ranging from 0.6-8.0 μm diameter (produced by a spinning top generator) were introduced. The aerosols were drawn through the replicate at constant inspiratory flow rates of 7, 15, 30 and 50 l min^{-1}. The deposition efficiencies for particle diameters ranging from 0.6-4.1 μm are shown in Fig. 4 (Swift, 1991); it can be seen that at any diameter, deposition efficiency increases with flow rate. By plotting the deposition efficiency as a function of the inertial parameter d_p^2Q, where d_p is particle diameter and Q is flow rate, the curves collapse to a single efficiency curve (Fig. 5). The sigmoid shape of this curve is characteristic of an idealized impactor, and is in good agreement with the averaged *in vivo* curve for a number of normal human subjects presented by the ICRP (1995).

Constant flow rate inspiratory deposition experiments were also carried out in the same nasal replicate with polydisperse di-2-ethyl hexyl sebacate (DEHS) aerosols by Kesevanathan and Swift (1995). Flow rates of 15, 20, 30 and 35 l min^{-1} were employed in these experiments. Average deposition efficiency values were plotted, as above, as a function of d_p^2Q (Fig. 6). The best fit equation which is shown in Fig. 6, is given by:

$$DE_{insp} = 100 - 1000\exp(-0.0001 d_p^2 Q) \quad (1)$$

In Vivo Nasal Deposition of Fine Particles

Results of past studies in normal human subjects were presented by ICRP (1995) and are shown in Fig. 7. Large inter-subject variability can be observed in this data set, but the average deposition curve is in good agreement with the deposition curve for the replicate model, Eq. 1. No nasal dimensional data is available for the subjects whose nasal deposition efficiencies are shown in Fig. 7. It should be noted that all the subjects reported in Fig. 7 are adult, healthy, male, non-smoking and European. It is likely that their nasal dimensions differed, but it is not known what differences existed during their nasal deposition efficiency measurements.

One of the goals of the nasal deposition studies being performed at our laboratory is to compare nasal deposition in individuals of different ethnic origins, with different measured nasal dimension, in order to see if a relationship can be established which indicates the range of depositions to be expected over a wide ethnic range and is predictive on the basis of nasal dimensions. As of yet, no single nasal dimension, such as the cross sectional area of the nasal valve, can be said to be determinative of nasal deposition efficiency. Results of preliminary studies are presented below.

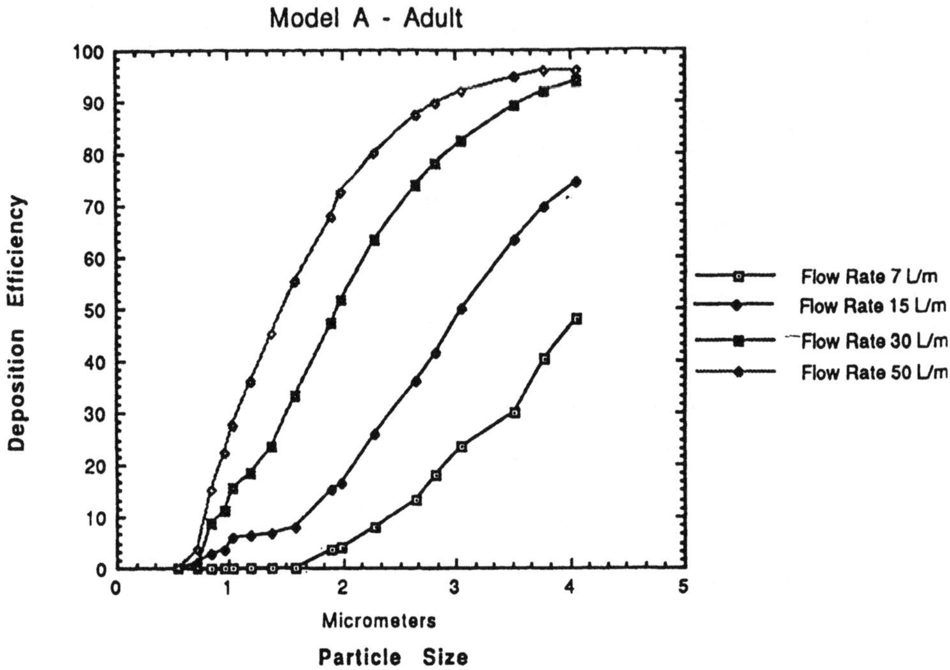

Fig. 4. Inspiratory Deposition Efficiencies in Nasal Replicate Model

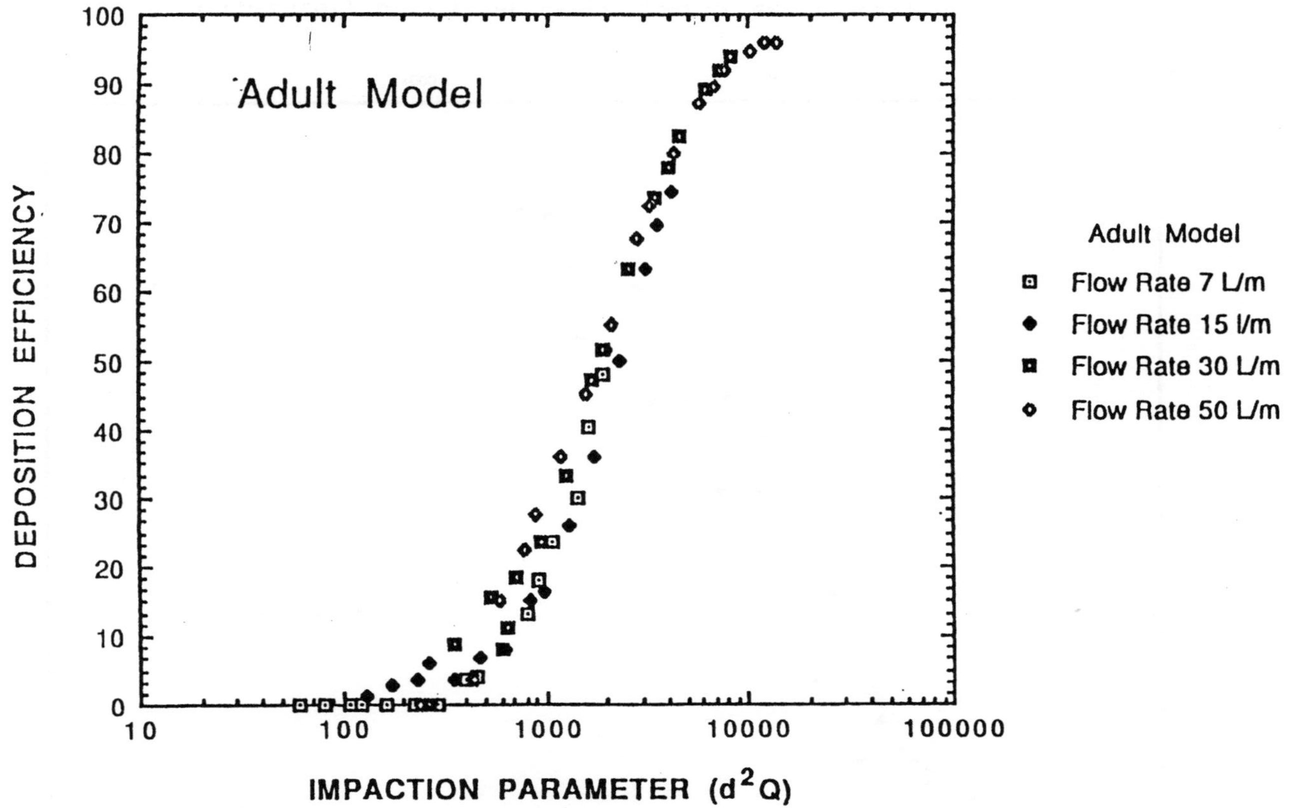

Fig. 5. Nasal Deposition Efficiencies vs. Inertial Parameter, d_p^2Q

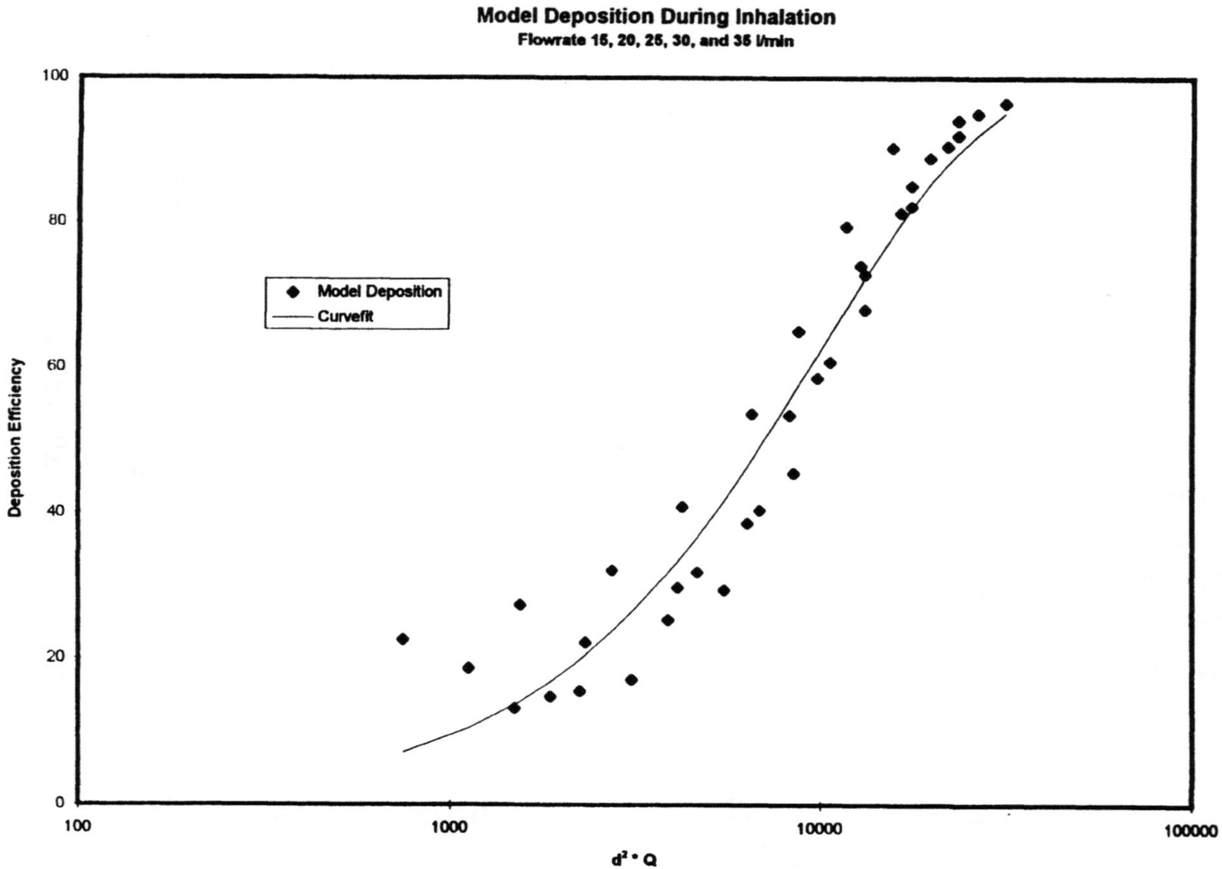

Fig. 6. Inspiratory Deposition Efficiency of Nasal Model Using Polydisperse DEHS Aerosol

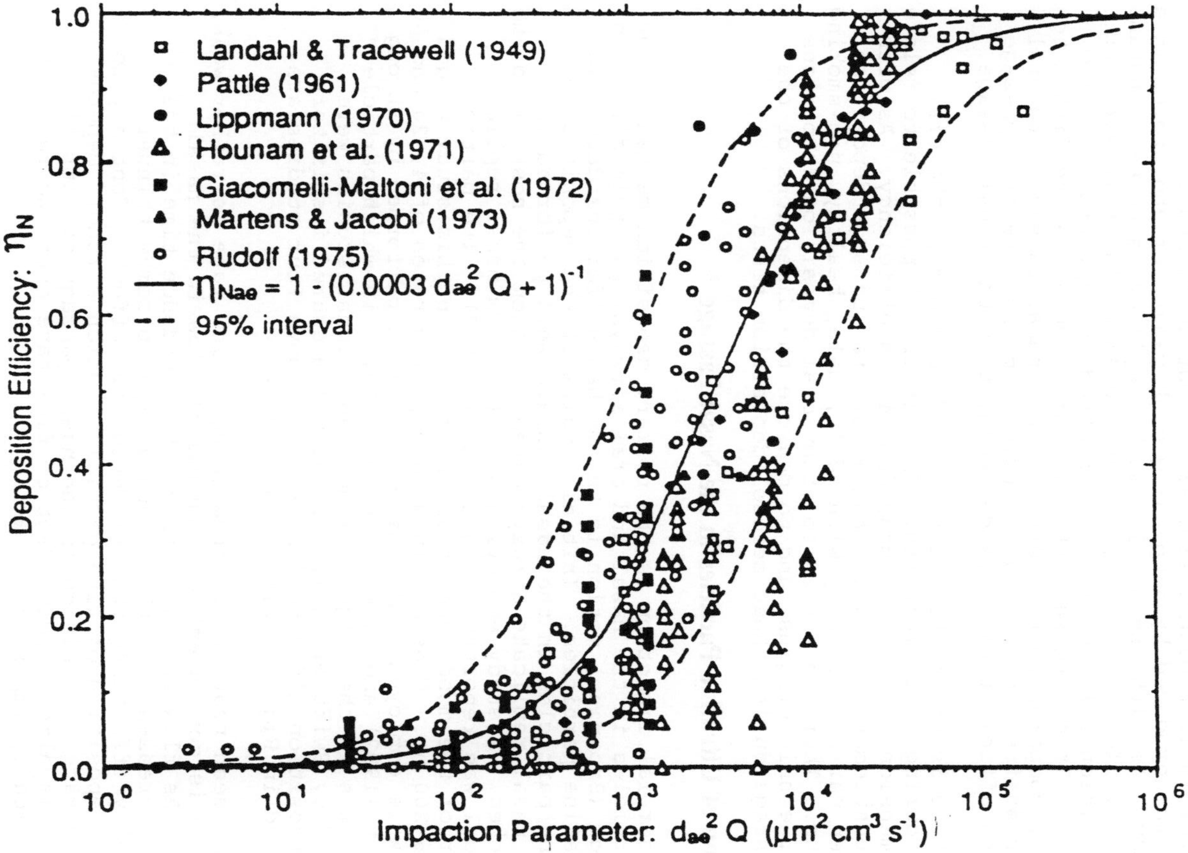

Fig. 7. Nasal Inspiratory Deposition Results from Past Studies from ICRP (1995)

Deposition of Fine Particles in Oral Passage Replicate Models

An oral passage replicate model was studied with monodisperse NaCl aerosols in a manner similar to that described above. The aerosol was produced by a spinning disk generator and stored in a 100 l box. As with the nasal passage replicate studies, constant inspiratory flow rates of 7, 15, 30 and 45 l min^{-1} were employed. The deposition efficiencies were plotted vs. d_p^2Q and are shown in Fig. 8. It can be seen that a sigmoid curve can be fitted to the data; compared to the curve for the nasal passage, the oral curve is shifted to the right, indicating that the oral passage is a less efficient particle collector than the nasal passage.

Very little is known about the exact dimensions of the oral passage during oral breathing. Thus it is not possible to relate oral dimensions to oral passage deposition efficiency. As with nasal measurements, *in vivo* studies of oral passage deposition have been carried out only with adult, male, healthy, non-smoking Europeans. There is a need to study oral deposition in persons of other ethnic origins, and to measure the dimensions of the oral passage during these aerosol deposition tests.

Deposition of Ultrafine Particles in the Nasal Passage

While inertial impaction is the primary deposition mechanism for fine particles in the nasal and oral airways, particle diffusion by random (Brownian) motion is the mechanism for the deposition of ultrafine particles. Initial studies of the deposition of ultrafine particles in the nasal passage were performed in several nasal replicate airways extending from the nostril to the mid-trachea. A study employing a *post mortem* nasal replicate cast model was reported by Cheng *et al* (1988). The particle diameter of the monodisperse aerosols used in this study ranged from 4.6-200 nm. Constant inspiratory flow rates of 4, 20 and 50 l min^{-1} were employed. Similar studies were performed by Strong and Swift (1992) and by Cheng *et al* (1991), in which particle diameter was in the range of 1 nm. Additional studies were performed by Swift *et al* (1992) and the measured inspiratory nasal deposition efficiencies from the above four studies are summarized in Fig. 9 (Swift *et al*, 1992).

It can be seen that the dependence upon flow rate is rather weak, $Q^{-0.125}$, and is consistent with turbulent flow in the nasal passage. As flow rate increases, the residence time in the passage decreases, but the boundary layer becomes thinner, practically offsetting the residence time effect. Contrary to inertial deposition of fine particles, which strongly depends upon flow rate, deposition in the nasal passage of ultrafine particles is almost independent of flow rate over the range of physiologically important flows.

The dependence of inspiratory deposition efficiency upon particle

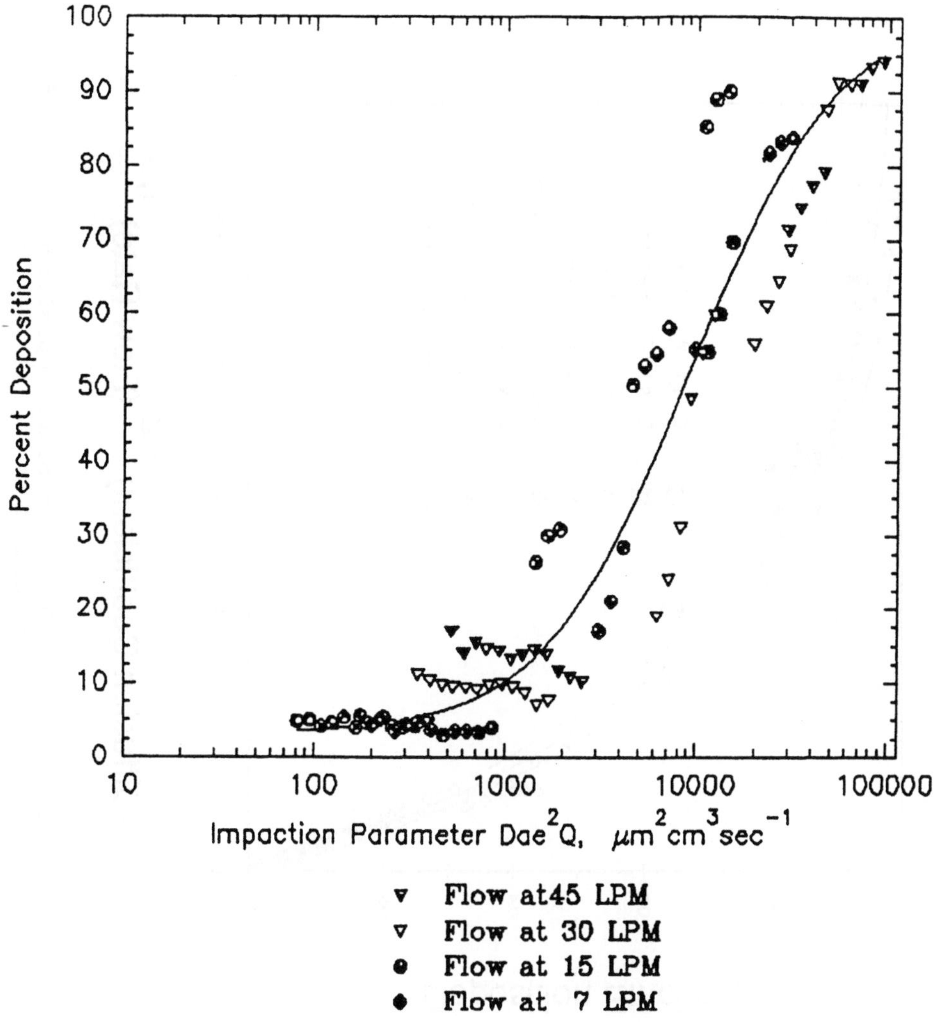

Fig. 8. Oral Passage Inspiratory Deposition Efficiency for Replicate Model vs. Inertial Parameter, d_p^2Q

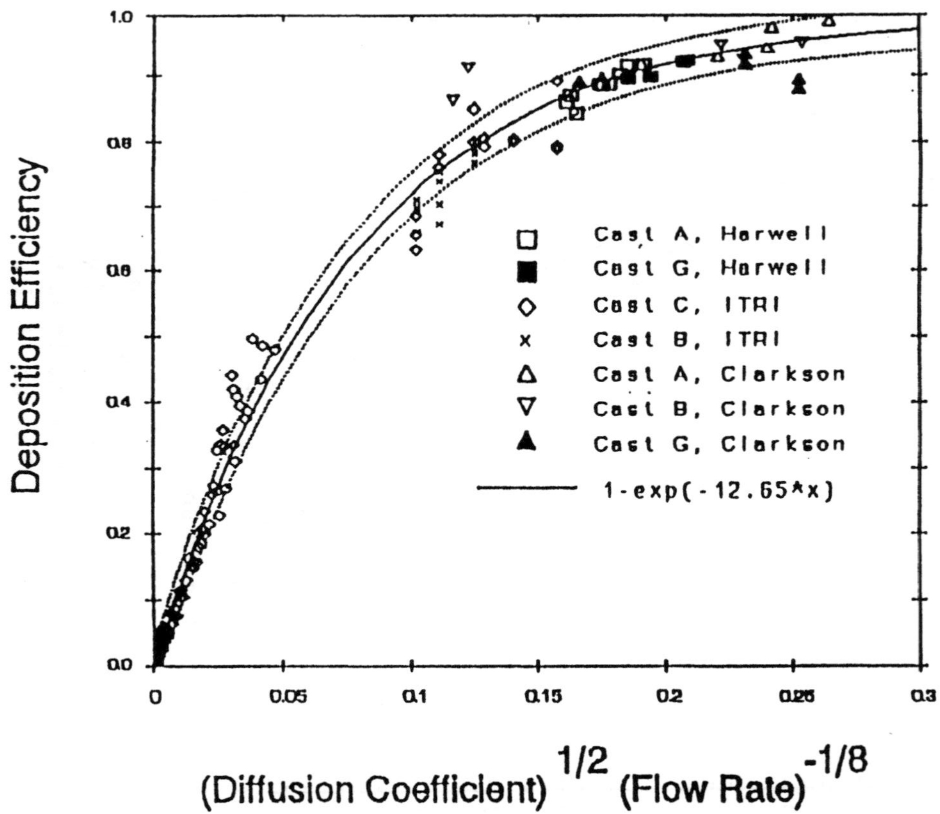

Fig. 9. Inspiratory Nasal Replicate Model Deposition of Ultrafine Aerosols from Four Studies

diffusivity, D cm^2 sec^{-1}, is much stronger. In the early work (Yamada *et al*, 1988), the particle diameters extended downward to 4.6 nm, and the best fit curve gave a dependence of $D^{2/3}$. However, with the addition of data down to 1 nm, the dependence of deposition upon particle diffusivity over the entire particle diameter range was found to be $D^{1/2}$. There is thus a significant effect of changing particle diameter upon diffusional deposition which is important because in many cases, ultrafine particles are dynamic with respect to diameter, growing by vapor uptake and by coagulation with similar and larger diameter particles.

The replicate models used in the studies reported in Swift *et al* (1992), Fig. 9, included both adult and small child (Cast G, 1.5 year old child) nasal airways. It can be seen that all models gave approximately the same inspiratory deposition efficiency, an important conclusion when considering the particle distribution in the whole population.

The local nasal passage diffusional deposition of 1.3 nm particles (^{212}Pb, the progeny of Rn-220 decay) was studied by Cheng *et al* (1993). A "sandwich" model of the nasal passage, composed of 3 mm sections fashioned from MRI scans, was used for these studies. After the aerosol exposure at a constant inspiratory flow rate of 20 l min^{-1}, the model was disassembled and each section was counted for ^{212}Pb deposition. Each section was characterized by a surface area obtained by the product of the section thickness and the airway perimeter. From this information and the particle diffusivity, the surface deposition density (% cm^{-2}) and the mass transfer coefficient (cm sec^{-1}) could be calculated for each section. These values are plotted in Fig. 10, where it can be seen that both parameters are maximum 3 cm distal to the nostril entrance, just beyond the nasal valve.

Deposition of Ultrafine Particles in the Oral Passage

The deposition of monodisperse ultrafine aerosols of diameter 5-200 nm in a human *post mortem* oral cast (denoted Model C) was measured by Cheng *et al* (1990) at both inspiratory and expiratory flow rates of 4, 20 and 40 l min^{-1}. As with the deposition studies in the nasal passage over this particle size range, the dependence of deposition upon flow was weak, $Q^{-0.125}$, while that of particle diffusivity was stronger, $D^{2/3}$. Empirical equations were presented for both inspiration and expiration. If the product $Q^{-0.125}D^{2/3}$ is denoted χ and the deposition efficiency is denoted η, the fitted equations were:

$$\eta_{insp} = 1 - \exp(-40.2\chi) \tag{2}$$

for inspiration, and:

$$\eta_{exp} = 1 - \exp(-34.3\chi) \tag{3}$$

for expiration. Subsequently, the deposition of smaller diameter

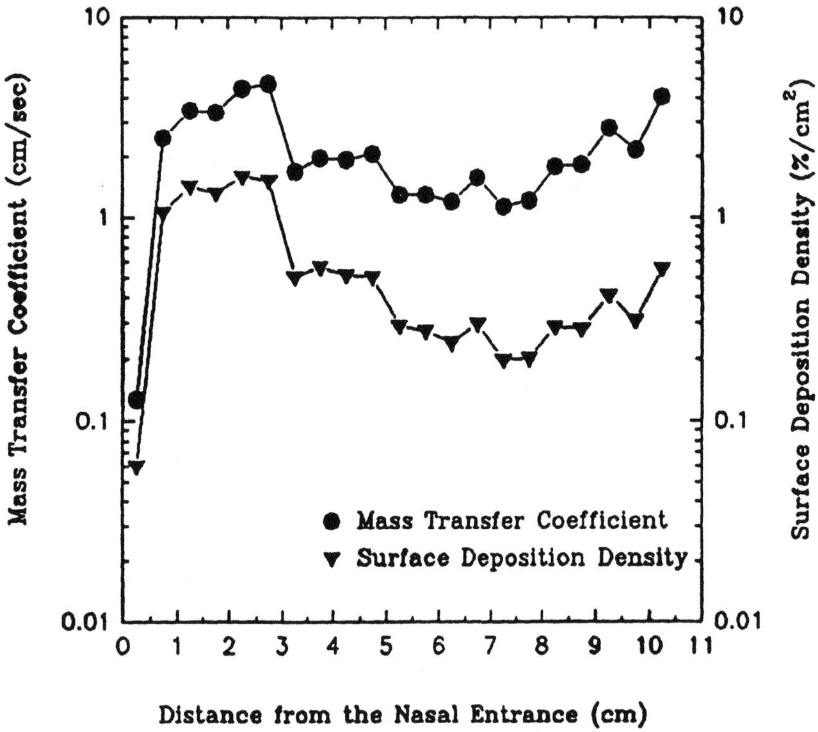

Fig. 10. Mass Transfer Coefficient and Surface Deposition Density of Ultrafine Aerosol Particles

ultrafine particles, down to 1 nm, was measured in this model and another adult oral passage model fashioned from an MRI scan, denoted Model B (Swift *et al*, 1994). As with the nasal studies, when the smaller diameter particles were studied, the dependence upon diffusivity changed from the ⅔ power to the ½ power. The empirical inspiratory deposition equation, obtained from these data shown in Fig. 11 (in which the parameter $X \equiv Q^{-0.125}D^{0.5}$) is:

$$\eta_{insp} = 1 - \exp(-9.925X) \qquad \textbf{(4)}$$

Inter-subject Variability of Upper Airway Particle Deposition

As noted above, it is known for the case of nasal passage fine particle deposition that significant inter-subject variability exists (ICRP, 1995). However, the studies reported on inter-subject variability have not included any dimensional information on the nasal passage with which differences may be correlated, nor have the studies included ethnic groups other than European. Recent studies have been undertaken in order to investigate the factors related to inter-subject variability by Kesevanathan and Swift (1995) for fine particles, and by Cheng (1995) for ultrafine particles.

The studies of fine particles were carried out with 40 subjects, among which were European-Americans and African-Americans. All the subjects were characterized by acoustic rhinometry (AR)(from which the minimum cross sectional area of the nasal passage was obtained), and the shape of the nostril entrance. From the outline of the nostril entrance, the degree of ellipticity could be derived, defined as the ratio of the major axis/minor axis, E. Deposition of fine particles from 1-10 μm diameter was measured in these subjects at a single flow rate of 25 l min^{-1}.

The data are shown in Fig. 12, including a best fit curve for all 40 subjects in whom the aerosol was drawn through both nostrils. An analysis of the best fit dependence of deposition efficiency upon particle diameter, d_p, minimum area, A_{min}, and nostril ratio, E, yielded the exponents which are shown in Fig. 12. Even when these dimensional factors are included in the plot of nasal inspiratory deposition, there remains a significant degree of inter-subject variability.

Ten normal adult subjects were studied with ultrafine aerosols by Cheng (1995). Particle diameters of 4, 8, 20 and 150 nm were studied at flow rates of 10 and 20 l min^{-1}. The subjects were characterized by MRI scans (done prior to the aerosol deposition experiments) and AR, performed at the time of the measurement. The deposition efficiencies were assumed to depend upon: (1) particle diffusivity, D, (2) flow rate, Q, (3) a shape factor, S_f, an MRI derived factor which expresses the degree of convolution of the turbinate region, and (4) the ratio A_s/A_{min}, where A_s is the total surface area of the nasal passage (derived from the MRI scan) and A_{min} is the minimum cross sectional area of

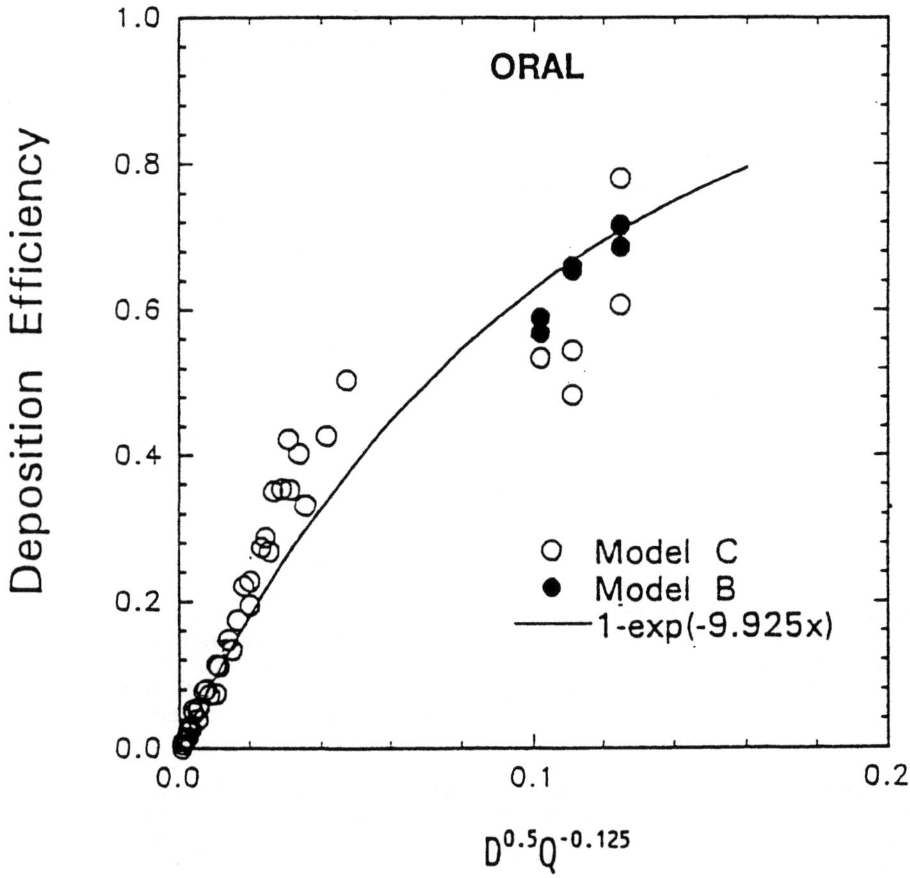

Fig. 11. Replicate Oral Passage Model Deposition of Ultrafine Aerosols

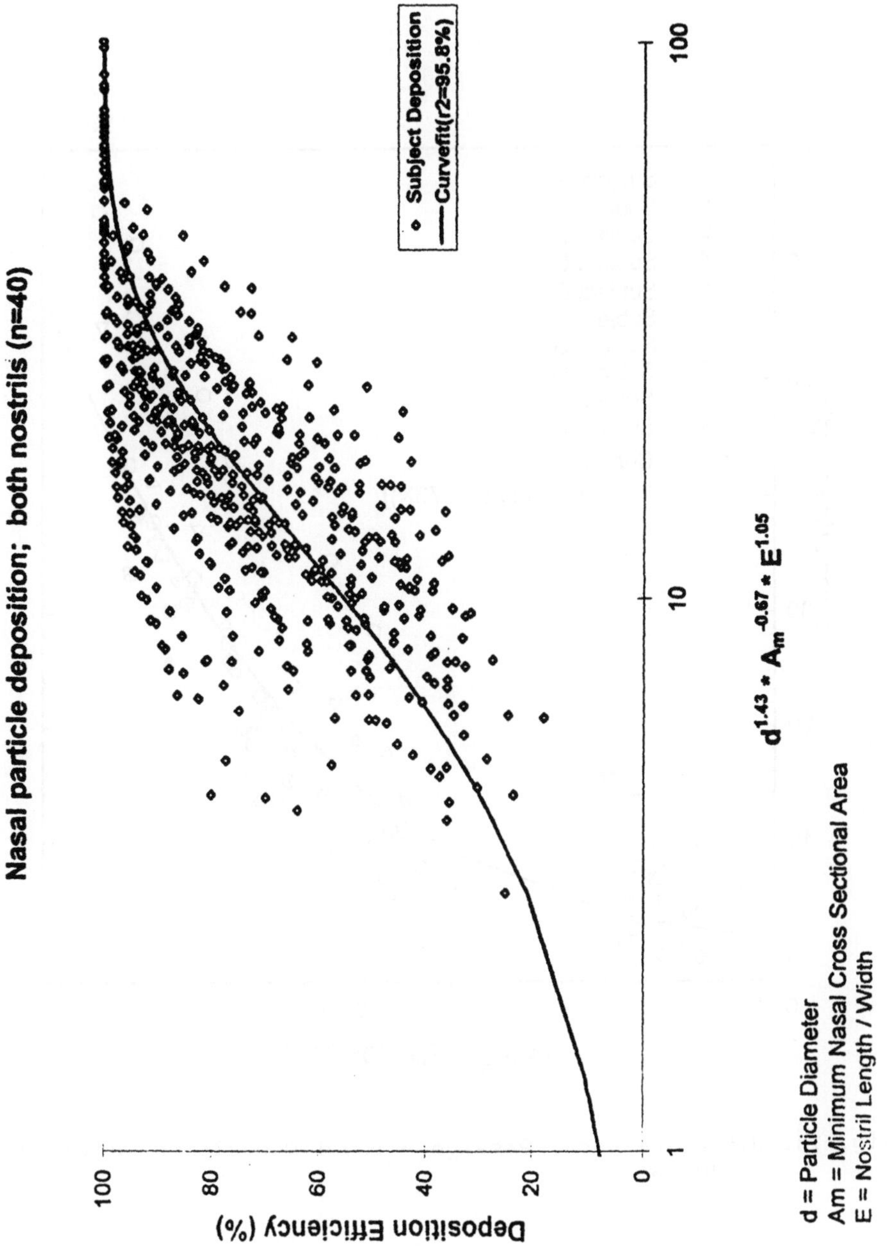

Fig. 12. Inter-subject Variability of Fine Particle Nasal Deposition in 40 Normal Subjects

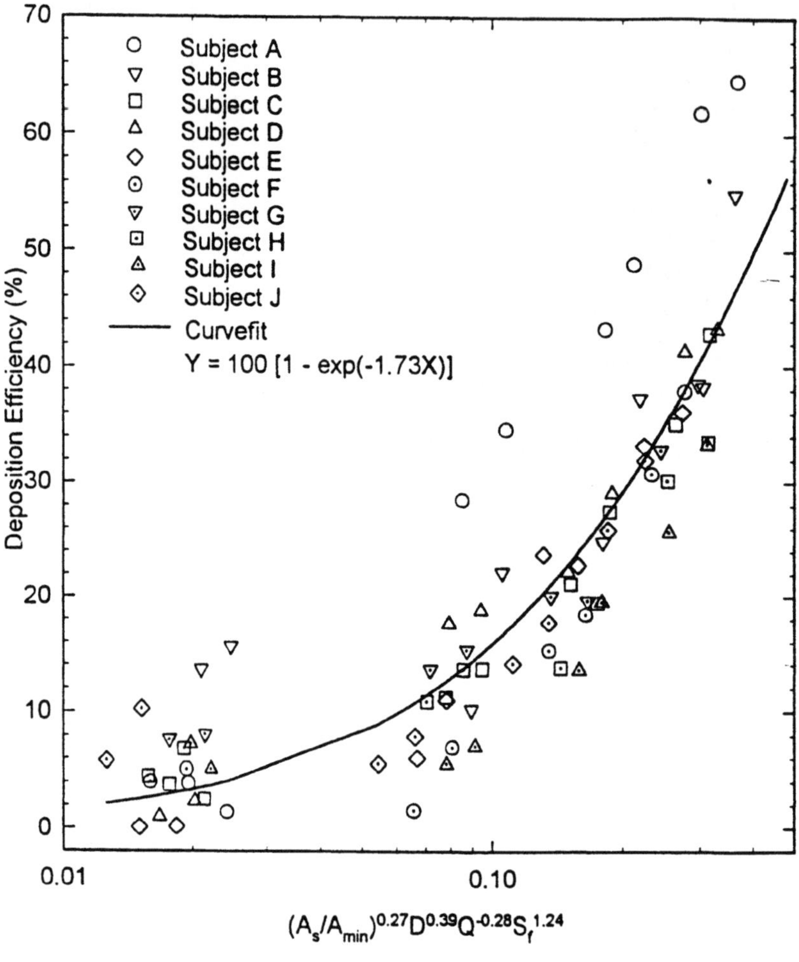

Fig. 13. Ultrafine Nasal Passage Aerosol Deposition in 10 Normal Subjects

the nasal airway (derived from AR).

A similar parameter fitting procedure was used to determine the parameter powers which gave the best fit curve for the data. The data are shown in Fig. 13. Most of the ten subject fall near the best fit curve, with the exception of Subject A at small particle diameters and Subject G at large particle diameters. This indicates that the choice of normalizing factors is still not able to bring all subjects into proximity of the best fit line, although the fit is better than that seen in plots of inspiratory nasal deposition vs. D^nQ^m, such as presented in ICRP (1995).

Conclusion

The upper airways are an important region of the respiratory tract when considering the regional deposition of particles ranging in diameter from 1 nm-20 μm. Deposition of inhaled particles in this region acts as a protective mechanism for the more distal airways, but such deposition may lead to toxic effects. In respiratory dose estimations for the whole population, it is important not only to know the average behavior of inhaled particles with respect to the upper airways, but the degree of inter-subject variability so that limits of exposure may be estimated.

The research efforts described above have shown that studies using replicate casts of the upper airways yield much useful information which is either difficult or impossible to obtain *in vivo*. Where studies in human subjects have been undertaken, they are found to be in generally good agreement with the replicate model results. The replicate studies are also valuable as range finding studies, allowing investigators to more carefully design *in vivo* studies.

Studies with ultrafine aerosols have yielded new information on the dependence of deposition upon particle diameter and flow rate, and have demonstrated that as particle size approaches 1 nm, the deposition in the nasal airway is essentially quantitative. Oral passage deposition is also high for these particle diameters, but is less than that for the nasal passage. Local deposition studies in replicates indicated the pattern of deposition for ultrafine particles.

Inter-subject variability is an issue which remains difficult to understand. Efforts continue to find those dimensional parameters which are critical in determining the deposition efficiency. It is important both for constructing population exposure models and for including different ethnic groups in the work place in terms of respiratory deposition patterns of inhaled particles. Increasing interest in the behavior of ultrafine particles means that further work on respiratory deposition patterns of these particles is needed.

Acknowledgement

Recent work at Johns Hopkins cited in this paper was carried out with Kuo-Hsi Cheng, Jana Kesevanathan, and Mark Bickert. MRI scans were provided by Raymond A. Guilmette of the Inhalation Toxicology Research Institute (ITRI), Albuquerque, NM. Collaborative studies were carried out with: (1) Yung-Sung Cheng, Hsu-Chi Yeh, Yin-Fong Su, and Yuichi Yamada at ITRI, (2) John C. Strong at AEA Technology, Harwell, UK, and (3) Philip K. Hopke, Natalie Montassier, and Kim Karpen-Hayes at Clarkson Univ., Potsdam, NY. The important contributions of these collaborators is gratefully acknowledged.

References

Cheng, K-H, *In Vivo Measurements of Nasal Airway Dimensions and Deposition of Ultrafine Aerosols in the Extrathoracic Airways*, Ph.D. Thesis, Johns Hopkins Univ. (1995)

Cheng, Y-S, Yamada, Y., Yeh, H-C, and Swift, D.L., Diffusional Deposition of Ultrafine Aerosols in a Human Nasal Cast, *J. Aerosol Sci.*, **19**, 741-751 (1988)

Cheng, Y-S, Yamada, Y., Yeh, H-C, and Swift, D.L., Deposition of Ultrafine Aerosols in a Human Oral Cast, *Aerosol Sci. and Tech.*, **12**, 1075-181 (1990)

Cheng, Y-S, Yeh, H-C, and Swift, D.L., Aerosol Deposition in Human Nasal Airways for Particles 1 nm to 20 μm: A Model Study, *Rad. Prot. Dosimetry*, **38**, 41-47 (1991)

Cheng, Y-S, Su, Y-F, Yeh, H-C, and Swift, D.L., Deposition of Thoron Progeny in Human Head Airways, *Aerosol Sci. and Tech.*, **18**, 359-375 (1993)

Christensen, J.V., Personal Communication (1988)

Hilberg, O., Jackson, A.C., Swift, D.L., and Pedersen, O.F., Acoustic Rhinometry: Evaluation of Nasal Cavity Geometry by Acoustic Reflection, *J. Appl. Physiology*, **66**, 295-303 (1989)

ICRP, *Human Respiratory Tract Model for Radiological Protection*, ICRP Rept. 66, Annexe D. Deposition of Inhaled Particles, pp.231-299, Oxford, Pergamon Press (1995)

Izundu, C.S., *Pressure, Velocity and Efficiency Measurements in a Model of Human Upper Airways*, M.S. Thesis, Johns Hopkins Univ. (1982)

Kesevanathan, J. and Swift, D.L., Unpublished data (1995)

Proctor, D.F. and Swift, D.L., Temperature and Water Vapor Adjustment, Chapter in *Respiratory Defense Mechanisms, Part 1*, pp. 95-124, ed. J.D. Brain, D.F. Proctor and L.M. Reid, New York, Marcel Dekker (1977)

Proetz, A.W., *Applied Physiology of the Nose, 2nd Ed.*, St.Louis, Annals Publishing Co. (1953)

Strong, J.C. and Swift, D.L., Deposition of "Unattached" Radon Daughters in Models of Human Nasal and Oral Airways, in *Indoor Radon and Lung Cancer: Reality or Myth, Part 1*, pp. 227-234, ed. F.T. Cross, Richland, WA, Battelle Press (1992)

Swift, D.L. and Proctor, D.F., Access of Air to the Respiratory Tract, Chapter in *Respiratory Defense Mechanisms, Part 1*, pp. 63-94, *Ibid.* (1977)

Swift, D.L., Experimental Studies of Inertial Aerosol Deposition in Human Upper Airway Casts: Implication for the Proposed NCRP Lung Model, *Rad. Prot. Dosimetry*, **38**, 29-34 (1991)

Swift, D.L., Montassier, N., Hopke, P.K., Karpen-Hayes, K., Cheng, Y-S, Su,Y-F, Yeh, H-C, and Strong, J.C., Inspiratory Deposition of Ultrafine Particles in Human Nasal Replicate Cast, *J. Aerosol Sci.*, **23**, 65-72 (1992)

Swift, D.L., Cheng, Y-S, Su, Y-F, and Yeh, H-C, Ultrafine Aerosol Deposition in the Human Nasal and Oral Passages, in *Inhaled Particles VII*, pp. 77-81, ed. J. Dodgson and R.A. McCallum, Oxford, Pergamon Press (1994)

A Deposition Model for Fibres in the Deep Parts of the Lung Based on Similarities of Bead Beds and Human Lungs

Y.Zhou, J.C.M.Marijnissen, S.M.Lemkowitz and B.H.Bibo

Sections Risk Management & Particle Technology
Faculty of Chemical Technology & Material Science
Delft University of Technology
Julianalaan 136, 2628 BL, Delft, The Netherlands

Keywords: Deposition, Fiber, Granular bed filter, Human lung, Lung model

Introduction

Fibrous particles are those with an elongated shape. They include glass fibres and asbestos. Asbestos has become very widely used because it is a low-cost material with very desirable chemical and physical properties, for instance, excellent heat and fire resistance. This unique combination makes the economical replacement of asbestos very difficult in many applications.

The mining and use of asbestos have always been associated with disease among the workers involved. Occupational diseases resulting to fibre inhalation are related to the deposition of those particles in the lung. Deposition of fibre particles in the lung is thought to be determined primarily by diameter and secondarily by the aspect ratio (Timbrell, 1965). Due to geometric differences, the deposition pattern of fibrous particles is different from that of spherical particles. It has long been hypothesized that fibre particles tend to align themselves with the direction of flow and are therefore more likely to penetrate deeper into the lung than spherical particles of the same aerodynamic diameter for aligned fibers.

However, this theory has never been proven by reliable experimental data. Toxicologists have until recently been unable to conduct experiments with inhalable fibres with well-defined lengths and diameters. Such fibre fractions have simply not been available due to the difficulties of preparing them directly or of fractionating fibres from mixtures on the basis of length and diameter.

Fibres which deposit in the lower regions of human lung are removed from the lung only very slowly, and therefore they are more dangerous than those deposited in the upper parts of the respiratory tract, where clearance occurs much more quickly. Occupational hygiene standards are usually based on alveolar deposition. From this point of view, this project is focused on investigating the deposition behaviour in the lower parts of the lung as a function of the physical parameters of fibres and the physical conditions existing in the human respiratory system during breathing.

The human lung respiratory tract

The human respiratory system can be classified into three regions (Hinds, 1982), each covering one or more anatomical regions. These regions differ markedly in structure, airflow patterns, function, and sensitivity to deposited particles. The first is the **head region**, which

J. C. M. Marijnissen and L. Gradoń (eds.), Aerosol Inhalation: Recent Research Frontiers, 127–142.

includes the nose, mouth and pharynx. Inhaled air is warmed and humidified in this region. The second is the **tracheobronchial region** (1 to 16 generation), which includes the airways from the larynx to the terminal bronchioles. This region resembles an inverted tree with a single trunk, the trachea, subdividing into smaller and smaller branches. Finally, beyond the terminal bronchioles is the **pulmonary or alveolar region** (17 to 23 generation), where gas exchange takes place.

A large variety of lung models exists. Over the years these lung models have become increasingly refined. A major improvement, and today's almost standard lung model, was developed by Weibel (1963, Weibel A model). It incorporates 23 generations of symmetrically dividing tubes of which the trachea is the 0th generation. The tubes of each generation end in the two tubes of the next generation. The length and diameter of the individual tubes can be described by simple mathematical relations. The lung structure dimensions based on Weibel model is shown in Table 1.

Table 1. Dimensions and average inspiratory flow conditions in the human lungs; Tidal volume = 1450 cm^3 (see list of symbols)

z (-)	N_z (-)	D_z (cm)	L_z (cm)	S_{lung} (cm^2)	V_{lung} (cm^3)	$A_{s,lung}$ (cm^2)	U_{lung} (cm/s)	$t_{r,lung}$ (s)
8	256	0.186	0.64	7.0	4.5	95.7	120	0.005
9	512	0.154	0.54	9.5	5.1	133.8	87.3	0.006
10	1024	0.130	0.46	13.5	6.2	192.4	61.5	0.007
11	2048	0.109	0.39	19.1	7.5	273.5	43.5	0.009
12	4096	0.095	0.33	29.1	9.6	403.4	28.7	0.012
13	8192	0.082	0.27	43.3	11.7	569.8	19.3	0.014
14	16.4e3	0.074	0.23	70.5	16.2	876.1	11.8	0.020
15	32.8e3	0.066	0.20	112	22.4	1358.9	7.42	0.027
16	65.5e3	0.060	0.165	185	30.5	2038.3	4.5	0.037
17	131e3	0.054	0.141	300	42.3	3135.3	2.78	0.051
18	262e3	0.050	0.117	514	60.1	4817.8	1.62	0.072
19	524e3	0.047	0.099	909	90	7664	0.916	0.108
20	1.05e6	0.045	0.083	1670	139	12304	0.5	0.166
21	2.10e6	0.043	0.070	3050	213	19831	0.273	0.256
22	4.19e6	0.041	0.059	5532	326	31875	0.150	0.394
23	8.39e6	0.041	0.050	11077	554	54025	0.075	0.666

The granular bed filter as a model for the lower parts of a human lung

In modelling particle deposition in the upper parts of the lung (Weibel (1963) generations 1 to 7), many researchers used glass bifurcations or airway casts from a real lung. However, as it is impossible to build a model of the lower parts of the respiratory tract (Weibel generations 8 to 23) with bifurcation or airway casts, the possibility of representing these parts by a granular bed filter, which is sometimes used for removing micron and sub-micron size particles from gases, was investigated. Deposition of particles in a lung is more often compared with filtration, because it relies on the same basic mechanisms that cause collection in a filter.

To simulate the human lung geometrically, the hydraulic diameter and length of the void per unit-cell of the granular bed filter should be equal to those of an airway channel in the lung. Internal surface area, cross section and volume in a granular bed filter layer should be equal to those in an airway generation. As the void per unit-cell is compared to an airway channel, the number of airway channels in each generation should be equal to the number of beads in the corresponding layer.

Especially important parameters (for the flow condition) calculating deposition probabilities are the airway residence time (in case of sedimentation and diffusion) and the airway velocity (in case of impaction and interception). The granular bed filter should therefore be designed to satisfy as many as possible of the following criteria:

- The hydraulic diameter of the voidspace per unit-cell is equal to the hydraulic diameter of an airway channel.
- The total cross section of the airspaces in a granular bed filter layer is equal to the total cross section of the corresponding airway generation.
- The total void volume in a layer is equal to the corresponding airway generation.
- Velocity and flow rate in each layer are the same as in the corresponding airway generation.

The calculated dimensions of each granular bed filter layer, representing an airway generation z (calculated with equations mentioned above), are shown in Table 2. Each airway generation is represented by one layer of beads.

Comparing Table 1 with Table 2, the residence times are seen to differ in the two systems. The ratios vary between 0.56 and 1.7. This is due to variation in volume, because residence time is flow rate divided by volume and the flow rate is equal in both systems. Variation in residence time will affect sedimentation and diffusion efficiencies.

Deposition in the last two generations of the lung, which is not completely filled by inhaled air and whose airway diameter increases during inspiration, is not very realistically simulated with filter deposition in a granular bed. The air in a filter flows through the whole system, and therefore the percentage of outside air in the last layers of the filter will be higher than in the last lung generations. In the granular bed filter transport in these regions occurs not only by diffusion but also by convection, causing an over-estimation for deposition in the last two generations according to our model.

Fluid dynamics of the two systems

It is assumed that laminar flow conditions generally prevail beyond the second airway generation under ordinary breathing conditions. Olson et al. (1970) state that it is unlikely that laminar flow is achieved in the bifurcating system. It is, however, generally assumed to be a good approximation in the lower airways. In Harris and Fraser's (1972) calculation for fiber deposition in lungs the flow is assumed to be distributed to an airway according to the

Table 2. Dimensions and average inspiratory flow conditions in a granular bed filter; Tidal volume = 1450 cm³

z	N_g	d_g	d_h	H_{layer}	S_{layer}	V_{vtot}	$A_{s,layer}$	U_{layer}	$t_{r,layer}$
(-)	(-)	(cm)	(cm)	(cm)	(cm^2)	(cm^3)	(cm^2)	(cm/s)	(s)
8	94	0.45	0.186	0.39	7.0	2.7	59.8	120	0.0033
9	178	0.38	0.154	0.33	9.5	3.2	80.7	87.3	0.0038
10	355	0.32	0.130	0.28	13.5	3.8	114.2	61.5	0.0046
11	699	0.27	0.109	0.23	19.1	4.4	160.1	43.5	0.0053
12	1491	0.23	0.095	0.20	29.1	5.8	247.8	28.7	0.0070
13	2864	0.20	0.082	0.17	43.3	7.4	360.0	19.3	0.0090
14	6033	0.18	0.074	0.16	70.5	11.3	614.1	11.8	0.014
15	12.0e3	0.16	0.066	0.14	112	15.7	965	7.42	0.019
16	22.1e3	0.15	0.060	0.13	185	24.1	1562	4.50	0.029
17	46.8e3	0.13	0.054	0.11	300	33.0	2485	2.78	0.040
18	92.7e3	0.12	0.050	0.10	514	51.4	4194	1.62	0.062
19	213e3	0.11	0.047	0.10	909	90.0	8097	0.916	0.11
20	391e3	0.11	0.045	0.10	1670	167	14863	0.500	0.20
21	714e3	0.11	0.043	0.10	3050	305	27141	0.273	0.37
22	1.55e6	0.10	0.041	0.09	5532	498	48695	0.150	0.60
23	3.10e6	0.10	0.041	0.09	11077	997	97389	0.075	1.2

percentage of the total lung volume subtended by that airway, so the airways are uniformly ventilated. They also assumed that inspiratory and expiratory flow rates are constant and do not vary cyclically with time during respiration. The assumptions made by Harris and Fraser do not cause considerable errors in calculating the deposition in lower regions of the lung. They do, however, cause errors in total deposition calculation in the upper parts.

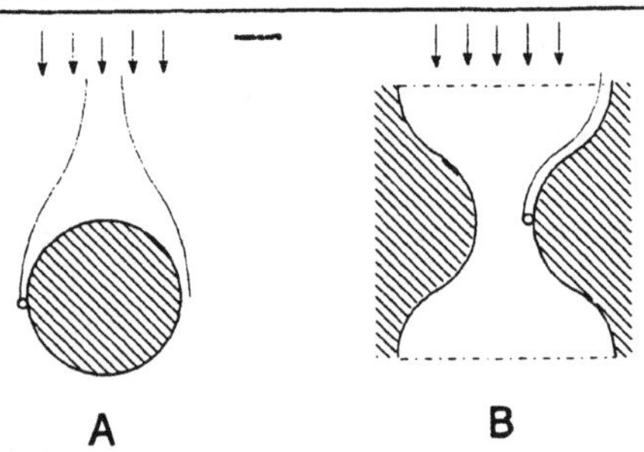

Figure 1. External (A) and internal (B) flow model for granular bed filters

Air follows a tortuous path in a granular bed filter, just as it does through the branching airways in the lung. The geometric structure of a granular bed filter is immensely complicated. Characterization of this complex structure with a simple geometric model is essential for the calculation of the flow within the bed. Generally, there are two models to describe the flow field in the granular bed filter, the external sphere-in-cell model and the internal constricted tube model, as shown in Figure 1. Most of the theoretical studies of granular bed filtration are based on the external model. The models of flow distribution in the granular bed filter assume that the distribution of flow over the granular bed filter is homogeneous, and the aerosol concentration is homogeneously distributed in each plane perpendicular to the flow direction, behind each layer of collectors.

Flow patterns in both the human lung and the granular bed filter are very complex and

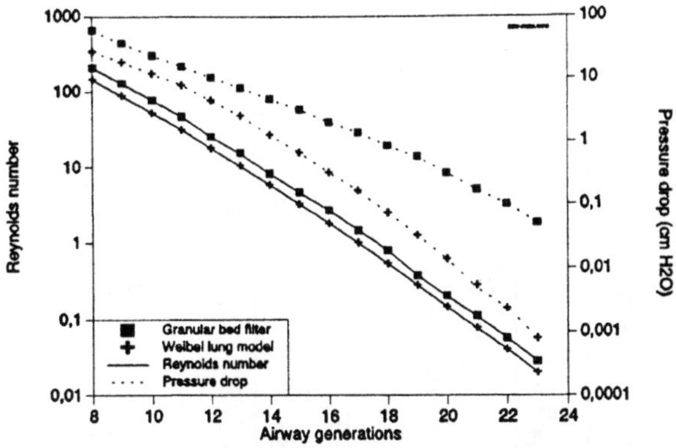

Figure 2. Comparison of the Reynolds number and the pressure drop in the two systems

certainly not comparable. In both systems different methods exist to calculate the flow patterns, and no method has as yet been proven to be valid. For this reason only a superficial comparison can be done. In Figure 2 the above assumptions are used to compare the Reynolds numbers and the pressure drops in both systems for the different generations. We see from Figure 2 that the Reynolds numbers are almost equal in the two systems, but pressure drops are different, especially in the deeper parts of the lung.

Particle deposition both in the human lung and a granular bed filter

Boulaud (1991) used granular bed filters as impactors in order to determine the size distribution of micrometer sized spherical aerosols. Four deposition mechanisms, sedimentation, diffusion, impaction and interception, are considered in his model. In order to predict *fiber* deposition in a granular bed filter, van der Put (1992) modified Boulaud's model using the equivalent diameter of fibers instead of particle diameter. Table 3 shows the essentials of the Boulaud/van der Put model. Note that van der Put used randomly orientated fibers in her model, in contrast with our assumption of fibers orientated parallel to the direction of the streamlines (as discussed later in this paper).

Inhaled particles deposit in the various regions of the respiratory system by the complex action of the same deposition mechanisms as in the granular bed filter. Particle deposition during respiration is separated into three processes, inspiration, expiration and pause. The relative contributions to total deposition from these three processes are presented in Figure 3. This figure shows that particle deposition during inspiration is the most significant contribution to total deposition, with the relative importance of inspiration increasing with particle size.

Mathematical models for the deposition of aerosol particles in the human respiratory tract have been developed to determine the probability of particle deposition during respiration. The Landahl (1950) and Taulbee (1978) models for spherical particles are widely used for calculating the regional particle deposition in lungs. Deposition equations determine the deposition probability of particles in an airway by sedimentation, diffusion, impaction and interception. In the Landahl and Taulbee models, interception is not included because it is

Table 3. Boulaud/van der Put deposition model

EQUATIOS.WPG

$$P = exp(\frac{-K_I(1-\epsilon)L_{bed}\eta}{d_g}) \qquad \eta = \eta_S+\eta_D+\eta_I+\eta_R$$

	Spheres	Random orientated fibers
$\eta_S=0.0375(Grv)^{0.5}+0.21(Grv)^{0.78}$	$Grv=\frac{v_s}{U_s}$	$v_s=\frac{\rho_p d_f^2 g}{8\mu(\frac{0.385}{ln(2\beta)-1/2}+\frac{1.230}{ln(2\beta)+1/2})}$
$\eta_D=B(Re)Sc^{f_1(Re)}Re^{f_2(Re)}$	$Sc=\frac{\nu}{D}$	$D=\frac{kT}{2\pi\mu d_f\beta(\frac{0.385}{ln(2\beta)-1/2}+\frac{1.230}{ln(2\beta)+1/2})}$
$\eta_I=\frac{(A(\epsilon,Re)Stk)^{3.55}}{1.67+(A(\epsilon,Re)Stk)^{3.55}}$	$Stk=\frac{\rho_p d_{ei}^2 U_s}{18\mu d_g}$	$d_{ei}=\frac{3}{2}d_f(\frac{0.385}{ln(2\beta)-1/2}+\frac{1.230}{ln(2\beta)+1/2})$
$\eta_R=H(\epsilon)(R_i)^2$	$R_i=\frac{d_p}{d_g}$	$R_i=\frac{l_f}{d_g}$

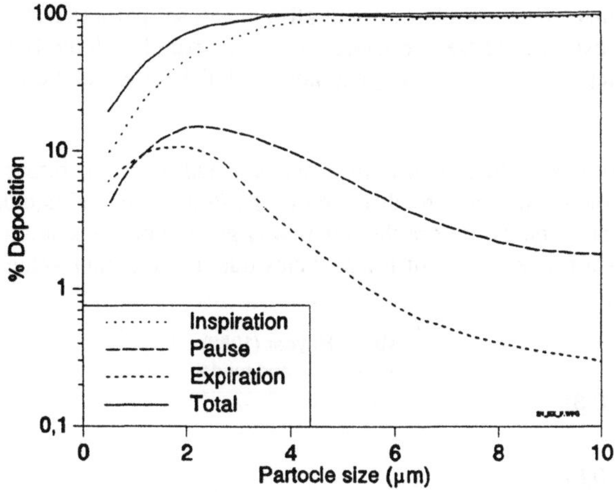

Figure 3. Relative contribution to total deposition from inspiration, breath pause and expiration as a function of particle size

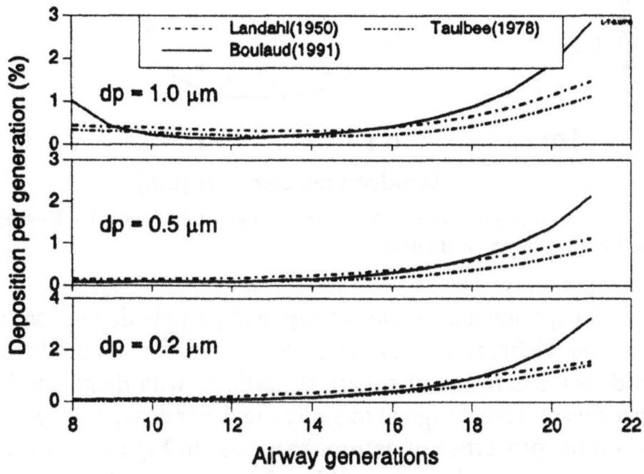

Figure 4. Regional particle deposition predicted by the three models under the tidal volume of 1450 cm_3

considered to be less important than the first three mechanisms (Hinds, 1982; Ferron et al., 1985). For the Landahl model, the sedimentation and diffusion depositions are calculated approximately in an airway, and the impaction deposition is based on the experimental data of Landahl and Herrmann (1949). Taulbee et al. (1978) uses equations for sedimentation (Thomas, 1958) and diffusion (Gormley and Kennedy, 1949) from a parabolic laminar airflow in a tube and an impaction equation based on experimental data of Johnston et al. (1977) and Schlesinger and Lippmann (1972). Landahl calculates 26% larger values for sedimentation compared to those of Taulbee et al. for deposition values below 10%. The data for impaction of both sets correspond well for small deposition values. Higher values for deposition by diffusion are found by Landahl for deposition probabilities below 1.5% and smaller ones above 1.5%. Taulbee et al. include impaction only during inhalation. The deposition in the alveoli is determined assuming that no air mixing occurs and that the aerosol fills an alveolus either entirely or not at all. An aerosol particle that does not deposit during respiration is exhaled. These two models were compared with the Boulaud model for spherical particles deposition in several generations at the same conditions, as shown in Figure 4.

From Figure 4 it can be seen that deposition in the Boulaud model increases in the last two generations because the contribution by diffusion in the Boulaud model is more important in this region than in other models. Before the 8th airway generation, deposition in the Boulaud model also increases for larger sizes of the particles due to the higher values for impaction in the Boulaud model.

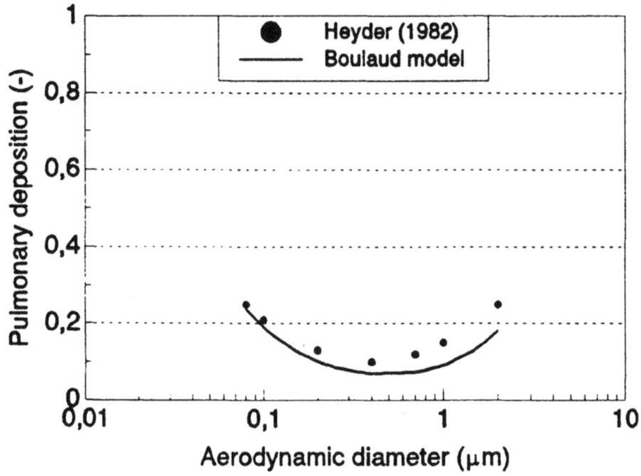

Figure 5. Comparison of experimental values for particle deposition found by Heyder and Gebhart with calculated values based on the theory of Boulaud

There are also several experimental studies of regional particle deposition in real lungs. The studies of Heyder and Gebhart (1982) were conducted for mouth-breathing and nose-breathing of monodisperse aerosols of spherical particles with diameters between 0.08 and 15 μm. These experiments were designed to study the dependence of deposition on breathing characteristics and on the properties of aerosol particles. In Figure 5 their experimental data of particle deposition in the alveolar region (17 to 23 generations) are compared with values predicted with the Boulaud model for a granular bed filter using the formula of Table 3 for spheres.

From Figure 4 and 5 it can be clearly seen that the particle deposition model in a granular bed filter is in good agreement with both mathematical models and experimental data for particle deposition in the lower parts of the lung. This result is very important as it makes credible the use a granular bed filter to represent the lower parts of the human lung.

Fiber orientation

The behaviour of fibrous particles in a shear flow under laminar conditions involves two competing mechanisms depending on the flow gradient and the mass ratio of the particle to fluid molecules. When the velocity gradient and mass ratio are large, the effect of Brownian motion is not significant. In this case the motion of the particle is dominated by the shear gradient. On the other hand, if the particle to fluid molecule mass ratio is small, Brownian motion can destroy the orderly behaviour of the particles caused by the shear flow. The particle has an equal probability of orienting in any direction in space. Both these situations are encountered in the flow of aerosol particles in the lung. In the upper airway generations, the velocity gradient is large, and the residence time of the particle is short; thus the particle orientation is dictated by the shear flow. Deep in the lung, however, velocity is greatly reduced because of an increased number of small airways. This in turn causes the velocity gradient to decrease, allowing the Brownian motion to alter the motion of small-sized particles (Asgharian, 1988).

For fully developed laminar flow, the orientation of fibers has been variously described by different authors as aligned parallel to the direction of the streamlines, or as exhibiting a periodic end-over-end rotation (Jeffery, 1922 and Harris, 1972). If fibers align with the streamline, the diameter is the significant parameter in the interception mechanism. In this case thin straight fibers of great length can easily be transported through the respiratory system, with deposition related only to the free falling speed, thus the fiber diameter, as is the case with compact particles. If, however, fibers are randomly orientated in the airstream, fiber length is the significant parameter in the interception mechanism, and interception may contribute a great deal to the total deposition. Note that fiber orientation for fully developed turbulent flow is assumed to be random.

In our research, the Reynolds number in the experimental system is small (max. 60). The fiber flow in the system is therefore assumed to be aligned with streamlines.

Experimental work for fiber deposition in the granular bed filter

The aim of our experiments was to investigate the deposition efficiency of a fibrous aerosol in a granular bed filter. We began the experimental program with measurements of deposition of spherical particles in a granular bed filter. The results were found to be in good agreement with the Boulaud model. After these initial measurements, the experiments were extended to include deposition of fibrous particles.

Polydispersed glass fibers were used in the experiments. We received glass fiber filters made by Warsaw University of Technology; these filters were made with fibers of constant diameter. At Delft University these fibers were sheared by a ball mill for 6 minutes at a speed of about 6,000 rpm. The diameter distributions of two samples of glass fibers are shown in Figure 6.

The experimental set-up and conditions are shown in Figure 7. Fibers were dispersed by a fluidized bed. The dispersed fibers were then neutralized by a radioactive source (Kr 85) before introducing them into the granular bed filter; this was done in order to avoid electrostatic effects on the particle collection. There are two sampling probes, one on the inlet and one on the outlet of the granular bed filter. A MIE Fiber Monitor (FM-7400, MIE Inc.) was used to measure the fiber concentration of the inlet and outlet of the granular bed as a function of fiber length in the fiber length range of 3 μm to 9 μm.

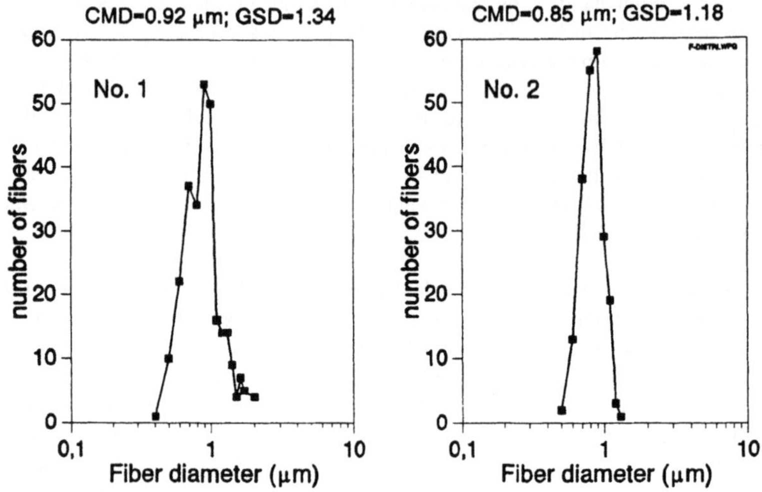

Figure 6. Diameter distribution of two glass fiber specimens

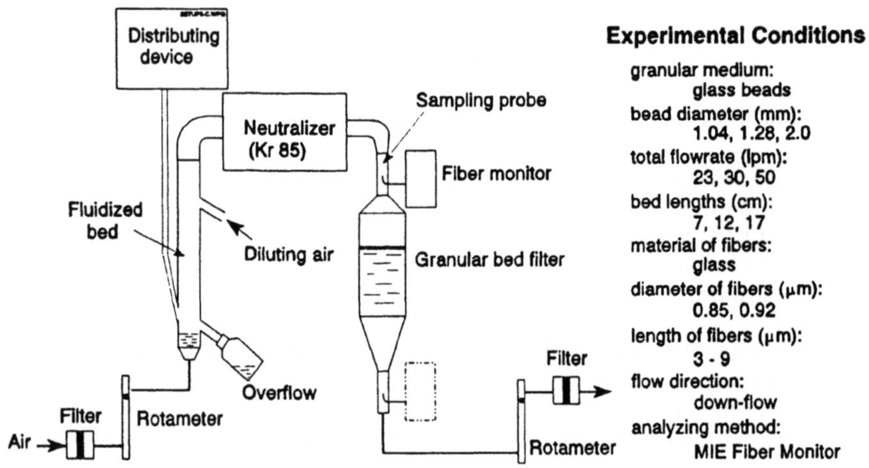

Figure 7. Experimental set-up and conditions

Results and modelling the fiber deposition in a granular bed filter

In order to discuss collection mechanisms in the granular bed filter, the total bed penetration, P, is converted to the single particle collection efficiency from Table 3 as follows:

$$\eta = -\frac{d_g}{K_1(1-\epsilon)L_{bed}}\ln P \qquad (1)$$

Formula 1 may be rewritten as a sum of efficiencies due to individual mechanisms, as shown in Table 3. We used the diffusion efficiency term of the Boulaud/van der Put model as part of our model. Because the fibers used in the experiments are not small enough to be significantly influenced by diffusion, deposition by diffusion was negligible in these experiments. Therefore Brownian motion will not be discussed further in this paper. For the new model presented here we use the same equation for diffusion efficiency as derived by Boulaud/van der Put (see equation for η_D in Table 3.

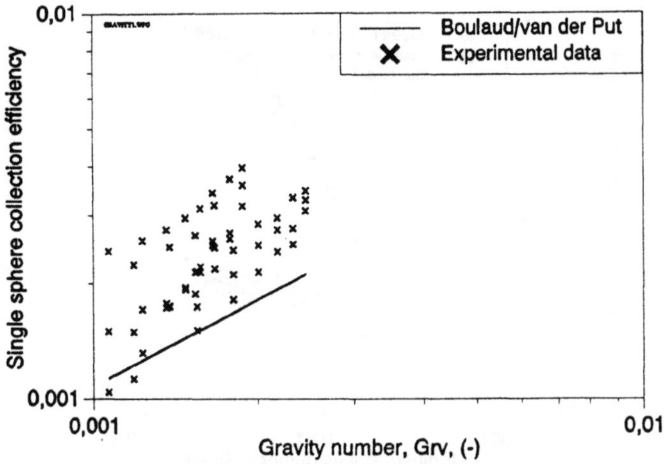

Figure 8. Single fiber efficiency vs. gravity parameter with different conditions

Gravitational settling becomes a predominant collection mechanism for large particles at low velocities. The single particle efficiency of large fibers is plotted in Figure 8 as a function of the gravity number, Grv, with different sizes of glass beads and fibers both for the Boulaud/van der Put model and our experimental results. It is assumed that the deposition mechanism in this case is sedimentation. The experimental data are in good agreement with the Boulaud/van der Put model. We can conclude that it is reasonable to use the Boulaud/van der Put model to predict the sedimentation effect for fiber deposition in a granular bed filter.

Inertial impaction is the controlling collection mechanisms for large particles at high velocities, and the other mechanisms are neglected in this case. Impaction efficiency is influenced by both Stokes number, Stk, and Reynolds number, Re. That is why in the calculation of Gal et al. (1985) the effective Stokes number, Stk_{eff}, was proposed as Eq. 2. By using the effective Stokes number, the single particle efficiencies are plotted in Figure 9. With linear regression the experimental data may be correlated by Eq. 3. In Figure 9 calculated points predicted by Boulaud/van der Put are shown for comparison. The present data differ significantly from the van der Put theory. The reason for this difference is fiber orientation. The fluid motion through the bed in this experimental set-up is always laminar. Therefore the fibers in the fluid are aligned in the same direction as the streamlines. The Boulaud model for spheres can not be used for this case (impaction), even not with a modification to account for the orientation of the fibers.

For spherical particle deposition in a granular bed filter, the effect of interception is usually

$$Stk_{eff} = (1.0 + \frac{1.75Re}{150(1-\epsilon)})Stk \tag{2}$$

$$\eta_I = \frac{Stk_{eff}^{0.52}}{71.47 + Stk_{eff}^{0.52}} \tag{3}$$

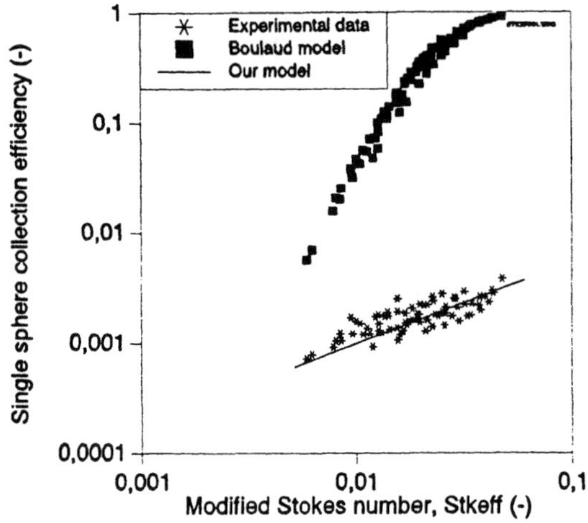

Figure 9. Inertial efficiency as a function of effective Stokes number

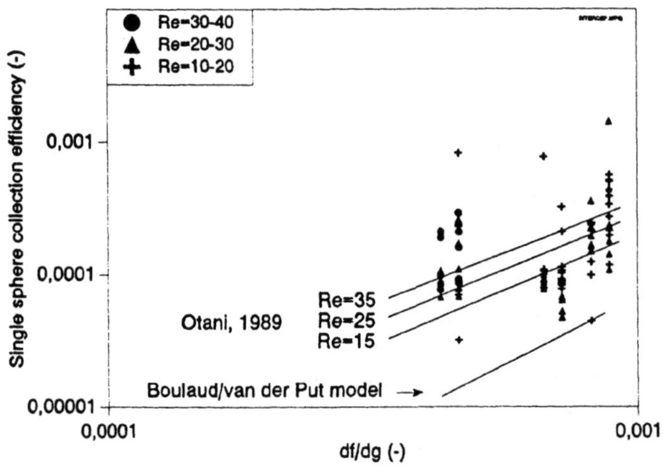

Figure 10. Interception efficiency as a function of R_i

small compared to other mechanisms. Interception becomes relatively important only in the transition regions of the other collection mechanisms. Therefore the interception efficiencies were obtained from the data of fibers in the transitional regions as the difference, after subtracting the sedimentation and impaction efficiencies from the total deposition efficiency.

For fibre deposition, interception also depends on the orientation of the fibres. As mentioned in the last paragraph, fibres are oriented parallel to one another in our experiments. The representative parameter of interception is R_i (= d_f/d_g). In this case, η_R is not the significant parameter for total fibre deposition in a granular bed filter. Figure 10 shows the residual interception efficiencies against R_i in the different range of Re numbers. Otani (1989) derived a formula for single sphere interception with R (= diameter of spheres / diameter of bed beads) as the parameter. By changing R to R_i, as done in formula 4, this equation is used and plotted together with our experimental results in Fig. 10. The thus calculated values correspond reasonably well with the measured ones.

$$\eta_R = 16R_i^{2-Re/(Re^{\frac{1}{3}}+1)^3} \tag{4}$$

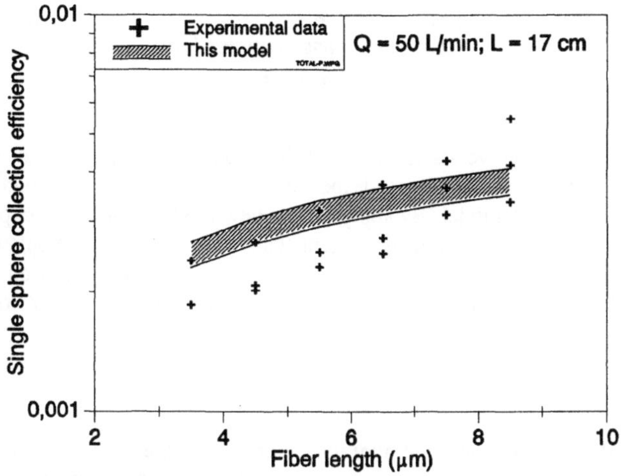

Figure 11. Comparison of the experimental data with the predicted theory, d_g=1.04 to 2.0 mm; d_f=0.92 μm

By using the equation for total penetration in Table 3, the collection efficiency due to the combination of the mechanisms can be predicted as a sum of the individual efficiencies. The predicted single sphere efficiencies are compared with the experimental data in Figure 11. It can be seen that the presented equations describe the experimental data well over the range of the fibre sizes as used in our experiments.

Our new model was also compared with the Asgharian (1988) model as shown in Figure 12. Asgharian developed his model for fibre deposition both in human lungs and rat lungs. He also compared it with several mathematical models for human lung deposition and experimental data for rat lung deposition. He proved it to be in good agreement with the data of rat lung deposition. Figure 12 shows that our model is in very good agreement with the Asgharian model for alveolar deposition. Our new model is also compared with the Harris and Fraser model. Here a big discrepancy between the two models is seen. This can be attributed mainly to the difference in sedimentation.

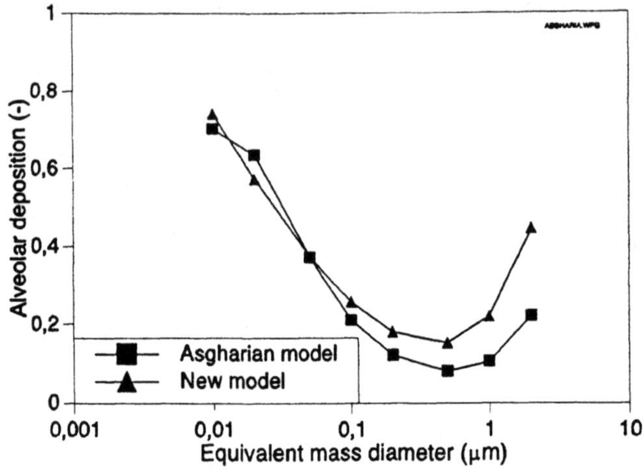

Figure 12. Comparison of the new model and the Asgharian model

To calculate fiber deposition we started with the Boulaud/van der Put model, which works well for spherical particles in bead beds. Experimental results, however, forced us to alter this model to a fiber-adapted model. By applying our new model to spheres, experiments and calculations seem to be in even better agreement than with the original Boulaud model. So it appears that our model is a general deposition model which can be used for different particle shapes. Further research is, however, necessary to validate or to falsify this statement.

Conclusions

From Tables 1 and 2 it is seen that the hydraulic diameter of the granular bed filter is equal to the diameter of the lung tube in each generation. The total cross section of air space and the average velocities in both systems are also equal. The residence times are different in the two systems. The variation in residence time will affect the diffusion efficiencies. In the lower parts of the lung, especially in the last two generations, this effect will be pronounced.

Flow patterns in both the human lung and the granular bed filter are very complex and not comparable. In both systems different methods exist to calculate the flow patterns, and no method has yet been proved to be valid. For this reason, it is difficult to make a definite conclusion about the difference in flow patterns between both systems. Therefore only a superficial comparison of the two systems can presently be carried out.

The Boulaud model, which is used to describe the particle deposition in a granular bed filter for spherical particles, can be compared with other models which are used to describe the particle deposition in the respiratory system (see Figs. 4 and 5).

Fiber orientation, which strongly influences the fiber deposition in a granular bed filter, depends on the flow situation. The fibers align themselves parallel, in the direction of the streamlines, when the fluid flow is laminar.

The Boulaud/van der Put model can not be used to predict the fiber deposition in a granular bed filter because the impaction and interception mechanisms can not be fit to the experimental data. Since the fiber orientation is parallel with the streamline direction, many large fibers can deposit in the lower parts of the lung which spherical particles of the same diameter can not reach.

The new model presented here is based on the Boulaud/van der Put model for the sedimentation and diffusion mechanisms. It has, however, not yet been proved to be valid for the diffusion mechanism. More experiments should be done with small fibers to investigate this mechanism. For impaction and interception other models than Boulaud/van der Put have been presented. The total model describes fiber deposition in a bead bed rather well.

List of symbols and dimensions

$A_{s,layer}$ internal surface area of a layer
$A_{s,lung}$ internal surface area of a lung generation
$A(\epsilon,Re)$ function of ϵ and Re
$B(Re)$ function of Re
Crv gravity number
D diffusion coefficient
D_z airway diameter
d_{ei} equivalent diameter for impaction
d_f diameter of fiber
d_g diameter of glass beads
d_h hydraulic diameter
d_p diameter of particles
$f_1(Re)$ function of Re
$f_2(Re)$ function of Re
g gravitational constant
H_{layer} layer height
$H(\epsilon)$ function of ϵ
k Boltzmann constant
K_1 constant
L_{bed} depth of bed
L_z length of airway
l_f length of fiber
N_g number of beads in a layer
N_z number of airways in generation z
P overall penetration
Re Reynolds number
R_i interception parameter
Sc Schmidt number
S_{layer} total cross section of air space in a layer
S_{lung} total cross section of air space in a airway generation
Stk Stokes number
T absolute temperature of the fluid
$t_{r,layer}$ residence time in a layer
$t_{r,lung}$ residence time in a airway gen.
U_{lung} average speed of a airway gen.
U_s velocity of the gas flow
U_{layer} average speed of a layer
V_{lung} total volume in generation z
v_s sedimentation velocity of sphere
V_{vtot} total void volume per layer
z airway generation
β aspect ratio
ϵ porosity of granular bed filter
η single-sphere collection eff.
η_D diffusion efficiency
η_I impaction efficiency
η_R interception efficiency
η_S sedimentation efficiency
μ dynamic viscosity of air
ν kinematic viscosity of the fluid
ρ_p particle density

References

Asgharian, B. (1988), The deposition of fibrous particles in the respiratory tract using a mathematical model, PhD dissertation, SUNY Buffalo, NY

Boulaud, D. (1991), Use of granular beds in the inertial impaction regime for aerosol size distribution measurement, J. Aerosol Sci., 22, 273-287

Ferron, A.G. et al. (1985), Comparison of experimental and calculated data for the total and regional deposition in the human lung, J. Aerosol Sci. 16, 133-143

Gal, E. et al. (1985), A study of inertial effects in granular bed filtration, AIChE J., 31, 1093-1104

Gormley, P.G. and Kennedy, M. (1949), Diffusion form a stream flowing through a cylindrical tube, Proc. Roy. Irish Acad. 52A, 163-169

Harris, R.L. (1972), A model for deposition of microscopic fibers in the human respiratory system, PhD Dissertation, Univ. of North Carolina at Chapel Hill.

Harris, R.L. and Fraser, D.A. (1976), A model for deposition of fibers in the human respiratory system, Am. Ind. Hyg. Assoc. J., 39, 73-89

Heyder, J. et al. (1982), Biological variability of particle deposition in the human respiratory tract during controlled and spontaneous mouth-breathing, Ann. Occup. Hyg., 26, 137-147

Hinds, W.C. (1982), Aerosol Technology, John Wiley & Sons, New York

Jeffery, G.B. (1922), The motion of ellipsoidal particles immersed in a viscous fluid, Proc.Roy.Soc., 102A, 161-179

Johnston, J.R. et al. (1977), Inhaled Particles IV, Vol. 1, Pergamon Press, Oxford

Landahl, H.D. (1950), On the removal of airborne droplets by the human respiratory tract: I. The lung, Bull. Math. Biophys. 12 (1950) 43-56

Landahl, H.D. and Herrmann, R. (1949), Sampling of liquid aerosols by wires, cylinders, and slides, and the efficiency of impaction of the droplets, J. Colloid Sci. 4, 103-136

Olson, D.E. et al. (1970), Pressure drop and fluid flow regime of air inspired into the human lung, J. Appl. Physiol. 28, 482-491

Otani, Y. et al. (1989) Experimental study of aerosol filtration by the granular bed over a wide range of Reynolds numbers, Aerosol Sci. Tech., 10, 463-474

Put S.J. van der (1992), The granular bed filter as a model of the lower parts of the human lung, Report of Delft University of Technology, The Netherlands

Schlesinger, R.B. and Lippmann, M. (1972), Particle deposition in casts of the human upper tracheobronchial tree, Am. Ind. Hyg. Ass. J. 33, 237

Taulbee, D.B. (1978), Aerosol transport in the human lung from analysis of single breaths, J. Appl. Physiol. 44, 803-810

Thomas, J.W. (1958), Gravity settling of particles in a horizontal tube, J. Air Pollut. Contr. Ass. 8, 32-34

Timbrell, V. (1965), The inhalation of fibrous dusts, Ann. N.Y. Acad. Sci., 132, 255-273

Weibel, E.R. (1963), Morphometry of the human lung. New York

Deposition of Hygroscopic Aerosol Particles in the Lungs

George A. Ferron, Bernhard Busch

GSF - Forschungszentrum für Umwelt und Gesundheit, Institut für Inhalation-biology, Neuherberg, D85764 Oberschleißheim, Germany

Keywords

Aerosol Particle; Particle Lung Deposition; Salt Particles; Drug Particles; Environmental Aerosols; Hygroscopic Particles; Particle Growth; Water Vapour Transport; Temperature Transport.

Introduction

A material such as a salt or a drug is hygroscopic if it absorbs water vapour from the air when the relative humidity (RH) of the air is larger than a certain value specific for this material. A material is non-hygroscopic, if it doesn't absorb water vapour at RH's below saturation. According to this definition a salt or drug is hygroscopic, that is all soluble in water. Hygroscopic materials are widely used in medicine for therapeutic and diagnostic purposes. Hygroscopic materials are present in atmospheric aerosol particles. High concentrations, mainly NaCl, are found near to the sea (Keith and Arons, 1954; Winkler and Junge, 1972). Continental aerosols also have a hygroscopic fraction (Hänel, 1976; Hicks and Megaw, 1985; McMurry and Stolzenburg, 1989; Busch *et al.*, 1994; Busch, 1995). Hygroscopic aerosols play an important role in the formation of clouds (Pruppacher and Klett, 1980).

Hygroscopic aerosol particles have a different behaviour in the lungs compared to non-hygroscopic particles. Dautrebande and Walkenhorst (1961) determined the minimal deposition for NaCl and coal aerosols as a function of particle size. The size for NaCl particles with minimum total deposition was about a factor of 7 lower than for coal particles. Experiments by Blanchard and Willeke (1984), Tu and Knutson (1984), Hicks *et al.* (1986), Gebhart *et al.* (1990) and Anselm *et al.* (1991) have demonstrated the differences in lung deposition between hygroscopic and non-hygroscopic particles. Models have been developed to estimate the deposition of hygroscopic particles in the human respiratory tract, which confirm most of the experimental data. Experimental and model results are limited to normal respiration conditions (average room temperature and humidity).

The aim of this contribution is to give an overview of our present knowledge and to apply this to some conditions found in the environment.

J. C. M. Marijnissen and L. Gradoń (eds.), Aerosol Inhalation: Recent Research Frontiers, 143–152.

Properties of hygroscopic aerosol particles

The size of an hygroscopic aerosol particle is determined by the dissolved materials in the droplet and the RH of the surrounding air. The dissolved materials reduce the water vapour pressure or water vapour concentration $c_{w,s}$ at the particle surface. This change can be described with the law of Raoult

$$\frac{c_{w,s}}{c_{w,o}} = 1 - \frac{in_s}{n_w} = 1 - \frac{im_sM_w}{m_wM_s} \tag{1}$$

where $c_{w,o}$ is the saturated water vapour concentration of pure water, i is the dissociation constant of the material, n_s and n_w are the number of salt and water molecules in the droplet, m_s and m_w are the salt and water mass in the droplet, M_s and M_w are the molecular weights, respectively. The dissociation constant i corresponds normally to the theoretical dissociation constant of the salt for a diluted solution (0.001 weight fraction of the salt or less). For larger concentrations, deviations up to a factor of 2 are found (Robinson and Stokes, 1968). So, for a precise determination of the water vapour concentration of a solution, experimental data have to be used. Data for pure salts can be found in Robinson and Stokes (1968) and the Handbook of Chemistry and Physics (Weast and Astle, 1979). Isotonic salt and drug concentrations are listed in The Merck Index (Budavari, 1989). The hygroscopic properties of salt mixtures are more complex than the properties of a pure salt (Tang, 1976). For salt mixtures hardly any data are available.

The vapour pressure of submicron particles has to be corrected for the Kelvin effect, which is caused by the curvature of the droplet surface

$$f_K = \exp\left(\frac{4\sigma M_w}{RT\rho d_{eq}}\right) = \exp\left(\frac{2.15 \cdot 10^{-3}}{d_{eq}}\right) \tag{2}$$

where σ is the surface tension of water, R is the general gas constant, T is the absolute temperature, ρ is the density of liquid water, and d_{eq} is the equivalent particle diameter (in μm). A correction of the Kelvin effect for the salt properties has been described by Nair and Vohra (1975).

The equilibrium size of an aerosol particle at a relative humidity RH is

$$RH = \frac{c_{w,s}}{c_{w,o}}f_K = \left(1 - \frac{im_sM_w}{m_wM_s}\right)\exp\left(\frac{4\sigma M_w}{RT\rho d_{eq}}\right) \tag{3}$$

The equilibrium size of initially dry salt particles is shown in Fig 1. The increase in size is significantly reduced for particles with an initial size below 1 μm.

The growth and evaporation of a droplet is derived from the law of Fick (1855). A slip correction for the discontinuity of a gas (Fuchs, 1962) has been introduced using the boundary layer theory. Then the mass flux I to a spherical aerosol particle is

$$I = 2\pi d_{eq}D_wC_w \tag{4}$$

where D_w is the diffusion constant of water vapour in air, and C_w is the slip correction. Information on the slip correction according to the boundary layer theory can be found in Fuchs (1962). Alternatives for the slip correction have been described by Fuchs and Sutugin (1971), Wagner (1982) and others.

The influence of the condensation of heat on the transport of water have

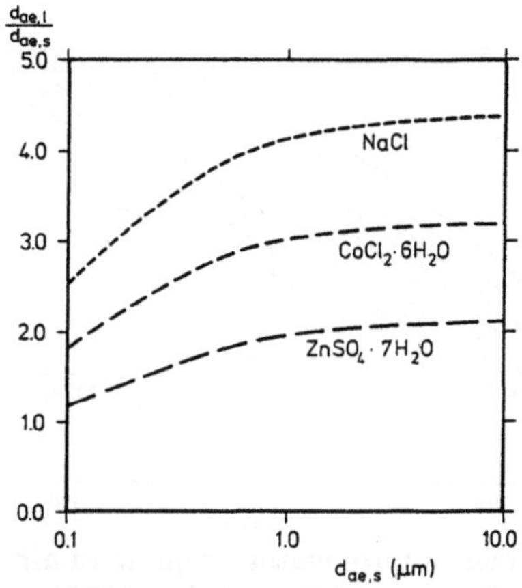

Figure 1. Ratio of the equilibrium aerodynamic particle size $d_{ae,l}$ and the aerodynamic diameter of the initially dry particle $d_{ae,s}$ in the human airways, assuming a relative humidity (RH) of 0.995 (Ferron *et al.*, 1988a). For copyright of this and other figures see Acknowledgements.

been estimated by Zebel (1956) and Fuchs (1962). They found a substantial reduction of the water vapour transport. Ferron (1977) used the boundary layer theory for both the water vapour and heat transport and solved the equations with an iterative method. This method can be applied to problems with time dependent boundary conditions which occur during respiration.

Ferron and Soderholm (1990) published information on times to evaporate or stabilize droplets. Fig. 2 shows the evaporation times of a pure water and several solution droplets with an initial particle size of 1 μm at a temperature of 20°C and a RH of 0.5. Evaporation of the water from a pure particle occurs within 13 ms. This time is also a good approximation for the evaporation for the solution

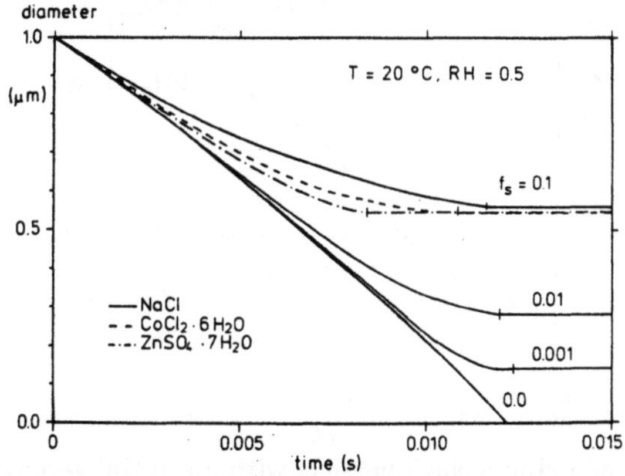

Figure 2. Evaporation of a pure water droplet and three droplets with 0.001, 0.01 and 0.1 weight fraction NaCl, and 0.1 weight fraction $CoCl_2.6H_2O$ or $ZnSO_4.7H_2O$ with initial aerodynamic diameter of 1 μm in air with a temperature of 20°C and a RH of 0.5 (Ferron and Soderholm, 1990).

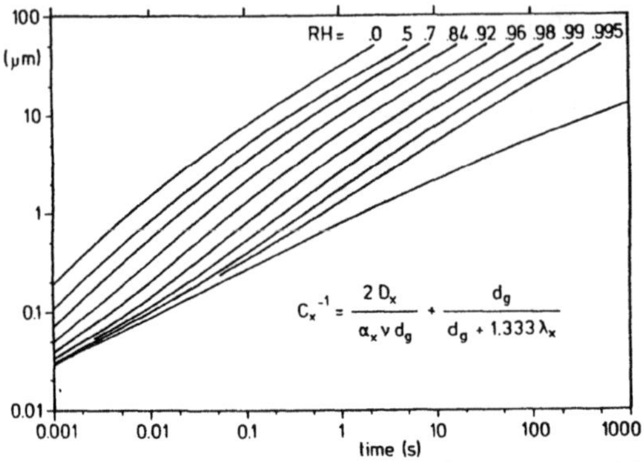

Figure 3. Evaporation times of pure water droplets in air with a temperature of 20°C and different RH's (Ferron and Soderholm, 1990).

droplets. Times for evaporation of pure water droplets of different size and at different RH's are shown in Fig. 3. Examples of the growth and stabilization of solution droplets at different RH's are given in Fig. 4. Here the small vertical line in each curve indicates the time a pure water droplet with size equal to the equilibrium size of the particle needs to evaporate at the given RH. The times of evaporation of pure water droplets are therefore a good estimate for the stabilization of solution droplets.

Models for deposition of hygroscopic particles in the lungs

Models to calculate the deposition of aerosol particles in the human

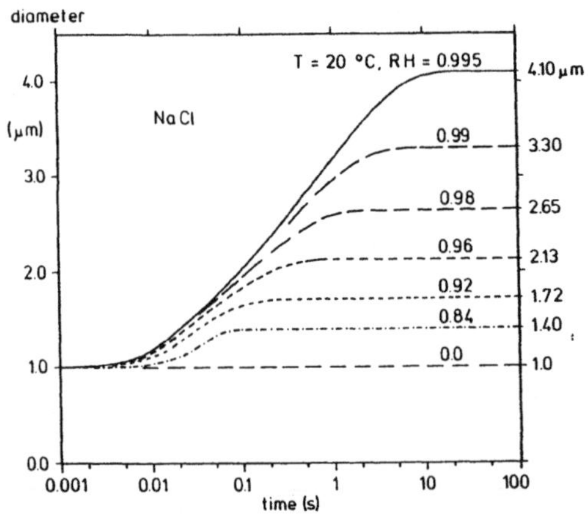

Figure 4. Growth of a single NaCl particle with an initial aerodynamic diameter of 1 μm in air with a temperature of 20°C and different RH's. The vertical bars in the curves indicate the time for a pure water droplet with an initial size equal to the final size of the solution droplet to evaporate in air with the same temperature and RH (Ferron and Soderholm, 1990).

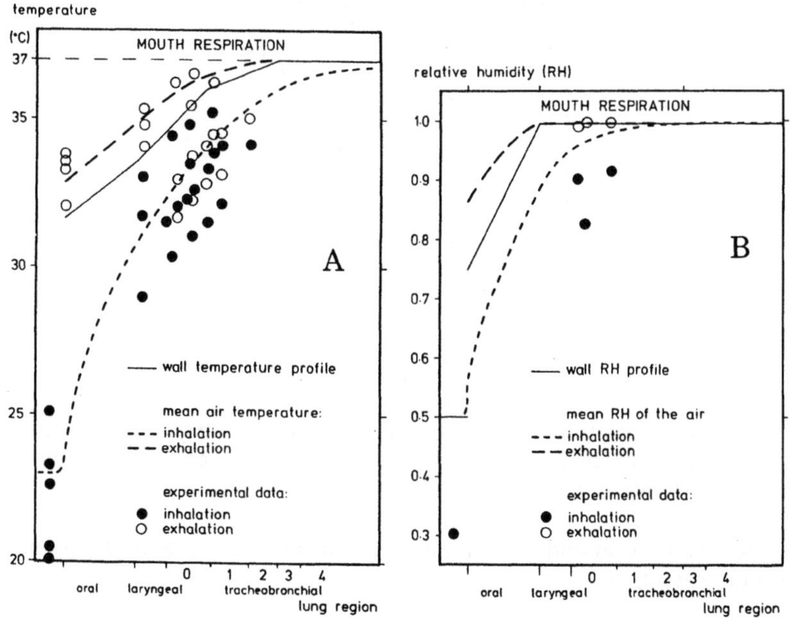

Figures 5a. Mean air and wall temperatures for inhalation and exhalation in the upper airways during mouth breathing (Ferron *et al.,* 1989).
Figure 5b. Same as Figure 5a, but now for RH.

respiratory tract are described by other contributors to this workshop. Considering hygroscopic properties, additional information is needed for the humidity and the temperature in the airways, both for inhalation and exhalation. Reviews on hygroscopic particle growth and deposition have been published by Morrow (1986), Hiller (1991) and James *et al.* (1994). The simultaneous transport of water vapour and heat in a tube has been solved by Ferron *et al.* (1988a) using finite difference methods for three and two dimensions for different conditions. These results were transformed to a human airways geometry using the model A of Weibel (1963) with a regular dichotomy extended with a nasal and oral cavity. The model boundaries included an increase of the temperature and the RH of the airway walls in the upper airways, which was assumed to be constant both for inhalation and exhalation, a boundary layer with a linear increase of the velocity with distance from the wall and a constant core airflow. An additional diffusivity increased the transport in the core flow to account for the instabilities in the air during respiration. Parameters were chosen to obtain a good approximation to the experimental data from the literature. Figs. 5a and b illustrate their results for mouth breathing. The total deposition of single monodisperse particles of three salts are given in Fig. 6. A substantial shift between the curve for the non-hygroscopic particles and the curves for salt particles are found. The aerodynamic diameters for the salts NaCl, $CoCl_2.6H_2O$ and $PbNO_3.7H_2O$ change by a factor 4.2, 3.2 and 2.1, respectively. Ferron *et al.* (1993) have shown that this factor is an adequate parameter to interpolate between the data for the different salts and the non-hygroscopic particles. This method can be applied to an arbitrary aerosol with

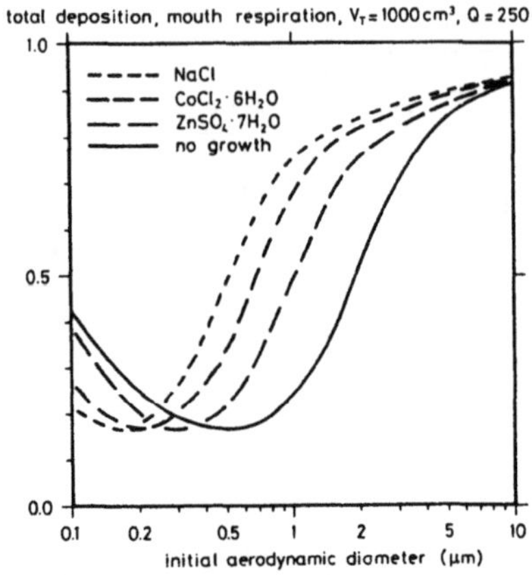

Figure 6. Total deposition for initially dry salt and non-hygroscopic (unit density) particles in the lungs for a tidal volume of 1000 cm^3 and inhalation and exhalation airflows of 250 cm^3s^{-1} (Ferron *et al.,* 1988b).

known size distribution and growth factor to estimate the total and regional deposition in the lungs.

Application to environmental conditions

Temperature and humidity are highly variable in the environment. At one side arctic air may have temperatures as low as -50°C and nearly zero water vapour content. At the other side temperatures as high as 50°C are found in the Sahara at low humidity and 40°C in India at RH's near to 1.0. Typical aerosol particle concentrations in an urban environment are 30,000 cm^{-3} with mass concentrations of 30 $\mu g\ m^{-3}$ (Seinfeld, 1986; Brand *et al.*, 1992). Extreme values up to a factor of 30 has been found.

Inhaling air causes the lungs to loose water and energy. The energy needed to accommodate the air with a temperature T to body temperature T_b (37°C) is $0.92(T_b - T)$ mJ cm^{-3}. The maximum energy to humidify dry air is 106 mJ cm^{-3}. The total energy needed to accommodate the inhaled air for different ambient temperatures and RH's are illustrated in Fig. 7. To accommodate air at room temperature 23°C and a RH of 0.5, 104 mJ cm^{-3} are needed. At freezing point the value is 141 mJ cm^{-3}, an increase of 35% with respect to a temperature of 23°C. So low air temperature have a moderate effect on the energy needed to accommodate the inhaled air.

Increase of the inhaled airflow has a strong influence on the air temperature (McFadden *et al.*, 1985). As a first approximation the difference between body temperature and the local temperature in the bronchial airways is proportional to the inhalation airflow.

Elevation is another parameter. At an elevation of about 5000m the air pressure is reduced by a factor of two. However, most people live at an elevation of below 500m, so the change in inhaled air volume is probably less than 7%.

Particle concentration has some influence on temperature and RH in the

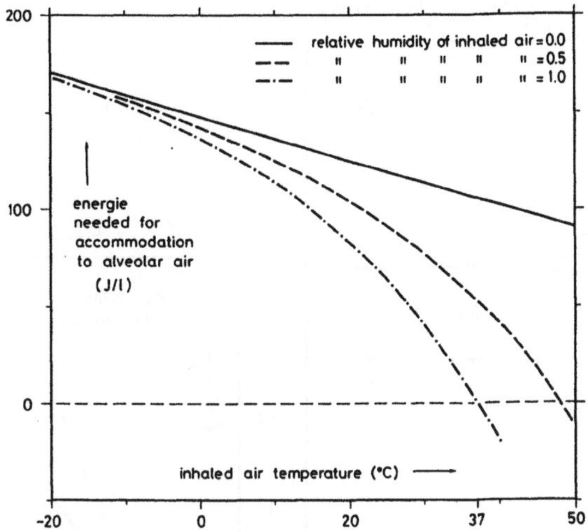

Figure 7. Energy needed to accommodate the inhaled air to a temperature of 37°C and a RH of 0.995 as a function of inhaled temperature and three RH's (Ferron *et al.,* 1988a).

upper airways as shown by Eisner *et al.* (1990) and Finlay and Stapleton (1995). A typical urban aerosol concentration is 30 µg m^{-3}. Assuming that all this material is NaCl, the aerosol will absorb about 3 mg m^{-3} or 0.003 µg cm^{-3} water in the lungs. This amount can be neglected if compared to the 44 µg cm^{-3} water vapour needed to humidify the air during respiration.

A change in temperature and RH of the inhaled air may change the RH in the upper airways. Ferron *et al.* (1984) calculated the RH of the air in the upper airways at different air temperatures. At temperatures below 20°C they found supersaturated air in the upper bronchial tree during inhalation (Fig. 8a). The size of an initially 0.3 µm dry NaCl particle is shown in Fig. 8b. Here maxima are found both for air temperatures of 10 and 0°C. A significant increase in total deposition was found for these air temperatures and particle sizes below 1 µm.

It is concluded that loss of energy from the lung is primarily proportional to the amount of absolute humidity the lungs have to supply per unit time. Less important parameters are the air temperature, elevation and particle concentration.

An estimation of the total deposition of atmospheric aerosols in the lungs is given by Hänel and Heyder (1980) based on experimental lung deposition data. A more detailed regional deposition for hygroscopic and non-hygroscopic aerosol particles can be calculated using an estimation method described by Ferron *et al.* (1993). This model is limited to particle sizes between 0.1 and 10 µm.

A large field of problems are found in the interaction of the inhalation of air with a low temperature and RH and the upper airways. Cold air may provoke asthma (Anderson *et al.*, 1985). Such effects are enhanced for lower air temperatures. Hardly any data are available on the interaction of the diseased lung and the environment.

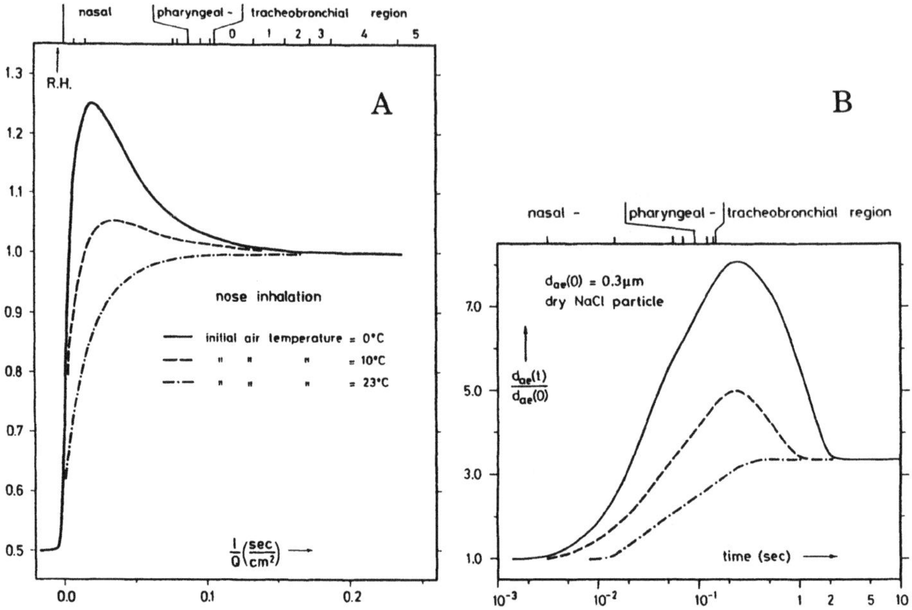

Figure 8a. Estimated RH of the air during inhalation of air with temperatures of 23, 10 and 0°C and RH of 0.5 (Ferron *et al.,* 1984).
Figure 8b. Size of a dry NaCl particle with an initial aerodynamic diameter of 0.3 µm during the inhalation of air with temperatures of 23, 10 and 0°C (Ferron *et al.,* 1984).

Discussion

Most of our knowledge concerning hygroscopic particle growth and deposition in the lungs is based on theory. Only a small number of experimental studies on total lung deposition (Dautrebande and Walkenhorst, 1962; Blanchard and Willeke, 1984; Tu and Knutson, 1984; Hicks *et al.*, 1986) or deposition of an aerosol bolus (Gebhart *et al.*, 1990; Anselm *et al.*, 1991) have been published. Data on regional deposition are not available. To obtain such data a technique using a Gamma Camera may be helpful (Flaming *et al.,* 1995). One study has determined the RH in the alveolar region using monodisperse NaCl particles and a breath-holding technique (Anselm *et al.,* 1991) to justify the widely used value for the RH of 0.995 in the alveolar region.

Some information is available on the hygroscopic properties of environmental aerosol particles. Winkler and Junge (1972) and Hänel (1976) studied the properties of sampled aerosols using a microbalance. Hicks and Megaw (1985) simulated the transport of the aerosols in the upper airways by passing the aerosol in a tube with dry or wet walls. McMurray and Stolzenburg (1989) used a Tandem Differential Mobility Analyser to measure the hygroscopic properties of selected particle size *in situ*. This method gives detailed information on particles with sizes less than 0.5 µm. A similar method is used by Busch *et al.* (1994). They found a hygroscopic and a non-hygroscopic fraction. The ratio of these two fractions is depending on the initial size. Additional measurements are needed to characterize different types of aerosols.

Acknowledgement. The work of Dr. Busch was supported by a grant of the Bavarian Ministry of Environment as part of the Bavarian Climate Research Program (BayFORKLIM). The authors are indebted to Dr. D. A. Edwards for his comments on the manuscript.
Copyrights: the reprint of the Figures 1, 2, 3, 4, 6, 7, 8 are with kind permission from Elsevier Science Ltd, The Boulevard, Langford Lane, Kidlington 0X5 1GB, UK, and the Figure 6 with kind permission from Mary Ann Liebert, Inc., 2 Madison Avenue, Latchmont, NY 10538, USA.

References

Anderson, S. D., Schoeffel, R. E., Black, J. L., Daviskas, E., *Eur. J. Respir. Dis.* 67 (1985) 20-30

Anselm, A., Heibel, T., Gebhart, J., Ferron, G., *J. Aerosol Sci.* 21 (1991) S427-430

Blanchard, J. D., Willeke, K., "Total deposition of ultrafine sodium chloride particles in the human lungs", *J. appl. Physiol.* 57 (1984) 1850-1856

Brand, P., Ruoss, K., Gebhart, J., *Atmosph. Environ.* 26A (1992) 2451-2457

Budavari, S., "Miscellaneous tables", in *The Merck Index*, pp. MISC:79-103, Merck & Co., Inc, Rahway, N. J., USA (1989)

Busch, B., Ferron, G., Karg, E., Solberg, A., Heyder, J., *J. Aerosol Sci.* 25 (1994) S143-144

Busch, B., Personal communication (1995)

Dautrebande, L., Walkenhorst, W., "Über die Retention von Kochsalzteilchen in den Atemwegen", pp. 110-119, in *Inhaled particles and vapours* I, Ed. Davies, C. N., Pergamon Press, Oxford, GB (1961)

Eisner, A. D., Graham, R. C., Martonen, T. B., *J. Aerosol Sci.* 21 (1990) 849-858

Ferron, G. A., *J. Aerosol Sci.* 8 (1977) 251-267

Ferron, G. A., Haider, B., Kreyling, W. G., *J. Aerosol Sci.* 15 (1984) 211-215

Ferron, G. A., Haider, B., Kreyling, W. G., *J. Aerosol Sci.* 16 (1988a) 343-363

Ferron, G. A., Karg, E., Peter, J. E., *J. Aerosol Sci.* 24 (1993) 655-670

Ferron, G. A., Kreyling, W. G., Haider, B., *J. Aerosol Sci.* 16 (1988b) 611-633

Ferron, G. A., Oberdörster, G., Henneberg, R., *J. Aerosol Med.* 2 (1989) 271-284

Ferron, G.A., Soderholm, S. C., *J. Aerosol Sci.* 21 (1990) 415-429

Fick, A., *Ann. Phys. Chem. Lpz.* 94 (1855) 59-86

Finlay, W. H., Stapleton, K. W., *J. Aerosol Sci.* 26 (1995) 655-670

Fleming, J. S., Nassim, M., Hashich, A. H., Bailey, A. G., Conway, J., Holgate, S., Halson, P., Moore, E., Martonen, T. B., *J. Aerosol Med.* 8 (1995) 66

Fuchs, N. A., "Evaporation and droplet growth in gaseous media", Pergamon Press, Oxford, GB (1962)

Fuchs, N. A., Sutugin, A. G., "High-disperse aerosols", pp. 1-60, in *Topics in current aerosol research*, Volume 2, Eds. Hidy, G. M., Brock, J. R., Pergamon Press, Oxford, GB (1971)

Gebhart, J., Anselm, A., Ferron, G., Heyder, J., Stahlhofen, W., "Experimental data on total deposition of hygroscopic particles in the human respiratory tract", pp. 1299-1302, in *Aerosol Science, Industry, Health and Environment,* Eds. Masuda, S., Takahashi, K., Pergamon Press, Oxford, GB (1990)

Hänel, G., *Adv. Geophys.* 19 (1972) 73-188

Hänel, G., Heyder, J., *Staub* 40 (1980) 9-13

Hicks, J. F., Megaw, W. J., *J. Aerosol Sci.* 16 (1985) 521-527

Hicks, J. F., Prichard, J. N., Black, A., Megaw, W. J., "Experimental evaluation of aerosol growth in the human respiratory tract", pp. 244-247, in *Aerosols: formation and reactivity,* Pergamon Press, Oxford, GB (1986)

Hiller, F. C., *J. Aerosol Med.* 4 (1991) 1-23

James, A. C., Stahlhofen, W., Rudolf, G., Köbrich R., Briant J. K., Egan, M. J., Nixon, W., Birchall, A., "Annex D. Deposition of inhaled particles", pp. 231-299, in *Human respiratory tract model for radiological protection,* ICRP Publication 66, Pergamon Press, Oxford, GB (1994)

Keith, C. H., Arons, A. B., *J. Atmosph. Sci.* 11 (1954) 173-184

McFadden, E. R. jr., Picurko, B. M., Bowman H. F., Ingenito, E., Burns, S., Dowling, N., Solway, J., *J. appl. Physiol.* 58 (1985) 564-570

McMurray, P. H., Stolzenburg, M. R., *Atmosph. Environm.* 23 (1989) 497-507

Morrow, P. E., "Factors determining hygroscopic aerosol deposition in airways", *Physiol. Rev.* 66 (1986) 330-376

Nair, P. V. N., Vohra, K. G., *J. Aerosol Sci.* 6 (1975) 256-271

Pruppacher, H. R., Klett, J. D., "Microphysics of clouds and precipitation", Riedel Publishing Company, Dordrecht, NL (1980)

Robinson, R. A., Stokes, R. H., "Electrolyte solutions", Butterworth, London, GB (1968)

Seinfeld, J. H., "Atmospheric chemistry and physics of air pollution", John Wiley and Sons, New York, USA (1986)

Tang, I. N., "Phase transformation and growth of aerosol particles composed of mixed salts", *J. Aerosol Sci.* 7 (1986) 361-371

Tu, K. W., Knutson, E. O., *Aerosol Sci. Technol.* 3 (1984) 453-465

Wagner, P. E., "Aerosol growth by condensation", pp. 129-178, in *Topics in current physics. Aerosol microphysics II* (Ed. Marlow, W. H.), Springer Verlag, Berlin, D (1982)

Weast, R. C., Astle, M. J., "Handbook of Chemistry and Physics", 45th edition, CRC Press, Boca Raton, Florida, USA (1979)

Weibel, E. R., "Morphometry of the human lung", Springer Verlag, Berlin, D (1963)

Winkler, P., Junge, C., *J. Rech. Atmosph.* 6 (1972) 617-638

Zebel, G., *Zeitschrift f. Aerosol-Forschung und -therapy,* 5 (1956) 263-288

The Influence of Charge on the Deposition Behavior of Aerosol Particles with Emphasis on Singly Charged Nanometer Sized Particles

Beverly S. Cohen, Judy Q. Xiong, Wei Li

Nelson Institute of Environmental Medicine, New York University School of Medicine, Long Meadow Road, Tuxedo, NY 10987, USA

Keywords

Charged Particle; Electrical Charge; Image Force; Ultrafine Particle Deposition; Lungs; Airways; Hollow Airway Casts

Abstract

The particle number concentration in ambient air is dominated by nanometer sized particles and most airborne particles carry a few charges. Our deposition studies in hollow airway casts demonstrate that electrical forces can result in significantly enhanced deposition of these particles in human airways. Deposition of highly charged airborne particles by electrical forces has been previously demonstrated experimentally in the respiratory tract in animals and people. Theory predicts that image forces will increase deposition of charged compared with neutral particles. However, modeling and experiments with particles whose deposition is dominated by inertial and gravitational forces suggest the effect is only discernible for highly charged particles. Our results show that the effect of single electrostatic charges carried by ultrafine particles cannot be neglected in modeling lung deposition.

Introduction

The electrical charge on an aerosol particle is a fundamental parameter that is important in governing particle deposition under many circumstances. Thus, the charge distribution on an aerosol can effect the dose of an inhaled toxicant by affecting the quantity and pattern of particle deposition in the human respiratory system. In the absence of charge, deposition of particles in the lung airways occurs by sedimentation, impaction, diffusion and interception. The first three mechanisms depend on the aerodynamic behavior of particles whereas interception depends merely on the geometry of particles and is negligible for spherical and other relatively compact particles. In the upper airways, in which air velocities are high, there is not sufficient time for particles to deposit by gravitational settling. The dominant deposition mechanisms are therefore diffusion for very small (d<100 nm) particles, and impaction for large particles (d >500 nm). Although sedimentation is negligible due to the high flow rates in the airways, it is a dominant process for deposition in the alveolar, or pulmonary, region of the lung. Electrical forces on charged particles may act in combination with any of the forces that cause deposition of neutral particles.

Background

Animal Studies

The potential for increased lung deposition of charged airborne particles by electrical forces has been recognized and explored both experimentally and theoretically in animals and people since an enhancement of lung deposition by particle charge was suggested by Wilson (1947). Longley (1960) and Longley and Berry (1961) investigated the deposition of charged aerosols in mice that were charged to a high potential (4 kV)

J. C. M. Marijnissen and L. Gradoń (eds.), Aerosol Inhalation: Recent Research Frontiers, 153–164.

compared to grounded (zero potential) mice. A median optical diameter of about 0.4 μm and $\sigma_g \approx 2$ can be calculated from their data for the chromium compound aerosols they used. They found that charges of either sign influenced the level of aerosol deposition in animal lung. Fraser (1966) exposed rabbits to eight dusts with median optical diameters between 0.11 (cobalt blend) and 3.5 μm (silica). He showed that the fraction of the dust that was retained by an animal increased as the charge per particle increased, and the total deposition in the respiratory tract doubled with a charge of about 1000e per particle, where e is the elementary unit of charge (1.6×10^{-19} Coulomb). No differences were observed based on either the polarity or type of dust.

The effect of charge on the deposition of Union International Centre le Cancer (UICC) amosite and crysotile fibers in rats' lungs was studied by Vincent et al. (1981) and Jones et al. (1983). The rats exposed to asbestos dust "normally" charged during the mechanical dispersing process received a lung burden of up to 40% greater for amosite, and 120% greater for crysotile, than those exposed to a similar concentration of neutralized fibers. The median number of charges per particle were 20 and 16 for normally dispersed crysotile and amosite respectively, and 1 and 8 for the neutralized dusts. The lungs were examined from 3 to 63 days post exposure and the ratio of the mass burden for charged fibers to that of neutralized fibers ranged from 1.24 to 2.26. Since early clearance was complete this must have resulted from deposition in the slow clearing part of the respiratory tract. It implies that the retention of fibers in the rat lung increases due to electrostatic charge on airborne fibers. There was no apparent effect on clearance rate. Ferin et al. (1983) found that when titanium dioxide powder dispersed by a Wright dust feed (MMAD = 1.0 μm and $\sigma_g = 2.3$) was charge neutralized with a ^{85}Kr source, the particle charge was reduced by a factor of 5.5 and deposition in exposed rats decreased by about 20%. Clearance rates were similar for charged or neutralized particles.

Human Studies

Experiments conducted *in vivo* in humans (Melandri et al., 1977; Tarroni et al., 1980; Melandri et al., 1983) have all shown an increase in respiratory tract deposition due to particle charge. Melandri et al. (1977, 1983) investigated the total deposition of 0.3, 0.6 and 1.0 μm diameter aerosol particles that had been charged to 50e-100e per particle. Both negative and positive charges were found to increase deposition by 15-30% as compared with that of neutralized aerosol particles. There is a linear increase in deposition with charge in human subjects similar to that seen in animals (Prodi and Mularoni, 1985). However, enhanced deposition by electrical forces was noted only if charge on the particle exceeded a threshold value q_c. The value of q_c depends on the particle size as well as the airway size (Yu, 1985; Prodi and Mularoni, 1985) but is normally greater than 30e. This is consistent with the results of Fry (1970) who found no significant effect on nasopharyngeal deposition of 4 μm particles with a median of 2-36 charges per particle.

Cast Studies

Further data on the effect of particle charge on deposition were provided by *in vitro* experiments with hollow lung airway casts. Chan et al. (1978) investigated the deposition of monodisperse charged particles with aerodynamic diameters of 2-7 μm, and observed significantly increased deposition of charged particles (360-1100 charges/particle) over uncharged particles. They also found that the enhancement was more prominent at low flowrate (15 and 30 LPM) than at high flowrate (60 LPM). This enhancement of deposition was observed in both larynx and trachea. They suggested that an even greater effect might be seen for submicrometer sized particles and potentially charged particles in cigarette smoke, mining atmospheres, and auto emissions should be studied. Since the airway casts used for these experiments were made of nonconducting plastic it is difficult to interpret their results. The material has a high relative dielectric constant, as does the oil used to wet the inner cast surface, which would reduce the magnitude of the image charge. The very high charges on the particles may explain why effects could be detected.

Theory of Charged Particle Deposition

There are two electrical forces which cause particles to deposit on the airway surfaces. One is due to interaction between the particle and the wall (image force) and the other is due to electrostatic scattering, the mutual repulsion between particles with like charge (space charge force). The image force is a single particle effect and increases as the particle nears the wall. The space charge force on the other hand depends on the particle concentration. When $n_oR^3 >> 100$, scattering causes more deposition than that resulting from image forces and when $n_oR^3 << 10$ image force is the predominant mechanism for deposition, where n_o is the initial concentration of charged particles and R is the airway, or alveolar, radius (Yu, 1977; Yu and Chandra, 1978). Due to low particle concentration deposition by electrostatic scattering is found to be much smaller than the image force in the lung.

In this discussion it is assumed that the inner surfaces of the lung airways are conducting. Although little is known about the electrical conductivity of the lining of the respiratory tract this assumption seems justified because of the high ion content of body fluids in general.

For the deposition efficiency due to image force alone (η_i), Yu (1977) derived an expression for deposition of particles uniformly distributed in a cylindrical tube:

$$\eta_i = 1 - \chi^2 \tag{1}$$

where χ is a critical radius beyond which all particles will deposit. It is related to the dimensionless residence time τ_e, by:

$$\tau_e = 4\,(\chi^{-1} - 2\ln\chi - \chi) \tag{2}$$

and

$$\tau_e = \frac{q^2 t B}{4\pi\varepsilon_o R^3} \tag{3}$$

in which q is the charge on the particle, t is the elapsed time in the airway, ε_o is the permittivity of air, R is the radius of airway, and B is the mobility of the particle.

$$B = \frac{C}{F} \tag{4}$$

where F is the drag per unit velocity, and C is the Cunningham correction factor, given by:

$$C = 1 + Kn\,(1.257 + 0.4\,e^{-1.1/Kn}) \tag{5}$$

Kn is the Knudson number given as:

$$Kn = \frac{2\lambda}{d} \tag{6}$$

in which d and λ are the particle diameter and the mean free path in air respectively.

For $\tau_e << 1$, Pich (1978) obtained the following simple results for the case where deposition by image forces is very small ($\eta_i \rightarrow 0$):

$$\eta_i = (6\tau_e)^{1/3} \tag{7}$$

and when deposition by image forces is large ($\eta_i \rightarrow 1$):

$$\eta_i = 1 - e^{-3\tau_e} \tag{8}$$

It is difficult to determine an exact expression for deposition efficiency in the lung because the motion of a particle is a result of the vector sum of many forces. Yu (1985) assumed a simple sum deposition model for the combined deposition efficiency, where total deposition efficiency is the sum of the efficiency of particle deposition of neutral particles by conventional mechanisms plus the contribution by particle charge due to the image force. For particles that deposit in the alveoli by sedimentation he derived a threshold charge below which no effect due to charge is detected. This model was in good agreement with the results of Tarroni et al. (1980) and Melandri et al. (1983) (Figure 1).

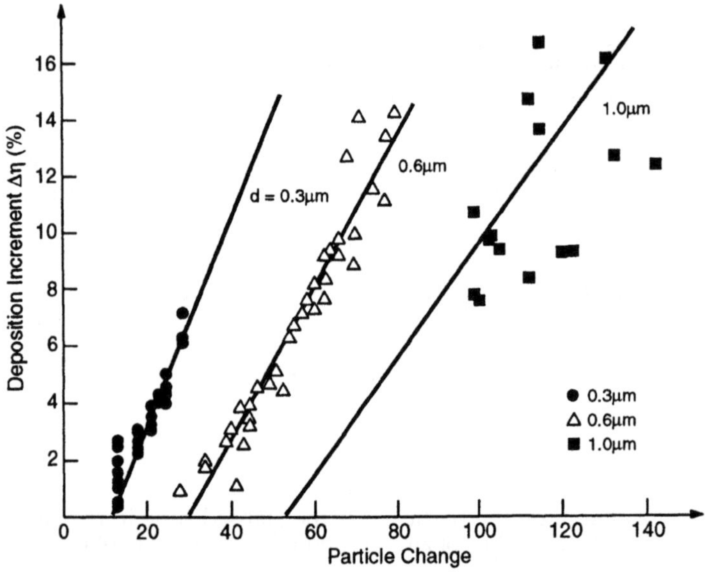

Figure 1. The increment in deposition $\Delta\eta$ vs number of particle charges carried by particles with aerodynamic diameters of 0.3, 0.6 and 1.0 μm. Points show experimental data of Tarroni et al (1980) and Melandri et al. (1983). Solid lines represent results calculated by Yu (1985). $\Delta\eta = (\eta - \eta_o)/(1-\eta_o)$ where η is the total deposition of charged particles and η_o is the deposition for the same particles without charge. (Adapted from Yu, C.P. 1985).

Cohen et al. (1995) recently examined the potential for increased deposition of ultrafine particles by image forces when diffusion is the primary mechanism of deposition. To determine the relative importance of diffusion and charge, they calculated the ratio,

$$\alpha = \eta_i/\eta_d \tag{9}$$

where η_i is given by equation (1) and η_d is the efficiency of deposition by diffusion. For η_d they used an the empirical equation derived from deposition data for ultrafine particles obtained in hollow airway casts (Cohen and Asgharian, 1990). Knowing that deposition fraction, η, depends only on Δ, the measured data of Cohen (1987) were fitted to

$$\eta = a\Delta^b \tag{10}$$

where a and b are found to be 2.97 and 0.568, respectively for $10^{-9} < \Delta < 10^{-4}$.

$$\Delta = \frac{\pi LD}{4Q} \tag{11}$$

in which L is the airway length, Q is the flow rate through the airway, and D is the diffusion coefficient given by:

$$D = kTB \tag{12}$$

where k is the Boltzmann constant, T is the absolute temperature, B is the mobility.

$$\alpha = \frac{\eta_d}{\eta_i} = 1.63\frac{\Delta^{0.568}}{\tau_e^{1/3}} \tag{13}$$

Results calculated for unit charge (unipolar) particles of diameters between 10 nm and 100 nm for flow rates from 300 cm^3s^{-1} to 600 cm^3s^{-1} show that α increases from 1.4 for d = 100 nm to about 4.4 for d = 10 nm. This implies that for this range of particle size deposition by diffusion and electrostatic charge effects are of the same order of magnitude. Although the effect of charge decreases with increasing diameter, its importance increases relative to deposition by diffusion. These results are only an approximation since deposition by diffusion and electrical forces are not independent processes and thus are not additive. Chen and Yu (1993) proposed an equation of the form $\eta_{12} = [\eta_1{}^2 + \eta_2{}^2 - (\eta_1\,\eta_2)^2]^{1/2}$, where η_1 and η_2 are respectively individual deposition efficiencies by mechanisms 1 and 2 and η_{12} is the efficiency for the combined mechanisms. The theoretical solutions they derived for combined electrostatic and diffusive deposition for highly charged particles and uniform flow compare well with numerical results previously derived by Yu and Chandra (1977, 1978). They predict that any additional lung deposition resulting from a single charge will be virtually indistinguishable from deposition by diffusion.

Measured Deposition of Ultrafine Particles in Airway Models

Ultrafine Particle Deposition in Surrogate Tracheas

We measured and compared the deposition of charged and neutral ultrafine particles in simple tracheal models (Cohen et al., 1995). For these experiments nebulized NaCl particles were size classified with Differential Mobility Analyzer (DMA, Model 3071, TSI, St. Paul, MN) to produce monodisperse, singly charged, particles with diameters from 15.5 to 95 nm. The deposition in cylindrical copper tubes was compared for charged and "neutralized" aerosols. The charge "neutralized" particles were produced by passing the aerosol through a bipolar charge region produced by a ^{210}Po source. Deposition differences were detectable for particles < 30 nm in 10, 23, and 30 cm long tubes, and for particles up to 95 nm in the longest tube. Those results are shown in Figure 2. The value of η_i calculated from equation (7) is also shown for comparison. The magnitude of the deposition by image forces, η_i is much greater than that predicted by equation (7) or by Chen and Yu (1993). Some of the difference may result from an underestimate of η_d by equation 2, so that although it is clear that charge will increase tracheal deposition, the absolute magnitude of the increase is uncertain. The value of η_d predicted by equation (2) that was used in these analyses is greater than that for parabolic flow but less than that predicted for uniform flow. If η_i is evaluated from the experimental data using η_d predicted for uniform flow, the excess deposition by image forces would be reduced, but a large excess would still be seen.

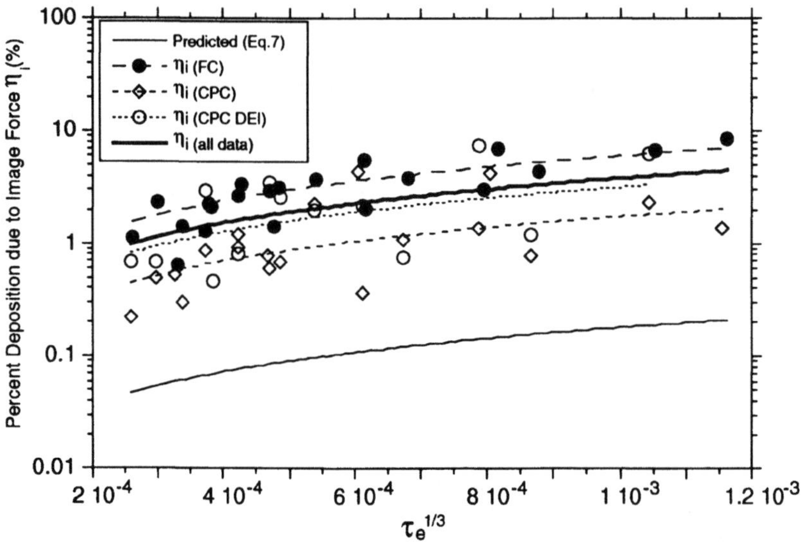

Figure 2. Deposition efficiency of singly charged particles in tracheal tubes due to image forces [$\eta_i(\%) = 100(\eta-\eta_d)/(1-\eta_d)$]. Data points are mean values for singly charged particles as measured with a condensation particle counter (CPC, open diamonds), Faraday Cup (closed circles) and the charged fraction of the "neutralized" particles (CPC-DEI, open circles). The latter were calculated from the values for deposition of charge neutralized particles measured with the CPC. The lines are a linear least squares fit to the data, weighted by standard error. The upper solid line is a single fit to all of the data points similarly weighted. The lower solid line represents values predicted from $\eta_i(\%) = 100\ [6^{1/3}\ \tau_e{}^{1/3}]$ (eq. 7). (Cohen et al., 1995. Reprinted from the Journal of Aerosol Science, Pergamon, with permission of the Editors-in-Chief.)

Ultrafine Particle Deposition in Hollow Airway Casts

In recent experiments we measured the deposition of charged, charge-neutralized, and zero-charge 20 nm and 125 nm particles in hollow-cast models of human airways. These particle sizes were selected because they are about where modal peaks occur for the activity of the short-lived radon progeny in indoor air. The original cast was prepared from a human lung obtained at autopsy, and replicates have been used in prior studies to measure detailed deposition of ultrafine charge-neutralized particles (40 nm < d < 200 nm) (Cohen, 1987). A new system was utilized for production of replicate conducting casts for the present study in order to simulate the electrical behavior of the human airways. A hollow plastic mold of the original cast was produced. The mold was made of many separate layers that could be lifted away to release a solid copy. The solid copy in turn was made by pouring a low melting point metal alloy into the hollow mold. After releasing and separating the solid metal cast, it was coated with many thin layers of silastic to produce a hollow cast for use in deposition experiments. Separation of the hollow central airway cast from the metal mold was accomplished by melting the metal out of the plastic cast. The resulting hollow cast retained a thin layer of conducting metal on the inner surface so that it could be used directly for charged particle deposition experiments. The conducting inner layer is currently achieved with a graphite impregnated silicone.

A schematic of the experimental setup is shown in Figure 3. Briefly, a Collison nebulizer is used to generate a polydisperse aerosol by atomizing a fluorescein-0.1N ammonium hydroxide solution. The dried polydisperse fluorescein aerosol is introduced into a DMA where it is classified within a narrow size range according to the particle electrical mobility. The extracted monodisperse aerosol is introduced into the lung cast either 1) directly, with each particle carrying a single positive charge, 2) after neutralization, for measuring the deposition efficiency of charge neutralized particles, or 3) after neutralization followed by total removal of charged particles. The zero-charge particles were produced by passing the charge neutralized monodisperse aerosol stream through an Electrical Aerosol Size Analyzer (Model 3030,TSI, St. Paul, MN) adapted to act as an electrostatic precipitator.

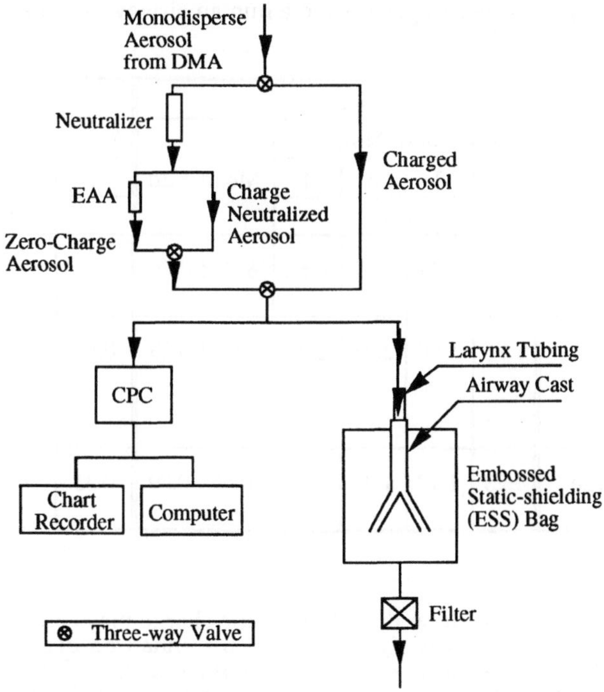

Figure 3. Schematic diagram of experimental system for measuring deposition efficiency of charged, charge-neutralized, and zero-charge particles in a hollow human airway cast.

To reduce electrostatic effects on the measurement the cast was enclosed in an electrically-conductive embossed static-shielding (ESS) bag. The entire system was charge neutralized by a 0.76 mCi ^{226}Ra, or a 200 mCi ^{241}Am source. The concentration of the aerosol was monitored with a Condensation Particle Counter (CPC, Model 3025, TSI, St. Paul, MN) and recorded by an interfacing personal computer.

The fluorescein was collected from the inner surface of the cast, the filter, filter holder, the ESS bag, connection tubing, and the outer surface of the cast by rinsing either with distilled water or 0.1 N NH_4OH solution. The recovery efficiency was 99% for most parts of the system, but only 90% for the outside of the cast. The masses of fluorescein were quantitated by fluorometry with a Luminescence Spectrometer (Model LS-5, Perkin-Elmer Co., Oak Brook, IL).

Results

Results are presented in Table 1. Mean deposition (± standard error) of singly charged 20 nm particles was 5.3±0.3 times greater than that of zero-charge, and 3.4±0.3 times that of charge-neutralized particles in the hollow airway casts. For 125 nm singly charged particles deposition averaged 6.2±0.7 times greater than that of zero-charge, and 2.3±0.3 times that of charge-neutralized particles.

Table 1. Measured deposition efficiency, η (%), of charged, charge-neutralized and zero-charge fluorescein particles in a human airway cast at an average flowrate of 5 liters/minute.

Particles	d = 20 nm					d = 125 nm			
	Experiment Number				Mean ± 2σ	Experiment Number			Mean ± 2σ
	1	2	3	4		1	2	3	
	η (%)					η (%)			
Singly Charged	6.78	7.44	5.75	6.52	6.62 ± 0.70	1.62	1.54	1.68	1.61± 0.08
Charge-Neutralized	1.96	1.58	2.03	2.28	1.96 ± 0.29	0.86	0.61	0.64	0.70 ± 0.16
Zero-Charge	1.23	1.17	1.33	-	1.24 ± 0.09	0.31	0.26	0.21	0.26 ± 0.06

Discussion

These studies all show that electrostatic charge plays an important role in lung deposition of particles. The significance of the enhanced deposition with respect to human health must be considered in view of the quantity of charge that is present on airborne particles and their size distribution. Aerosol particles generated from specific sources by mechanical processes tend to be larger and more highly charged than ordinary ambient particles. Published measurements for workplace aerosols have been reviewed by Johnston et al., 1985. They concluded that there is a wide variety in electrical characteristics of aerosols depending on the charging process. They were able to fit their field measurements of the median magnitude of charge per particle, q_m, together with a variety of measurements reported for laboratory dispersed dusts, to an equation of the form $q_m/e=Kd^n$ where d is particle diameter in micrometers, and e is the charge on an electron. K then represents the median magnitude of charge carried by a 1 μm particle. They report K values from 3 to 86 for a variety of 9 laboratory dispersed dusts, and from 2.1 to 49.8 for about 28 workplace processes in the United Kingdom. For the laboratory dusts n ranged from 1.0 to 1.6, for the workplace the range was 0.72 to 1.91.

In reviewing the relevance of highly charged industrial aerosols, John and Vincent (1985) considered that important effects are only seen for highly charged particles, and high numbers of charges are carried by large particles and long fibers. Large particles deposit with high efficiency in the respiratory tract, principally on tracheobronchial airways, as a result of inertial forces, thus any increase that results from charge would be minimal. For particles that deposit primarily in the alveolar region, very high charges are unlikely based on the reported measurements. They concluded therefore that "...for isometric dusts, in normal circumstances.....the electric charge effects will have no significant effect on lung deposition and, in turn, on health risk." Thus the increase that results from electrical forces is not important for particle sizes and lung regions where deposition efficiency is high since all deposition mechanisms act on the same particles. This was not the case however for fine fibrous dusts like asbestos for which they conclude "that electrostatic effects may contribute substantially to lung deposition and therefore, possibly to health risk."

Indoor air environments are usually dominated by nanometer sized particles (Owen et al., 1990; Kamens et al., 1991), and most outdoor airborne particles are in the ultrafine particle mode. This differs from a workplace where dust particles in the micrometer size range are typical because of mechanical processes. Infiltration of outdoor aerosol is an important source of the indoor aerosol. Early studies of ambient airborne particle size distributions reported over 90% of all airborne particles in nuclei less than 50 nm in diameter (NRC, 1977; Whitby et al., 1972) The mode varies depending on atmospheric conditions, and the median size of particles reported for eight typical atmospheric size distributions ranged from 15 to 38 nm with an average of 29±7 nm in the nuclei mode, and a range of 180 to 360 nm with an average of 290±60 nm for the accumulation mode (John, 1993; Whitby and Sverdrup, 1980). Recent more detailed measurements including 1987 data from downtown Los Angeles, an urban site at Long Beach California, and a clean background site in the Rocky Mountains, Colorado, summarized by Hering (1995, personal communication) indicate that the geometric mean diameter and fraction of particle number in the nuclei mode, and the relationship between particle number and particle volume vary with site and atmospheric conditions, but numerical values are in the same general range. Many of these small aerosol particles carry one or more electrical charges (Bricard et al., 1966).

Ambient airborne particles are charged by free ions in the air that are produced by cosmic rays and by natural radioactivity. The particle charge distribution can be estimated by a Boltzmann distribution for equilibrium conditions. The predicted charge as a function of particle size is shown in Figure 4. For particles with d = 20 nm more than 10% are expected to be charged and the number rises to over 50% for d > 80 nm, with equal numbers of particles of either sign. When charge distributions are measured equilibrium conditions are not normally met; a larger fraction of charged particles and an excess of charge of one sign is common. In addition, the Boltzmann distribution does not apply for very small particles. The number of charged particles in test aerosols with diameters from 4-40 nm exceed the predictions based on Boltzmann equilibrium, and a ratio of 0.35:0.65 was observed for positively to negatively charged particles (Kousaka et al., 1983). One set of measurements of charge has been reported for the short-lived radon progeny. Those measurements in a ^{222}Rn test chamber (d ≈ 100 nm) suggested that 32% of the ^{218}Po and 91% of its offspring, ^{214}Pb, were charged (Maiello and Harley, 1989).

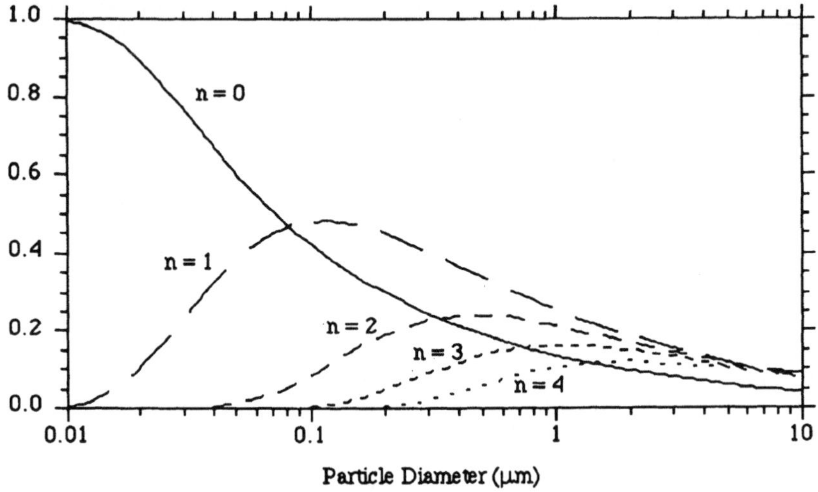

Figure 4. Fraction of aerosol particles carrying n units of charge at equilibrium based on the Boltzmann distribution.

We monitored the aerosol charge distribution in our laboratory (Lanza Labs, NYU, Tuxedo, NY) and found that the fractions of charged particles were about 50% - 100% higher than predicted for a Boltzmann distribution in the size range 10 nm - 500 nm. The fraction of naturally charged particles with electrical mobility equal to a singly charged 125 nm particle for example was about 60% higher than that of a Boltzmann distribution.

Ambient atmospheres are dominated by particles in the nanometer size range and a significant proportion of them are normally charged. The results of our cast deposition studies show significant differences in deposition between singly charged, charge neutralized, and neutral (zero-charge) ultrafine particles. This is in contrast to theoretical predictions of lung deposition based on the models presented above which suggest that increased deposition of nanometer sized particles under normal ambient conditions would be insignificant. The additional deposition that results from particle charge, must be evaluated at higher flowrates that are more applicable to normal breathing patterns. But it is clear that the charge effect can be very important. This suggests that the effect of electrostatic charge on lung deposition of ultrafine particles cannot be neglected as has been common in dosimetry models.

There is little doubt that enhanced deposition of singly charged nanometer sized particles would also be observed in the nasal and alveolar regions of the respiratory tract. Deposition of ultrafine particles is very low in the upper airways compared with total respiratory tract deposition, but because a high proportion of lung cancers are bronchogenic, this small fraction is clearly important for toxicants that act before they are removed by normal clearance. It is particularly important for radon progeny in homes and mines, but also affects dose predictions for other agents that evoke responses based on airway deposition (e.g., acid aerosols and environmental tobacco smoke).

Conclusions

Electrical charge on aerosol particles increases their deposition in the respiratory tract because of image forces. The significance of this increase depends on the size of the particle and the number of charges it carries. For large particles that deposit with high efficiency on tracheobronchial airways, any increase that results from charge would be minimal. For particles that deposit primarily in the alveolar region by sedimentation, effects are also minimal because few carry very high charges based on the measurements reported for workplace aerosols. Fibrous particles differ, however, and charge effects can be significant. The importance of electric charge increases for small particles that deposit primarily by diffusion. Ambient atmospheres are dominated by nanometer sized particles and a significant proportion of them carry only one, or a few, charges. The results of our cast deposition studies show significant differences in deposition between singly charged, charge neutralized, and neutral (zero-charge) ultrafine particles. This differs from predictions based on theoretical lung models. The additional deposition that results from particle charge, can be very important where deposition is ordinarily low, because in such cases even a small enhancement factor can cause a significant increase in the delivered dose. Thus the effect of electrostatic charge on lung deposition of ultrafine particles should not be neglected in dosimetry models used to evaluate the health effects of ambient aerosols.

Acknowledgments

This research was supported in part by the U.S. Department of Energy, Grant No. ER60592.

References

Bricard, J., and Pradel, J. (1966) "Electric charge and radioactivity of naturally occurring aerosols", pp. 87-109 in Aerosol *Sciences*, Ed. Davis, C.N., Academic Press, New York, NY

Chan, T.L., Lippmann, M., Cohen, V.R., and Schlesinger, R.B. (1978) "Effect of electrostatic charges on particle deposition in a hollow cast of human larynx-tracheobronchial tree", *J. Aerosol Sci.* **9**: 463-468

Chen, Y.K., and Yu, C.P. (1993) "Particle deposition from duct flows by combined mechanisms", *Aerosol Sci. Technol.* **19**: 389-395

Cohen, B.S. (1987) "Deposition of ultrafine particles in the human tracheobronchial tree: A determinant of the dose from radon daughters", pp. 475-486 in *Radon and Its Decay Products: Occurrence, Properties and Health Effects*, Ed. Hopke, P., American Chemical Society, Washington, DC 20036

Cohen, B.S., and Asgharian, B. (1990) "Deposition of ultrafine particles in the upper airways: An empirical analysis", *J. Aerosol Sci.* **21**: 789-797

Cohen, B.S., Xiong, J.Q., Asgharian, B., and Ayres, L. (1995) "Deposition of inhaled charged ultrafine particles in a simple tracheal model", *J. Aerosol Sci.* , in press

Ferin, J., Mercer, T.T., and Leach, L.J. (1983) "The effect of aerosol charge on the deposition and clearance of TiO_2 particles in rats", *Environ. Res.* **31**: 148-151

Fraser, D.A. (1966) "The deposition of unipolar charged particles in the lungs of animals", *Arch. Environ. Health* **2**: 535-543

Fry, F.A. (1970) "Charge distribution on polystyrene aerosols and deposition in the human nose", *Aerosol Sci.* **1**: 135-146.
Hering, S. 1995 Personal Communication
John, W., and Vincent, J.H. (1985) "Static electrification of workplace aerosols: A perspective", *Ann. Occup. Hyg.* 29: 285-288
John, W. (1993) "The characteristics of environmental and laboratory-generated aerosols", pp. 54-76 in Aerosol Measurement: Principles, Techniques and Applications", Eds. Willeke, K. and Baron, P.A., Wiley, New York, NY
Johnston, A.M., Vincent, J.H., and Jones, A.D. (1985) "Measurements of electric charge for workplace aerosols", *Ann. Occup. Hyg.* **29**: 271-284
Jones, A.D., Vincent, J.H., Johnston, A.M., and McMillan, C.H. (1988) "Effects of electrostatic charge on the pulmonary deposition of mineral dust aerosols inhaled by rats", *J. Aerosol Sci.* **19**: 565-575
Kamens, R., Lee, C.T., Wiener, R., and Leith, D. (1991) "A study to characterize indoor particles in three non-smoking homes", *Atmos. Environ.* **25**: 939-948
Kousaka, Y., Adachi, M., Okuyama, K., Kitada, N., and Motouchi, T. (1983) "Bipolar charging of ultrafine aerosol particles", *Aerosol Sci. Technol.* **2**: 421-427
Longley, M.Y. (1960) "Pulmonary deposition of dust as affected by electric charges on the body", *Am. Ind. Hyg. Assoc. J.* **21**: 187-194
Longley, M.Y., and Berry, C.M. (1961) "Pulmonary deposition of aerosols", *Arch. Environ. Health* **2**: 535-543
Maiello, M., and Harley, N.H. (1989) Determining the charged fractions of ^{218}Po and ^{214}Pb using an environmental γ-ray and Rn detector", *Health Phys.* **57**: 51-59
Melandri, C., Prodi, V., Tarroni, G., Formignani, M., DeZaiacomo, T., Bompane, G.F., and Maestri, G. (1977) "On the deposition of unipolar charged particles in the human respiratory tract", pp. 193-200 in Inhaled Particles IV, Ed. Walton, W.H., Pergamon Press, Oxford
Melandri, C., Tarroni, G., Prodi, V., DeZaiacomo, T.G., Formignani, M., and Lombardi, C.C. (1983) "Deposition of charged particles in the human airways", *J. Aerosol Sci.* **14**: 657-669
National Research Council (1977) Airborne Particles, National Academy of Sciences
Owen, M.K., Ensor, D.S., Hovis, L.S., Tucker, W.G., and Sparks, L.E. (1990) "Particle size distributions for an office aerosol", *Aerosol Sci. Technol.* **13**: 486-492
Pich, J. (1978) "Comments on the paper: C.P. Yu's precipitation of unipolarly charged particles in cylindrical and spherical vessels", *J. Aerosol Sci.* **9**: 275-278
Prodi, V., and Mularoni, A. (1985) "Electrostatic lung deposition experiments with humans and animals", *Ann. Occup. Hyg.* **29**: 229-240
Tarroni, G., Melandri, C., Prodi, V., DeZaiacomo, T., Formignani, M., and Bassi, P. (1980) "An indication on the biological variability of aerosol total deposition in humans", *Am. Ind. Hyg. Assoc. J.* **41**: 826-831
Vincent, J.H., Johnson, W.B., Jones, A.D., and Johnston, A.M. (1981) "Static electrification of airborne asbestos: a study of its cause , assessment and effects of airborne asbestos: A study of its cause, assessment and effects on deposition in the lungs of rats", *Am. Ind. Hyg. Assoc. J.* **42**: 711-721
Whitby, K.T., Husar, R.B., and Liu, B.Y.H. (1972) "The aerosol size distribution of Los Angeles smog", *J. Colloid Interface Sci.* **39**: 177-204
Whitby, K.T., and Sverdrup, G.M. (1973) "California aerosols: Their physical and chemical characteristics", p. 495 in *The Character and Origins of Smog Aerosols*, Eds. Hidy, G.M., et al., Wiley, New York
Wilson, L.B. (1947) "The deposition of charged particles in tubes, with reference to the retention of therapeutic aerosols in the human lung", *J. Colloid Sci.* **2**: 271-276
Yu, C.P. (1977) "Precipitation of unipolar charged particles in cylindrical and spherical vessels", *J. Aerosol Sci.* **8**: 237-241
Yu, C.P. (1985) "Theories of electrostatic lung deposition of inhaled aerosols", *Ann. Occup. Hyg.* **29**: 219-227
Yu, C.P., and Chandra, K. (1977) "Precipitation of submicron charged particles in human lung airways", *Bull. Math. Biol.* **39**: 471-478
Yu, C.P., and Chandra, K. (1978) "Deposition of charged particles from laminar flows in rectangular and cylindrical channels by image force", *J. Aerosol Sci.* **9**: 175-180

Effects of Ultrafine Particles in the Lung and Potential Relevance to Environmental Particles

G. Oberdörster
The University of Rochester
Department of Environmental Medicine
Rochester, NY 14642

ABSTRACT

A number of epidemiological studies has been published over the past 5 years which reported an association between low ambient particle concentration (< 100 $\mu g/m^3$) and acute morbidity and even mortality in the elderly. Since larger particles above ~0.3 μm cannot explain such dramatic effects we hypothesize that ultrafine particles (< 50 nm) present in the urban atmosphere at high number concentrations may be causally involved. We have evaluated this hypothesis by using model ultrafine particles of Teflon® fume with diameters of 10-26 nm. These particles when inhaled by rats proved to be extremely toxic, leading to severe hemorrhagic pulmonary edema and even death at short-exposure durations (10-30 minutes) and particle number concentration of 1-5x10^5 particles/cm^3. The respective calculated mass concentrations are in the range of 30-60 $\mu g/m^3$. We conclude that certain ultrafine particles when inhaled as singlet particles can be highly toxic and that there is a need to study more closely environmentally occurring ultrafine particles.

KEY WORDS

Ultrafine particles; pulmonary toxicity; inhalation; deposition

INTRODUCTION

Although evidence from a number of epidemiological studies indicates that acute morbidity and acute mortality seen in those studies is best correlated with ambient particulate levels at very low concentrations (< 100 $\mu g/m^3$) (e.g., Schwartz et al., 1994; Dockery et al., 1993) this association does not imply a causal relationship of the observed effects and exposure to ambient particles. Indeed, it is difficult to find a plausible explanation how such extremely low concentrations of particles can induce these adverse acute effects. Several hypotheses have been proposed. One is that acidic particles or a coating of particles with acid (e.g., H_2SO_4) could be responsible. Laboratory data in guinea pigs by Chen et al. (1992) indicate that low levels of H_2SO_4 coated onto carrier particles resulted in effects in the conducting airways of the animals. Another suggestion is that surface coating of ambient particles with transition metals (e.g., iron) will lead to an increased inflammatory reaction after phagocytosis of these particles due to the generation of more toxic oxygen-derived radicals via a Fenton and Haber-Weiss reaction (Costa et al., 1994; Ghio et al., 1992). Yet another theory suggests that opsonized particles leads to amplified inflammatory response

J. C. M. Marijnissen and L. Gradoń (eds.), Aerosol Inhalation: Recent Research Frontiers, 165–173.

upon uptake by cells in the lung which may also induce such reactions in the neighboring cells. Indeed, Kobzik et al. (1990, 1993) could show that only opsonized particles lead to a significantly increased respiratory burst in macrophages whereas non-opsonized particles after phagocytosis caused actually a decrease in the oxidative burst; similar results were most recently reported by Hill et al. (1995) for opsonized fibrous particles. Still another suggestion is that highly insoluble ultrafine particles (<50 nm) which occur in the ambient air can be highly reactive and cause subsequent adverse reactions in the respiratory tract (Oberdörster et al., 1995; Seaton et al., 1995). Since no studies have been performed with relevant ultrafine ambient particles this hypothesis is primarily based on laboratory findings with model ultrafine particles which will be summarized in the following paragraphs. Perhaps a combination of these theories should be developed which may best explain observed epidemiological data.

Toxicity of ultrafine particles: A plausible hypothesis?

In order for any of these hypotheses to be plausible the following criteria have to be met: Adverse effects should be observed at low inhaled mass concentrations (<100-50 $\mu g/m^3$); particles should be present indoors as well as outdoors, i.e., should be able to penetrate easily; and the particles should have greater effects in people with compromised respiratory systems (COPD). Since smaller particles below 0.5 μm are more affected by diffusive processes, and especially ultrafine particles behave more like gases these are more likely to penetrate into indoor spaces. Certain ultrafine particles which have been tested so far (see below) also show very high acute pulmonary toxicity at very low inhaled mass concentrations, whereas larger-sized particles (>0.2 μm) have not been shown to elicit such effects. Only if concentrations in the mg/m^3 range are administered for such larger particles were effects seen; however, these were of a chronic nature, i.e., after repeated long-term exposures. It appears that under such high exposure conditions any particle type of low cytotoxicity and low solubility will induce effects in laboratory rodents, specifically in rats. These effects include pulmonary fibrosis and lung tumors which are then associated with a phenomenon of lung particle overload (Morrow, 1988; Muhle et al., 1990). Previous experiments with aggregated ultrafine particles such as TiO_2 and gallium oxide (Oberdörster et al., 1992; Wolff et al., 1988) have shown that these particles are significantly more toxic to the lung than larger-sized particles of the same chemical nature, although exposure concentrations were still high (mg/m^3) and exposures had to be repeated over an extended period of time. It was also found that inhaled aggregated ultrafine particles of TiO_2 penetrated more easily across alveolar epithelial barriers into the pulmonary interstitium (Takenaka et al., 1986; Ferin et al., 1992) in contrast to singlet ultrafine particles which may behave very differently and thus cause effects already at lower concentrations, especially if these particles are more reactive.

The hypothesis that freshly-generated ultrafine particles when inhaled as singlets at very low mass concentrations can be highly toxic to the lung is supported by their high deposition efficiency in the lower respiratory tract, their large numbers per unit mass, and their increased surface areas available for interactions with cells. In addition, particle-cell interactions could be potentially amplified by the presence of radicals on these surfaces. Based on the predictive lung deposition model for inhaled particles published by ICRP (1994) ultrafine particles inhaled as singlets of a size of about 20 nm have the highest deposition efficiency in the alveolar region of the lung in humans. Up to 50% of inhaled ultrafine particles of this size are deposited there, and assuming a resting minute ventilation of 8 l/min and an exposure concentration of 1 x 10^5 particles per cm^3, about 80 particles would be

TABLE 1

Numbers and surface areas of particles of unit density of different sizes at a mass concentration of 10 μg/m^3

Particle Diameter μm	Particle Number 1/cm^3	Particle Surface Area μm^2/cm^3
0.02	2,400,000	3016
0.1	19,100	600
0.5	153	120
1.0	19	60
2.5	1.2	24

TABLE 2

Research Needs, Ultrafine Particles

- Physico-chemical characterization of ambient u.f. particle phase
- Behavior under specific meteorological conditions
- Outdoor/indoor relationships
- Toxicological evaluation of relevant ultrafine particles
  (carbon; metal-oxides; O_3, NOx - olefinic reactions)
- Cellular and molecular mechanisms of u.f. particle effects
- Study particle effects in relevant animal model with cardio-respiratory disease

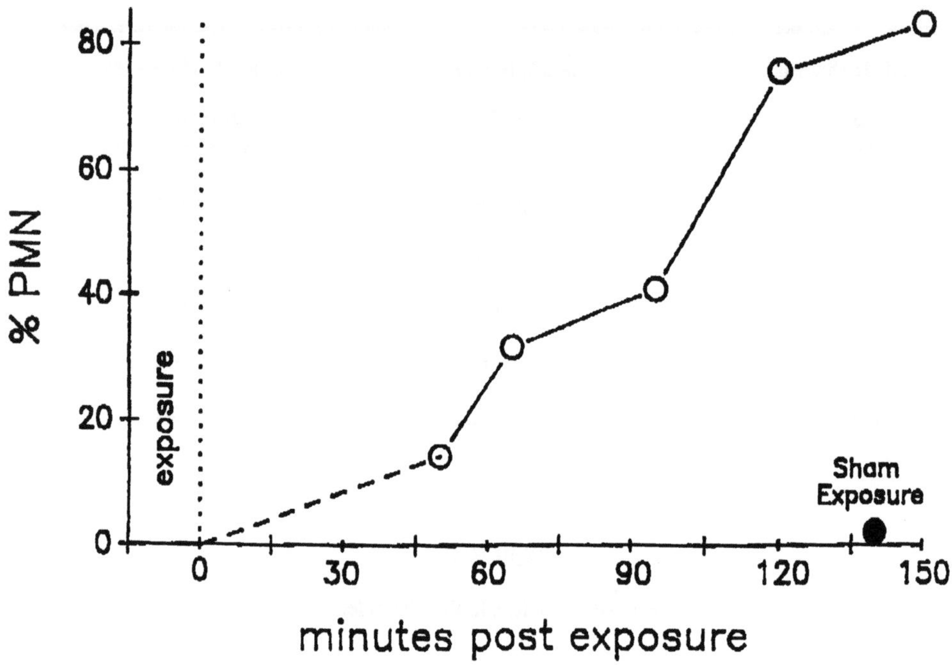

Figure 1: Percent of PMN in lung lavage at different times after a 15-min. exposure to 5×10^5 PTFE particles/cm3 generated in an argon atmosphere.

deposited in one hour per alveolus in a human lung assuming 3 x 10^8 alveoli per lung. This is obviously a very high number; however, in terms of particle mass, this is very low as is demonstrated in Table 1 in which particle numbers and surface areas for different particle diameters are compared resulting in the same mass concentration. For 20 nm particles of unit density, 2.4 x 10^6 particles/cm^3 are required to achieve 10 $\mu g/m^3$, whereas for 2.5 μm particles only about 1 particle/cm^3 gives the same mass concentration.

Studies with model ultrafine particles.

A number of studies have been performed in the past using fumes generated by heating of polytetrafluoroethylene (PTFE, Teflon®) which showed that these fumes are highly toxic to different animal species (Waritz and Kwon, 1968). It was suggested early on that the toxicity is related to the particle phase in these fumes rather than toxic gas phase compounds; it was shown more recently that the particles consist of ultrafine particles ranging between 10 and 30 nm in diameter (Seidel et al., 1991; Warheit et al., 1990; Oberdörster et al., 1995). Concentrations of 5 x 10^5 particles/cm^3 inhaled for 15-30 mins. by rats resulted in the development of acute pulmonary hemorrhagic edema with a high influx of inflammatory PMNs in the lung. About 10^5 particles/cm^3 of these PTFE fume particles is equivalent to a mass concentration of 9 $\mu g/m^3$. To exclude the possibility that oxygen-derived radicals on the surface of these particles may cause the highly toxic response, particles were also generated by heating Teflon® in an argon gas atmosphere which resulted in the same high toxicity. Figure 1 shows an example of the time course of the post-exposure influx of PMN into the alveolar space as measured by bronchoalveolar lavage of animals exposed to 5 x 10^5 particles/cm^3 for 15 mins. to ultrafine Teflon® particles generated in an argon gas atmosphere. The highly inflammatory response was accompanied by severe hemorrhagic pulmonary edema with damage to epithelial and endothelial cells. Earlier studies as well as newer studies have shown that this toxicity can be prevented by filtering out the particles which implies that the high pulmonary toxicity is associated with the particle phase (Waritz and Kwon, 1968; Coleman et al, 1968a,b; Warheit et al., 1990, Lee and Seidel, 1991).

It appears that deposited ultrafine particles penetrate rapidly into the pulmonary interstitium, thereby interacting with epithelial cells and subsequently with interstitial and endothelial cells. Elicitation of PMNs and their activation results in the production of reactive oxygen species which could lead to damage of both the epithelial and endothelial barrier function. This in turn leads to the development of interstitial as well as alveolar edema and also the exudation of erythrocytes as seen in the acute PTFE-fume experiments in rats. Several pro-inflammatory cytokines such as TNFa, IL-1 and IL-6 have been found to be up-regulated in lungs after ultrafine PTFE particle exposures (Oberdörster et al., 1995). In addition, message levels for antioxidants such as MnSOD and metallothionein increased also after ultrafine PTFE particle fume exposures. Moreover, initial data indicated that rats with elastase-induced pulmonary emphysema were more sensitive to these particles than healthy rats (Oberdörster et al., 1995).

Ambient ultrafine particles.

Obviously, PTFE-fume particles do not occur in the ambient atmosphere and the question is as to whether ultrafine particles in general are highly reactive in the lung because of their high numbers and high surface area. Measurements of particle size distribution in the urban atmosphere have shown that it consists principally of three particle modes: The coarse particle mode with a median size of about 5 μm; the accumulation mode with a median size of about 0.2 μm; and the nucleation mode with a median size of about 20 nm (ultrafine particles) (Whitby and Cantrell, 1976; Wilson et al., 1977). Whereas the coarse particles and accumulation mode particles contribute most to the overall mass concentration, it is the nucleation or ultrafine particle mode that by far exceeds the other modes in terms of particle numbers. However, the ultrafine particles are not very stable in that they coagulate and aggregate onto the accumulation mode particles, but they are constantly regenerated from several sources and thus available as freshly-generated ultrafine particles. These particles consist of carbon black, combustion nuclei, oil smokes, metallurgical dusts and fumes and are most prevalent under smog conditions (NRC, 1979). Both inorganic and organic ultrafine particles are generated. Inorganic particles from combustion processes result in carbon and metal oxides; for example, platinum oxide in tailpipe emissions of catalyst-equipped cars. Organic particles are generated from gas to particle conversions in the atmosphere; for example, via reactions of ozone and NOx with olefinic compounds which are emitted from anthropogenic and natural sources (NRC, 1979).

Concentrations of ambient ultrafine particles are generally extremely low by mass, but can be very high by number. For example, Brand et al. (1991; 1992) determined ambient ultrafine particle levels up to about 300,000 particles/cm^3 in an urban atmosphere in Germany. They found a significant increase in ultrafine particle numbers when a warm front was moving through the area. Grosjean and Friedlander (1980) measured cyclic olefins which are present in gasolines and car exhaust to be at a level at 2-10 ppb in the Los Angeles atmosphere. They calculated that via gas to particle conversions 10 ppb of cyclic olefins could generate 10 μg/m^3 of an organic aerosol. Recent data by Castellani (1993) of seasonal atmospheric measurements in Bologna, Italy, show the presence of ultrafine particles in the urban atmosphere at levels ranging from about 3 x 10^4 to 3 x 10^5 particles/cm^3. No studies have been performed yet to determine the toxicity potential of these ambient ultrafine particles, however, it is conceivable that these particles could be highly reactive depending on the process of generation, e.g., reactive platinum oxide in tailpipe emissions (Rosner and Merget, 1990; Schlögl et al., 1987).

Outlook

If indeed ambient ultrafine particles can induce adverse responses in the respiratory tract, especially in sensitive subpopulations with an existing compromised cardiorespiratory system, the question arises as to whether the right particle types are regulated using the definition of PM_{10} or $PM_{2.5}$. Ambient ultrafine particles contribute only very little to the overall mass concentration and even a significant increase in ultrafine particle numbers may not be detectable by determining PM_{10} or $PM_{2.5}$ using the traditional measure of mass concentrations. Further research needs include the physico-chemical characterization of ambient ultrafine particles and their behavior under specific meteorological conditions as well as their outdoor/indoor relationships. Most importantly, a toxicological evaluation of relevant ambient ultrafine particles is needed. With respect to effects in the respiratory tract, research should focus on specific cellular and molecular mechanisms of the effects induced by ultrafine particles. Specifically, studies should be performed in relevant animal models which mimic sensitive subpopulations, i.e., animal models incorporating a compromised cardio-respiratory tract (Table 2). Until such studies have been performed, the ultrafine particle theory is just one of several which needs to be rigorously tested.

ACKNOWLEDGEMENT

This paper was supported in part by NIH Grants ESO1247 and RO1 ESO4872 and NASA Grant NAGW-2356.

REFERENCES

Brand, P., Gebhart, J., Below, M., Georgi, B., Heyder, J. Characterization of environmental aerosols on Heligoland Island. Atmospheric Environment 25A(No. 3/4): 581-585, 1991.

Brand, P., Ruob, K., Gebhart, J. Performance of a mobile aerosol spectrometer for an in situ characterization of environmental aerosols in Frankfurt City. Atmospheric Environment 26A(No. 13): 2451-2457, 1992.

Castellani, C.M. Characterization of the atmospheric aerosol in an urban area and the vicinity of a main highway. In: ENEA Environment Department meetings and seminars held in Bologna and Rome, INTO 7(93): 35-55, 1993.

Chen, L.C., Miller, P.D., Amdur, M.O., Gordon, T. Airway hyperresponsiveness in guinea pigs exposed to acid-coated ultrafine particles. J. Toxicol. & Environm. Health 35: 165-174, 1992.

Coleman, W.E., Scheel, L.D., Kupel, R.E. and Larkin, R.L. The identification of toxic compounds in the pyrolysis products of polytetrafluoroethylene (PTFE). Amer. Ind. Hyg. Assoc. J. 29: 33-40, 1968a.

Coleman, W.E., Scheel, L.D., Kupel, R.E. and Gorski, C.H. The particles resulting from polytetrafluoroethylene (PTFE) pyrolysis in air. Amer. Ind. Hyg. Assoc. J. 29: 54-60, 1968b.

Costa, D.L., Tepper, J.S., Lehmann, J.R., Winsett, D.W., Dreher, K., and Ghio, A.J.: Surface complexed iron (Fe+3) on particles: Its role in the induction of lung inflammation and hyperreactivity. Presented at the Colloquium on Particulate Air Pollution and Human Mortality and Morbidity, January 24-25, 1994, Irvine, CA.

Dockery, D.W., Pope, C.A., Xu, X., Spengler, J.D., Ware, J.H., Fay, M.E., Ferris, B.G. and Speizer, F.E. An association between air pollution and mortality in six U.S. cities. The New England Journal of Medicine 329 (24): 1753-1759, 1993.

Ferin, J., Oberdörster, G., and Penney, D.P. Pulmonary retention of ultrafine and fine particles in rats. Am. J. Resp. Cell and Mol. Biol. 6: 535-542, 1992.

Ghio, A.J., Kennedy, T.P., Whorton, A.R., Crumbliss, A.L., Hatch, G.E., Hoidal, J.R.: Role of surface complexed iron in oxidant generation and lung inflammation induced by silicated. American Journal of Physiology 263: L511-L518, 1992.

Grosjean, D. and Friedlander, S.K. Formation of organic aerosols from cyclic olefins and diolefins. Adv. Environ. Sci. Technol. 10: 435, 1980.

Hill, I.M., Beswick, P.H., and Donaldson, K. Differential release of superoxide anions by macrophages treated with long and short fibre amosite asbestos is a consequence of differential affinity for opsonin. Occup.and Environ. Med. 52: 92-96, 1995.

ICRP: (International Commission on Radiological Protection) (1994): Human Respiratory Tract Model for Radiological Protection. A Report of Committee 2 of the ICRP. Pergamon Press, Oxford, UK.
Kobzik, L., Godleski, J.J., Brain, J.D. Selective down-regulation of alveolar macrophage oxidative response to opsonin-independent phagocytosis. J. Immunology 144 (No. 11): 4312-4319, 1990.

Kobzik, L., Huang, S., Paulauskis, J.D., Godleski, J.J.: Particle opsonization and lung macrophage cytokine response. In vitro and in vivo analysis. The Journal of Immunology. 151, No. 5:2753-2759, September 1, 1993.

Lee, K.P., Seidel, W.C. Pulmonary response to perfluoropolymer fume and particles generated under various exposure conditions. Fundamental and Applied Toxicology. 17:254-269, 1991.

Morrow, P.E. Possible mechanisms to explain dust overloading of the lungs. Fundam. Appl. Toxicol. 10, 369-384, 1988.

Muhle, H., Creutzenberg, O., Bellmann, B., Heinrich, U., and Mermelstein, R. Dust overloading of lungs: Investigations of various materials, species differences and irreversibility of effects. J. Aerosol Med. 3: Suppl. 1, S111-
S128, 1990.

NRC, Airborne Particles, Subcommittee on Airborne Particles, Committee on Medical and Biologic Effects of Environmental Pollutants, University Park Press, 1979.
Oberdörster et al. (1992) p-6146

Oberdörster, G., Gelein, R.M., Ferin, J. and Weiss, B. Association of particulate air pollution and acute mortality: Involvement of ultrafine particles? Inhalation Toxicology 7: 111-124, 1995.

Rosner, G. and Merget, R. Allergenic potential of platinum compounds. In: Immunotoxicity of Metals and Immunotoxicology, (ed., A.D.Dayan et al.), Plenum Press, New York, pp. 93- , 1990.

Schlögl, V.R., Indlekofer, G. and Oelhafen, P. Mikropartikelemissionen von Verbrennungsmotoren mit Abgasreinigung - Röntgen-Photoelektronenspektroskopie in der Umweltanalytik. Angew. Chem. 99: 312-322, 1987.

Schwartz, J. Air pollution and daily mortality: A review and meta analysis. Environ. Res. 64: 36-52, 1994.

Seaton, A, MacNee, W., Donaldson, K., Godden, D. Particulate air pollution and acute health effects. The Lancet 345: 176-178, 1995.

Seidel, W.C., Scherer, K.V., Jr., Cline, D., Jr., Olson, A.H., Bonesteel, J.K., Church, D.F., Nuggehalli, S., Pryor, W.A. Chemical, physical, and toxicological characterization of fumes produced by heating tetrafluoroethene homopolymer and its copolymers with hexafluoropropene and perfluoro(propyl vinyl ether). Chemical Research in Toxicology. 4: 229-236, 1991.

Takenaka, S., Dornhöfer,-Takenaka, H., and Muhle, H. Alveolar distribution of fly ash and of TiO2 after long-term inhalation by Wistar rats. J. Aerosol Sci. 17: 361-364, 1986.

Warheit, D.B., Seidel, W.C., Carakostas, M.C., Hartsky, M.A. Attenuation of perfluoropolymer fume pulmonary toxicity: Effect of filters, combustion method and aerosol age. Exper. and Mol. Path. 52: 309-329, 1990.

Waritz, R.S., Kwon, B.K.: The inhalation toxicity of pyrolysis products of polytetrafluoroethylene heated below 500 degrees centigrade. American Industrial Hygiene Association Journal. Jan-Feb., pgs. 19-26, 1968.

Wolff, R. K., Henderson, R., Eidson, A., Pickrell, J., Hahn, F, and Rothenberg, S. Toxicity of gallium oxide particles following a 4-week inhalation exposure. J. Applied Toxicology 8: 191-199, 1988.

Whitby, K.T. and Cantrell, B. Atmospheric aerosols – characteristics and measurements. In Session 29, Fine Particles. In International Conference on Environmental Sensing and Assessment, held in Las Vegas, Nevada, September 14-17, 1975. New York: Institute of Electrical and Electronics Engineers, 1976.

Wilson, W.E., Spiller, L.L., Ellestad, T.G., Lamothe, P.J., Dzubay, T.G., Stevens, R.K., Macias, E.S., Fletcher, R.A., Husar, J.D., Husar, R.B., Whitby, K.T., Kittelson, D.B. and Cantrell, B.K. General Motors sulfate dispersion experiment: Summary of EPA Measurements. JAPCA 27(1): 46-51, 1977.

ATMOSPHERIC PARTICULATE POLLUTANTS AND ENVIRONMETAL HEALTH
A Physico-Chemical Standpoint

K.R.SPURNY
Aerosol Chemist, Eichenweg 6, 57392 SCHMALLENBERG, Germany

ABSTRACT

Several chemical and epidemiological investigations have been made during the last decade showing that correlations exist between ambient air concentrations of aerodispersed (particulate) pollutants and the health risk for the general population. Since then, changes in air particulate pollution definitions, measurements, analyses, and air quality standards have been proposed. In this review we review the "chemical standpoint" of the problem and its impact on the measurement strategy and on the air quality standard assessment.

KEYWORDS

aerodispersed (particulate) pollutants, air particulate pollution, air quality standard assessment, air quality standards, ambient air concentrations, analyses, chemical investigations, chemical standpoint, correlations, definitions, epidemiological investigations, health risk, measurements, measurement strategy.

INTRODUCTION

Recent epidemiological studies indicate health effects on the general population when air particulate mass concentrations lay below existing air quality standards. These studies indicate that increases in human mortality and morbidity are associated with levels of air particulate pollution significantly lower than were previously thought to affect human health. Despite some uncertainties, these new epidemiological data are considered sufficiently coherent to warrant the re-evaluation of the existing air particulate pollution philosophy and strategy. In the majority of published papers, concentrations of the total aerosol (TSP-total suspended particulates) and the PM-10 fraction (air suspended particulate with an aerodynamic particle diameter less than 10 μm) were correlated to observed health effects. These papers did not consider the detailed chemical composition of the TSP or PM-10 and are therefore significantly deficient. In fact, the air particulate mass of both TSP and PM-l0, are insuficiently defined and too obscure to serve as standards to measure pollution. It is only the ambient air quality standard that is not chemically specific. These important statements dealing with aggravated health effects observations and the non-specificity of ambient air particulate pollutants have been published by Friedlander and Lippmann 1994 (1). The consequences of these statements are expected to include new definitions, better measurement methodology, and new air quality standards for particulate air pollutants.

J. C. M. Marijnissen and L. Gradoń (eds.), Aerosol Inhalation: Recent Research Frontiers, 175–185.

Table 1 : Summarized results of the most important epidemiological studies (7).

Combined Effect Estimates[1)] of Daily Mean Exposure to Particulate Matter

Health endpoint	# of studies	Attributive Risk per 10 μg/m³ PM10 [95% CI]	
Premature mortality:			
- total deaths	8	1.0%	[0.7% ; 1.4%]
- respiratory deaths	4	3.3%	[1.5% ; 5.0%]
- cardiovascular deaths	4	1.4%	[0.7% ; 2.0%]
Increase in hospital use (all respiratory):			
- admissions	4	1.5%	[0.2% ; 2.8%]
- emergency department visits	3	1.8%	[0.0% ; 3.6%]
Exacerbation of asthma[2)]:			
- asthmatic attacks	3	4.4%	[-2.0% ; 11%]
- bronchodilator	3	7.0%	[0.6% ; 13%]
- emergency department visits	1	3.4%	[0.9% ; 6.9%]
- hospital admissions	3	1.9%	[0.8% ; 3.1%]
Increase in respiratory systems reports[2)]:			
- lower respiratory	8	3.8%	[0.3% ; 7.1%]
- upper respiratory	8	2.0%	[-0.1% ; 4.1%]
- cough	7	5.4%	[-1.1% ; 12%]
Decrease in lung function[2)]:			
- FEV_1	3	0.20%	[0.02% ; 0.38%]
- PEF	7	0.12%	[0.05% ; 0.19%]

1) Adapted from: Dockery, D.W. en C.A. Pope III. Acute respiratory effects of particulate air pollution. Annu Rev Public Health (1994) 15; 107-132

2) among primary school children

Air Particulate Epidemiology and Toxicology

Over the past few years a number of epidemiological studies have concluded that ambient air particulate exposure is associated with increased mortality and morbidity among the general population. The health effects were more aggravated among children, the elderly, and other vulnerable populations (2-8). These studies demonstrate positive correlations between particulate air pollution concentrations (mainly PM-10) and various health effects. These effects were observed at particulate mass concentrations as small as 10 $\mu g/m^3$. Examples of such results are shown in Fig.1 and Table 1.

The mechanisms responsible for these adverse effects are nevertheless practically unknown. There is toxicological evidence (animal inhalation experiments) for adverse health effects from polluted air (9); and animal inhalation experiments with atmospheric particulates are in preparation (10). Numerous controlled toxicological investigations of individual chemical species have clearly shown that specific constituents of ambient air particulate matter are associated with adverse biological effects, including carcinogenicity (11-14). The atmospheric particulates are heterogeneous mixtures that vary in constituent particle sizes and chemical composition, depending on geographical location, meteorology and source emissions. Complex chemical mixtures present difficult problems for toxicological studies and risk assessment. It is probably not possible to predict the nature of any interaction solely on the basis of stated exposure conditions (6,15).

Air Particulate Chemistry

The atmospheric anthropogenic aerosol (AAA) is an aerodispersed system of solid and liquid particles with different sizes, particle forms, and chemical composition. Only a few parts of these particles, mainly the primary ones, are single inorganic or organic compounds. The majority of them represent chemical mixtures. Also the chemical composition of single particles can be anisotropic: this means that the chemical composition is different on the surface from the inside of particles. Furthermore, the very fine particles form agglomerates (Fig.2) and can adsorb or absorb volatile substances or gases because of relatively high specific surfaces. The AAA are produced in the atmosphere and undergo several physical, physico-chemical and chemical processes during their residence time. Their resulting product is a polydisperse system of chemically heterogeneous particles with a complex toxic and carcinogenic potential. Biochemically active components may have been present at the particle surface or inside the particles. The AAA can contain short-lived chemical species, including free radicals and other metastable components (16-22). The AAA are formed by two basic mechanisms: dispersion and condensation (including chemical reactions). The total AAA is a mixture of the primary and secondary aerosol and will have a bimodal mass size distribution. Because both components are of a different origin, they also have different chemical compositions. The idealized bimodal mass size distribution is shown in Fig. 3 (23). The particle size dependent chemical composition of the AAA has been described in several papers recently published (24-26). The results of these publications have confirmed that the fine fraction of the AAA (particle sizes less than about 2 μm) is associated with the most toxic, inorganic and organic air pollutants. This AAA fraction could also be designated as the 'Lung Toxic Mode Fraction' and could serve as a possible marker for the dosage of toxic and carcinogenic components of particulate air pollutants.

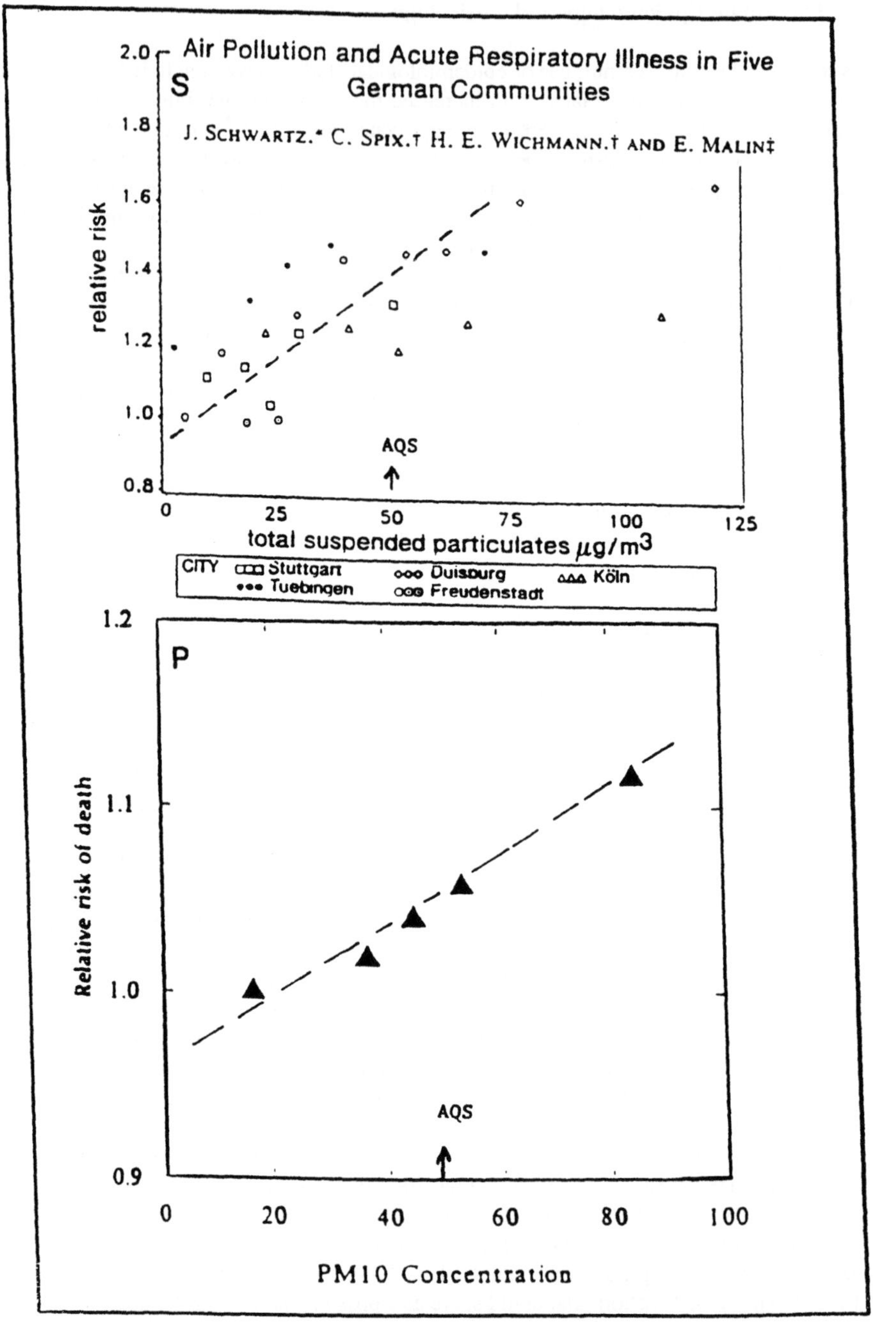

Fig. 1 : Relative risk of croup syndrome at different levels of air particulate concentrations in Germany (above)(12). Relative risk of death correlated to the PM 10 particulate concentrations in the Utah Valley (low)(14).

AAA- Definition, Sampling, Markers, and Monitoring

The mechanisms by which inhaled particulates induce adverse health effects depend initially on their depth of penetration, deposition and retention in the lung. The response to deposited material has been shown to be markedly greater when the tracheobronchial and alveolar regions of the respiratory tract, as opposed to the extrathoracic regions, were involved (2,6). Since it has been demonstrated that particles with an aerodynamic diameter of 10 μm or less enter the thoracic region (the bronchus and bronchioles) the term "Thoracic Particles" was proposed to describe such particulates. This fraction was measured as a mass concentration in $\mu g/m^3$ and was designated as "PM-10 ".

As already mentioned, this definition neglects the particulate chemistry (Table 2) and toxicity. For this reason, a standardized sampling of an additional mass fraction will be necessary: e.g. "PM-2.5".

The definition and measurement of the "Fine Particulate Fraction" (FPF) seems to be important even without the knowledge of the chemical composition. Measurements in Germany (34) have shown that the mass concentrations of the fine AAA fraction are continuously increasing, while the TSP and PM-10 mass concentrations have been rapidly decreasing during the last two decades.

The sampling procedures of the PM-10 mass fraction have already been well developed, and standardized equipment is commercially available (27,28). The necessary know-how and equipment for the FPF sampling is also available (27), but needs to be standardized. The cut-off point of 2.5 μm (aerodynamic diameter) seems to be scientifically justified as well (27). There exist considerable variations in the size distributions of ambient particles; however, there is generally a clear separation into fine and coarse modes, with a dividing point between 1.0 and 2.5 μm where the mass of the two modes is at a minimum. Condensation aerosols normally do not grow above 2 μm, and significant concentrations of dispersion aerosols are not normally found below 2 μm.

A very complete and accurate chemical analysis of the AAA samples is now possible and useful for several basic studies in the atmospheric environment and in toxicological research. Nevertheless, practical applications of such results in health risk assessment is difficult. Furthermore, the chemical analyses are usually conducted long after the aerosol has been sampled by filters or other sampling devices. Short-lived chemical species in the gas and/or aerosol phase that may be biochemically more active than the aerosol components measured later are not considered (1,16-18). For routine measurements and the evaluation of the FPF, the analysis or identification of some special markers could be a better and more useful approach for solving this problem. In addition to the total mass of the FPF, a continuous, in situ and on-line real time monitoring of carbon black and PAH-aerosols has been recommended recently in Germany (29,30). Commercially available equipment is able to monitor the ambient air concentration of the carbon black particulates (aethalometer) and of the PAH-concentration (photoelectric aerosol sensor) at the same time. Another possibility could be the determination of free radical (hydroxyl radical) activity associated with the PM-10 and PM-2.5 particulate fractions. This can estimate the marker ability of the AAA to produce free radical damages to DNA (18,21). Methods for measuring the free radical activity of particulate samples were already well developed (31).

Last but not least, the "in vitro" toxicity (mutagenicity) tests represent another tool for evaluating the general and integrated biological activity of the PM-10 and PM-2.5 particulate samples (32).

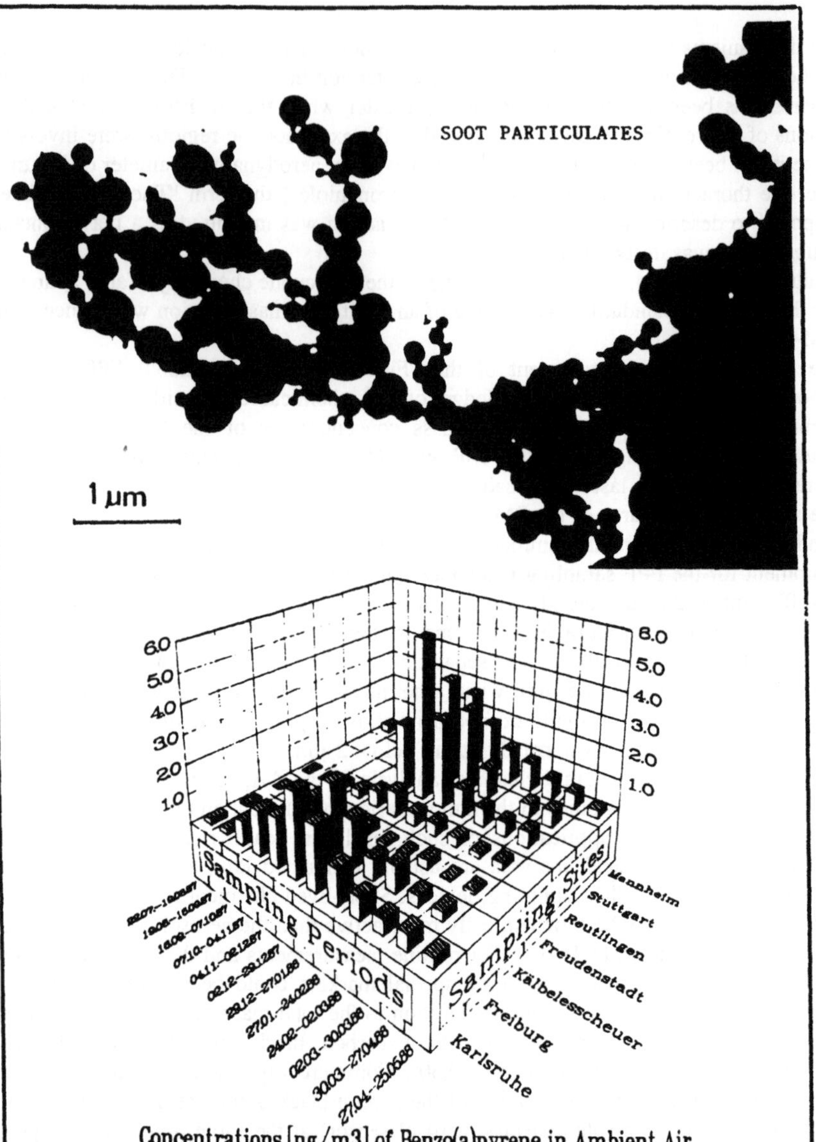

Fig. 2 : Electron micrograph (above) of an carbon black (soot) particle agglomerate as a vehicle for PAH; and the distribution of BaP concentrations on seven sampling sites in western Germany, measured between July and April (3).

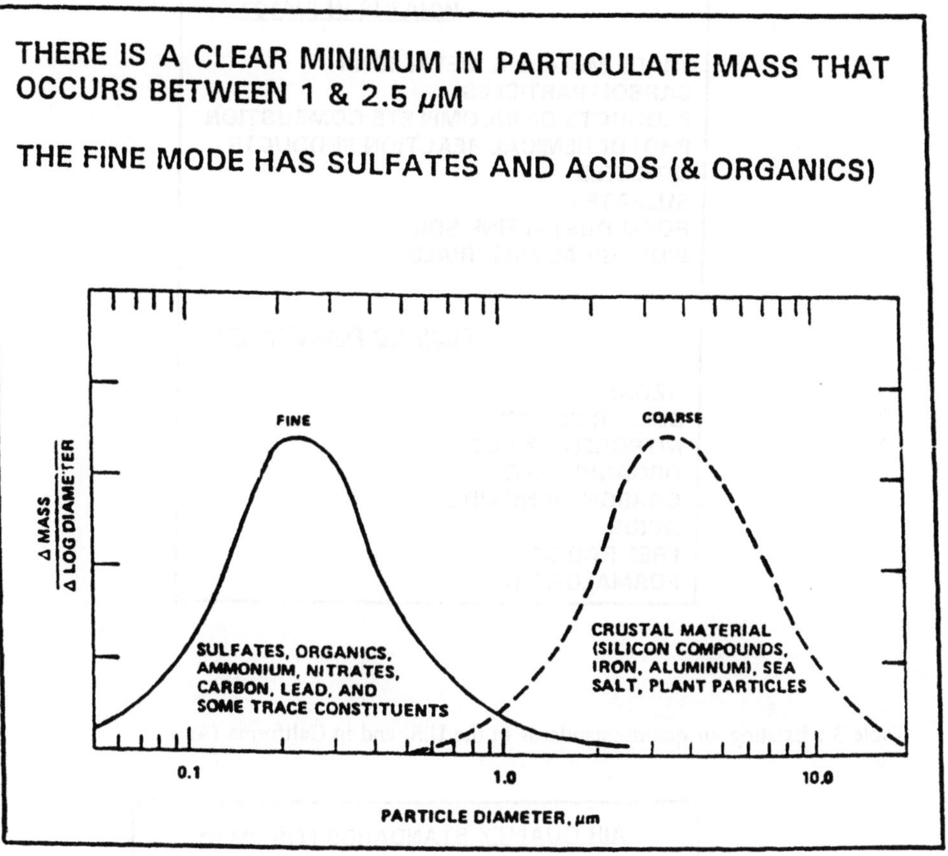

Fig. 3 : Idealized particle mass size distribution for urban aerosols (23).

Table 2 : The most important components of the ambient air particulate pollutants (4).

WHAT IS IN PM10?
TRACE METALS & METAL IONS
CARBON PARTICLES
PRODUCTS OF INCOMPLETE COMBUSTION
PHOTOCHEMICAL REACTION PRODUCTS
NITRATES
SULFATES
ROAD DUST & FINE SOIL
BIOLOGICAL MATERIALS
PLUS CO-POLLUTANTS
OZONE
SULFUR DIOXIDE
NITROGEN OXIDES
ORGANIC VAPORS
CARBON MONOXIDE
ACIDS
FREE RADICALS
FORMALDEHYDE

Table 3 : Existing air quality standards in the U.S. and in California (4).

AIR QUALITY STANDARDS FOR PM10		
NATIONAL AMBIENT AIR QUALITY STANDARDS (NAAQS)		
24 HR	PM10	150 $\mu G/M^3$
ANNUAL	PM10	50 $\mu G/M^3$
CALIFORNIA AMBIENT AIR QUALITY STANDARDS		
24 HR	PM10	50 $\mu G/M^3$
ANNUAL	PM10	30 $\mu G/M^3$

Particulate Air Quality Standards

The existing air quality standards (AQS) were settled for the PM-10 particulate fraction. An example of these standards is shown in Table 3. Similar AQS also exist in other highly developed countries. A value of 75 $\mu g/m^3$ (annual mean) is used in Germany (33).
These standards will need to be revised and an additional standard for the FPF (e.g. PM-2.5) settled. Based on the existing results of recent epidemiological studies, the AQS for PM-10 probably lies in the range of 10 $\mu g/m^3$ (annual mean), and the AQS for PM-2.5 should probably be less than 5 $\mu g/m^3$ (annual mean). The additional AQS for e.g. carbon black aerosols could be in the range of 3 $\mu g/m^3$ (annual mean); and for the PAH-aerosol an annual value less than 50 ng/m^3 (and or less than 5 ng/m^3 for BaP) could be adequate. Nevertheless, such proposals and their legislation will need still more scientific support, motivation and economic consideration.

Conclusions

The results of several recent epidemiological studies report increases in human mortality and morbidity associated with levels of particulate air pollution significantly lower than previously thought to affect human health. Results of these studies suggest the necessity to revise existing AQS for PM-lO, and to set new AQS for FPF (e.g. PM-2.5) and for some markers of the AAA, such as PAH, carbon black, free radical formation activity, and genotoxic/mutagenic activities. At the same time, the AAA should be newly defined and sampling and measurement strategies for the PM-2.5 should be developed and standardized.

REFERENCES

1. Friedlander, S.K. and Lippmann, M. (1994). Revising the particulate ambient air quality standards. Environ. Sci. Techol. 28, 148 A150 A.

2. Withey, J.R. (1989) A critical review of the health effects of atmospheric particulates. Toxicol and Industrial Health 5, 519-554.

3. Spurny, K.R. (1993) Atmospheric anthropogenic aerosol and its toxic and carcinogenic components. Wissenschaft und Umwelt, ISU (Aachen) 2, 139-151.

4. Phalen, R.F. (1994) PM-10 health effects: Scientific issues and uncertainties. 4th Int. Aerosol Conference, Los Angeles, pp.964-965.

5. Perkins, M.S., Phalen, R.F., Kleinman, M.T., Bhalla, D.K., Mautz, W.J. and Baker, D.B. (1994) A bibliography on particulate air pollution and human health. Colloquium on Particulate Air Pollution and Human Mortality and Morbidity, University of California, Irvine, (USA) pp.1-46.

6. Schlesinger, R.B. (1995) Toxicological evidence for health effects from inhaled particulate pollution.Inhalation Toxicology 7, 99-109.

7. Lebret, E., Houthuijs, D. and Dusseldorp, A. (1994) Accute and chronic studies in air pollution epidemiology. Int. Symposium "Environmental Health Hazards in Central and Eastern Europe", WHO, Sosnowiec, Poland.

8. Reichhard, T. (1995) Weighing the health risks of airborne particulates. Environ. Sci. Technol. 29, 360A-364A.

9. Saldiva, P.H.N., King, M., Delmonte, V.L.C., Macchione, M., Parada, M.A.C., Daliberto, M.L., Sakae, R.S., Criado, P.M.P., Silveira, P.L.P., Zin, W.A. and Bohm, G.M. (1992) Respiratory alterations due to urban air pollution. Environ. Res. 57, 19-33.

10. Sioutas, C., Koutrakis, P. and Burton, R.M. (1995) A technique to expose animals to concentrated fine ambient aerosols. Environmental Health Perspectives 103, 172-177.

11. Schwartz, J. (1991) Particulate air pollution and daily mortality in Detroit. Environ. Res. 56, 204-213.

12. Schwartz, J., Spix, C., Wichmann, H.E. and Malin, E. (1991) Air pollution and acute respiratory illness in five German communities. Environ. Res. 56, 1-14.

13. Schwartz, J. (1993) Particulate air pollution and chronic respiratory disease. Environ. Res. 62, 7-13.

14. Pope, C.A., Schwartz, J. and Ransom, M.R. (1992) Daily mortality and PM-10 pollution in Utah. Arch. Environ. Health 47, 211-217.

15. Lewtas, J. (1993) Complex mixtures of air pollutants. Environ. Health Perspectives 100, 211-218.

16. Kao, A.S. and Friedlander, S.K. (1994) Chemical signatures of the Los Angeles aerosol. Aerosol. Sci. Technol. 21, 283-293.

17. Kao, A.S. and Friedlander, S.K. (1995) Frequency distribution of PM-10 chemical components and their sources. Environ. Sci. Technol. 29, 19-28.

18. Kao, A.S. and Friedlander, S.K. (1995) Temporal variations of particulate air pollution: A matter for free radical dosage and adverse health effects? Inhalation Toxicology 7, 149-156.

19. Atkinson, R. and Arey, J. (1994) Atmospheric chemistry of gasphase polycyclic aromatic hydrocarbons: Formation of atmospheric mutagens. Environmental Health Perspectives 102, Suppl.4, 117-126.

20. Bayona, J.M., Casellas, M., Fernandez, P., Solanas, A.M. and Albiges, J. (1994) Sources and seasonal variability of mutagenic agents in the Barcelona city aerosol. Chemosphere 29, 441-450.

21. Donaldson, K. and Gilmour, P.S. (1995) Free radical activity associated with the surface of PM-10 material, ultrafine TiO_2, asbestos and man-made fibres. 5th European Meeting of Environmental Hygiene, Prague.

22. Fan, Z., Chen, D., Birla, P. and Kamens, R.M. (1995) Modelling of nitro-polycyclic aromatic hydrocarbons formation and their decay in the atmosphere. Atmospheric Environment 29, 1171-1181.

23. Hidy, G.M. (1975) Summary of the California aerosol characterization experiment. J. Air Poll. Control Assoc. 25, 1106-1114.

24. Sweet, C.W. and Vermette, S.J. (1993) Sources of toxic trace elements in urbain air in Illinois. Environ. Sci. Toxicol. 27, 2502-2510.

25. Kaplan, I.R. and Gordon, R.J. (1994) Non-fosil-fuel fine-particle organic carbon aerosol in southern California. Aerosol Sci. Technol. 21, 343-359.

26. Chow, J.C., Watson, J.G., Fujita, E.M., Lu, Z. and Lawson, D.R. (1994) Temporal and spatial variations of PM-2.5 and PM-10 in the southern California air quality study. Atmospheric Environment 28, 2061-1080.

27. Willeke, K. and Baron, P.A. (Eds.) (1993) Aerosol Measurement. Van Nostrand Reinhold, New York, pp.1-876.

28. Lundgreen, D.A. and Burton, R.M. (1994) The effect of particle size distribution on the cut point between fine and coarse ambient mass fraction. 4th Int. Aerosol Conference, Los Angeles. p.630.

29. Burtscher, H. (1994) Charakterizierung von Aerosolen mit dem photoelektrischen Aerosolsensor(PAS). Methoden der Aerosolmeßtechnik. GIV-Seminar, Frankfurt/M, pp.1-5.

30. Hansen, A.D.A. (1994) The development of the aethalometer. Methoden der Aerosolmeßtechnik. GIV-Seminar. Frankfurt/M, pp.1-19.

31. Brown, R.C., Hoskins, J.A. and Johnson, N.F. (1991) Mechanisms in fibre carcinogenesis. Ser. A. Life Sci. NATO ASI Ser. 223, Plenum Press, London, pp.1-589.

32. Hadnagy, W. and Seemayer, N.H. (1987) Comparative investigation of the genotoxicity of city smog and automobile exhaust. J. Aerosol Sci. 18, 679-699.

33. VDI-Reinhalt. Luft (1992) VDI-Richtlinie 2310. Maximale Immissions-Konzentrationen fur Schwebstaub. Beuth-Verlag, Berlin.

34. Winkler, P. and Kaminski, U. (1988) Increesing submicron particle mass concentration in Hamburg. I,II. Atmospheric Environment 22, 2871-1883.

Flow-induced Mixing at Low Reynolds Number in an Expanding and Contracting Alveolus Model

Yoshio Otani [1], *James P. Butler* [2], *Hitoshi Emi* [1] *and Akira Tsuda* [2]

[1]Department of Chemistry and Chemical Engineering, Kanazawa University, 2-40-20 Kodatsuno, Kanazawa 920, Japan, and [2]Physiology Program, Harvard School of Public Health, 665 Huntington Ave., Boston, MA 02115, USA

Keywords

Mechanical Mixing; Low Reynolds Number Flow; Expanding Acinus; Flow Reversibility; Model Acinus.

Introduction

A recent prospective cohort study (Dockery, 1993) has shown that particulate pollutants below a few micrometers pose the greatest health hazard. Aerosols of this size range penetrate deep into the distal region of the lung and are likely to deposit in the lung periphery. The mechanisms involved in mixing and deposition of these particles in the alveolar region, however, are still not fully understood. In prior studies it has been assumed that convective (flow-induced) mixing between inhaled particles and acinar residual gas is insignificant since the nature of acinar flow is fully viscous and reversible (Davis, 1972, Ultman, 1978). Reversibility implies that the expired gas retraces the motion of inspired gas. However, experimental data contradict this widely-accepted assertion (e.g. Heyder, et al., 1988, Anderson, et al., 1989). In studies on single breath bolus dispersion in human lung, for example, Heyder, et al. (1988) showed appreciable bolus dispersion in the lung periphery which could not be accounted for solely by the particles' intrinsic motion (Brownian motion and gravitational sedimentation). This suggested the existence of convective mixing at the level of acinus. Although several mechanisms (Cinkotai, 1974, Yu, et al., 1977, Taulbee, et al., 1978, Otani et al., 1990, 1991, Rosental, 1992, Miki, et al., 1993) have been postulated to resolve this question, none of these mechanisms has been tested fully.

Under normal circumstances, airflow in the alveolus is governed by the balance of pressure and viscous forces and determined by wall boundary conditions. Therefore, reversibility of alveolar wall motion may determine whether the flow is kinematically reversible in that region. Wall motion reversibility, however, can be defined in several different ways (Fig.1).

A) The wall can expand and contract through a series of geometrically similar shapes (Cinkotai, 1974). (Geometrically similar reversible wall motion)

B) The wall shape can change during an expansion-contraction cycle, but during deflation every boundary point must retrace its path it took during inflation in both space and time (Watson, 1974). (Reversible wall motion in space and time)

C) The wall shape can change during an expansion-contraction cycle, but during deflation every boundary point must retrace only its spatial path it took during inflation. The boundary points do not need to retrace their path in time. (Reversible wall motion in space but not in time)

D) The wall shape must come back to the original shape at the end of cycle, but the motion of boundary points do not have any restriction in space or time. (wall motion with reversible configuration)

An important and currently unsolved question is which one of these definitions of wall motion reversibility describes the conditions which are sufficient to allow kinematically reversible flow. The objective of this present study is to address this question in a simple experimental system, which allows us to examine systematically various combinations of wall motions.

J. C. M. Marijnissen and L. Gradoń (eds.), Aerosol Inhalation: Recent Research Frontiers, 187–193.

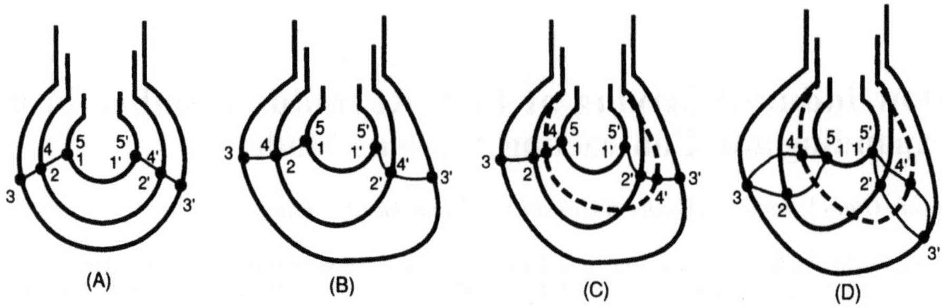

Fig.1 Definitions of wall motion reversibility. 1:t=0, 2:t=τ/4, 3:t=τ/2, 4:t=3τ/4, 5:t=τ.

Methods

Apparatus: A large-scale single alveolus model (TEE junction) was constructed, which consisted of a section of circular pipe (diameter: 20mm) with pistons at each end, connected to a vertical column (diameter: 20mm) in the middle (Fig.2). The TEE junction system was filled with glycerin (density 1250 kg/m^3, viscosity 1.4 Pa s, and incompressible) and flow was generated in the system by pistons' motion. Fluid motion associated with movement of the two pistons and the free boundary in the vertical column simulated gas flow in the expanding and contracting alveolus and the alveolar duct. Each piston was independently driven by a combination of automatic stage (LTS-100X, Sigma Koki, Ltd.) and stepping motor assembly (Model PH556-A, Oriental Motor Co., Ltd.), which was operated by a personal computer (NEC PC9801VX). The position and motion of each piston were precisely controlled at the accuracy of 1μm. The dimensionless parameters defining fluid motion in the system were set in the range of those occurring in the human alveolus. In particular, the Reynolds number based on duct diameter was kept much less than unity (Re=0.0089) and the Womersely parameter (α, ratio of unsteady inertial effects to viscous force) was less than 1.

Procedures: Various combinations of piston motion (MODE-1~5 defined below) were studied. In each case, the two pistons were initially located 60 mm away from the center of the vertical column and colored glycerin (red poster color dissolved into glycerin) was passively introduced

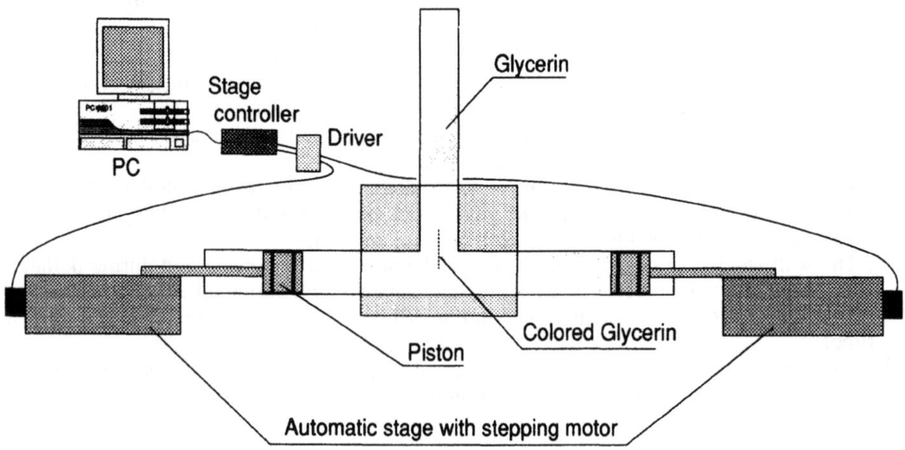

Fig.2 A large-scale single alveolus model.

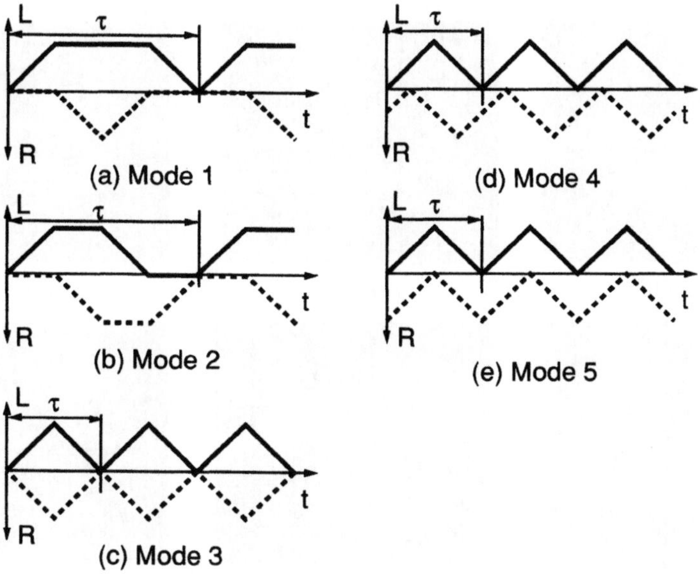

Fig.3 Patterns of pistons' movements.

manually with a syringe at the middle of the vertical column. The pistons' motions were set as follows (see Fig.3). In MODE-1 and 2, only one piston moved at one time. In MODE-1, during inspiration, the left wall expanded first, followed by right wall expansion. During expiration, the right wall contracted first, then the left wall contracted. In MODE-2, during inspiration, the order of walls' expansion were exactly the same as MODE-1, but during expiration, the order of walls' contraction was reversed, i.e., the left wall contracted first, followed by the right wall. In MODE-3, 4, and 5, both pistons were in motion at the same time. Each piston performed a simple back-and-forth periodic motion with a constant speed of 0.05mm/s but there was a phase difference (θ) between the pistons' motions. In MODE-3, no phase difference (θ=0) was set between the two pistons' motion so that both left and right walls expanded and contracted synchronously. In MODE-4, the two pistons' motions were $\pi/2$ out of phase ($\theta=\pi/2$), i.e., the left wall started to expand a quarter of cycle earlier than the right wall. In MODE-5, two pistons' motions were totally out of phase ($\theta=\pi$), i.e., the left wall started to expand when the right wall started to contract and vice versa.

The motion of tracer was monitored by a video camera (Toshiba IK-636) and recorded on S-VHS video tape for later analysis.

Results

The reversibility (or irreversibility) of low-Reynolds number flows largely depended on how the wall motion exhibited its reversibility.

Reversible wall motion in space and time: MODE-1 motion pattern of our model represented spatially and temporally reversible wall motion with only one piston moving at one time (Fig.4; oscillation period; τ=40s, each piston displacement; L=5mm). Changes in tracer pattern during the first cycle are shown in Fig.4a-e. The tracer was initially introduced in the vertical channel (Fig.4a). At time $t=\tau/4$, when the left wall fully expanded, the tracer was slightly displaced from the source duct and stretched toward the left channel (Fig.4b). At $t=\tau/2$, when the right wall was fully expanded, the tracer was largely stretched and deformed into the right channel (Fig.4c).

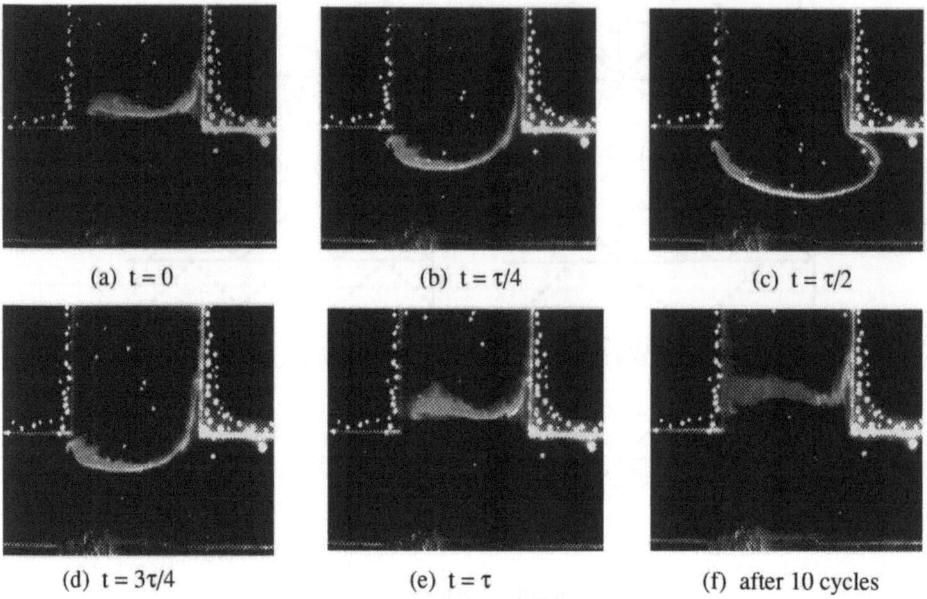

Fig.4 Evolution of tracer pattern in Mode 1. Oscillation period τ=40s, dispacement of piston L=5mm, Re=0.0089, α=0.24) .

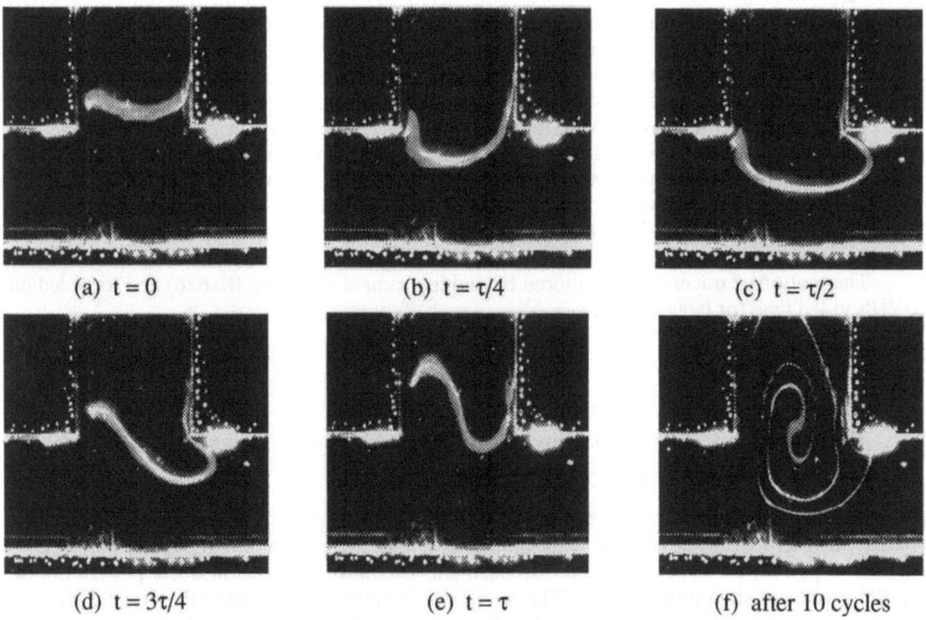

Fig.5 Evolution of tracer pattern in Mode 2. Oscillation period τ=40s, dispacement of piston L=5mm, Re=0.0089, α=0.24).

During expiration at t=3τ/4, when the right wall first contracted to the original position, the tracer (Fig.4d) returned to the same position as in Fig.4b. At t=τ, when the left wall contracted after a cycle, the tracer (Fig.4e) returned to exactly the same initial position (Fig.4a). MODE-1 was repeated for up to 10 cycles, and the tracer pattern showed no difference in shape (Fig.4f). In MODE-3 (θ=0, no phase difference between the two pistons' motion) and MODE-5 (θ=π, the two pistons' motions totally out of phase), our model represented spatially and temporally reversible wall motion with both pistons in motion at the same time. In all these modes, the tracer motions were perfectly reversible (data not shown).

Reversible wall motion in space but not in time: MODE-2 motion pattern of our model represents spatially but not temporally reversible wall motion; i.e., the walls' motion was exactly the same as in MODE-1 during inspiration, but the walls did not move in the reverse order during expiration (Fig.5). Changes in tracer pattern during the first cycle are shown in Fig.5a-e. Similar to the case in MODE-1, the tracer was initially introduced in the parent duct (Fig.5a). During inspiration (t=0~τ/2), the tracer motion (Fig.5a-c) was the same as those in MODE-1 (Fig.4a-c). However, during expiration, the tracer no longer reversed its pattern during inspiration. At t=3τ/4 when the left wall contracted to the original position, the left portion of the tracer was pushed into the source channel (Fig.5d). By the end of the first cycle (t=τ), the tracer pattern was very different from its original shape (Fig.5a). It was stretched and exhibited an S-shape wavy pattern. As the cycle number increased, the tracer was progressively stretched and curved. At the end of the 10th (Fig.5f), the tracer was highly stretched, thinned, and formed into a very well-ordered swirling shape. This tracer pattern is particularly interesting for transport of Brownian particles. Suppose the tracer is made of Brownian particles. This swirling pattern has two effects on diffusion. It progressively increases the surface area of the tracer, and simultaneously, it increases the concentration gradients by reducing the distance between the swirling tracer striation. Therefore, we expect that this convectively mixed pattern exponentially increases the diffusion process. In MODE-4 (θ=π/2, the two pistons' motions out of phase), our model also represented spatially but not temporally reversible wall motion with both pistons in motion at the same time. In this mode, the tracer motion was again not reversible (data not shown).

Discussion

When the kinematic reversibility of low-Reynolds number flows is studied, the reversibility of wall motion must be closely examined since those flows are presumably in a (quasi-) Stokes flow regime and chiefly governed by the wall motion. In past analyses, however, the definition of wall motion reversibility was often ambiguous and thus caused great confusion. In the present study, we specify and use four different definitions of wall motion reversibility (see Introduction) and examine the specific effects of two of these on flow reversibility.

The most strict definition of wall motion reversibility is "geometrically similar reversible wall motion (Definition-A)", i.e., the shape of the wall must be maintained in a self-similar shape throughout the cycle. Cinkotai (1974), performing simple physical model experiments using a large-scale latex balloon alveolar model, argued that this strict condition is necessary for the interior flow to be kinematically reversible. In our present experiments, however, we found many examples contradicting Cinkotai's conclusion. For instance, in MODE-1, 3, and 5, the flow was perfectly reversible even though the system did not undergo a self-similar expanding and contracting motion. We conclude that "geometrically similar reversible wall motion" is not a necessary condition for the flow to be kinematically reversible.

A slightly relaxed definition of wall motion reversibility is the "reversible wall motion in space and time (Definition-B)". This was embodied in our in-vitro mechanical model system as MODE-1, 3, and 5. In these modes, the tracer pattern at the end of every cycle (even up to 10 cycles) returned to exactly the same shape as its initial pattern, indicating that the flow in the system was kinematically reversible. This finding may be best explained as follows. Since during deflation every wall point retraced its path in step with one another, the sequence of the wall shapes appears as if a movie was running backward. Since the flow motion faithfully follows the wall motion in a Stokes flow regime, the flow must reverse its motion exactly in space and time.

Logically, therefore, the "reversible wall motion in space and time" results in a perfectly reversible flow (Watson, 1974).

From these data, the conclusion is that spatially and temporally reversible wall motion is sufficient to make the flow reversible. But is this condition necessary for flow reversibility? This is an important question because in the lungs the expansion and contraction of alveolar walls is known to exhibit small but systematic temporal hysteresis (Miki, et al, 1993). We examined the effects of "spatially but not temporally reversible wall motion (Definition-C)" on flow reversibility. This wall motion was generated in MODE-2. In this mode, the order of the two walls' contractions did not occur in the reverse order of their expansions, but every wall point still exactly retraced its spatial path it took during inspiration (since every point on the piston only moved on a simple back- &-forth one-dimensional path). In this case (Fig.5), the tracer pattern was no longer reversible. The tracer was stretched and swirled in every cycle and its pattern progressively became more convoluted as the cycle number increased (Fig.5f). Based on these experiments, therefore, we conclude that wall motion that is only spatially reversible (Definition-C) is not sufficient to make the flow kinematically reversible, whereas wall motion that is both spatially and temporally reversible (Definition-B) is sufficient to make the flow kinematically reversible.

This conclusion was further tested by performing the following experiments. Since the flow is expected to be reversible if the wall motion is reversible in both space and time, we tested whether the flow motion in MODE-2 (solely spatially reversible boundaries) could be made reversible if we used an additional cycle with wall motions representing the temporal "mirror image" of the first cycle. For example, the sequence of walls motion would be *left then right wall expansion during inflation and left then right wall contraction during deflation, followed by the temporal "mirror image" right then left wall expansion followed by right then left wall contraction.* We tested this case and found that the tracer motion indeed became perfectly reversible, further supporting our conclusion that the interior flow was reversible. From these results it follows that our fourth definition (Definition-D) of wall motion reversibility, which only requires wall motion with reversible configuration is certainly not sufficient for flow reversibility.

Conclusions

Through the experiments with a large-scale alveolar model at low Reynolds number, the following conclusions were obtained on the reversibility of alveolar airflow;

1) Geometrically similar reversible wall motion is not a necessary condition for the flow to be kinematically reversible.
2) The flow is kinematically reversible if the wall motion is reversible in both space and time.
3) Spatially but not temporally reversible wall motion leads to convective mixing (irreversible flow). The convective mixing may have significant effects on axial dispersion in the whole acinar level, gravitational sedimentation, and Brownian diffusion of inhaled aerosol particles.
4) Wall motion with reversible configuration is not sufficient for flow reversibility.

Further investigation is necessary in this area in order to understand the basic behavior of aerosol mixing and deposition in the lung periphery.

Acknowledgement

This study was supported by National Heart, Lung, and Blood Institute Grants HL47428 and HL54885

References

Anderson, P.J., Blanchard, J.D., Brain, J.D., Feldman, H.A., Mcnamara, J.J., and Heyder, J., Effect of cystic fibrosis on inhaled aerosol boluses, *Am. Rev. Respir. Dis.* **140**(1989)1317- 1324

Cinkotai, F.F., Fluid flow in a model alveolar sac, *J. Appl. Physiol.*, **37**(2)(1974)249-251

Davies, C.N., Breathing of half-micron aerosols: II Interpretation of experimental results, *J. Appl.*

Physiol., **35**(1972)605-611

Dockery, D.W., Pope III, C.A., Xu, X, Spengler, J.D., Ware, J.H., Fay, M.E., Ferris, B.G., and Speizer, F.E., An Association between air pollution and mortality in six U.S. cities, *New England J. Med.*, **329**(1993)1753-1759

Heyder, J., Blanchard, J.D., Feldman, H.A. and Brain, J.D., Convective mixing in human respiratory tract: estimates with aerosol boli, *J. Appl. Physiol.*, **64**(3)(1988)1273-1278

Miki, H., Butler, J.P., Rogers, R.A., and Lehr, J., Geometric hysteresis in pulmonary surface to volume ratio during tidal breathing, *J. Appl. Physiol.*, **75**(4)(1993)1630-1636

Otani, Y., Emi, H. and Tanaka, T., Mixing and deposition of Brownian particles in model alveolus, *J. Chem. Eng. Japan*, **23**(1990)326-331

Otani, Y., Emi, H., Tanaka, T. and Kamide, K., Application of mixing and deposition data of Brownian particles in a model alveolus to human alveoli, *J. Chem. Eng. Japan*, **24**(1991)154-159

Rosental, F.S., Blanchard, J.D., and Anderson, P.J., Aerosol bolus dispersion and convective mixing in human and dog lungs and physical models., *J. Appl. Physiol.*, **73**(3)(1992)862-873

Taulbee, D.B., Yu, C.P., and Heyder, J., Aerosol transport in the human lung from analysis of single breaths, *J. Appl. Physiol.*, **44**(5)(1978)803-812

Ultman, J.S., Gas transport in the conducting airways, In: L.A. Engel and M.Paiva (ed.), Gas mixing and distribution in the lung, ch.3, Lung Biology in Health and Disease, vol. 25, Marcel Dekker, Inc, N.Y. (1985)

Watson, E.J., Fluid flow in a model alveolar sac (Appendix), *J. Appl. Physiol.*, **37**(2)(1974)249-251

Yu, C.P., Rajaram, S. and Taulbee, D.B., Air mixing in a model lung alveolus, An ASME Publication, 77-Bio-4(1977)1-8

The Influence of Normal and Pathological Inhomogeneities of the Lung Tissues on Particle Transport and Deposition

DARIUSZ S. ORLICKI * and **LEON GRADON**

Department of Chemical Engineering, State University of New York at Buffalo, Amherst, NY 14260

Department of Chemical and Process Engineering, Warsaw University of Technology, 00-645 Warsaw, Poland

August, 1995

Abstract

A single alveolus is used to simulate a breathing cycle in human lungs. An alveolus elastic wall lined with liquid is deformed to accommodate the intrapleural pressure changes. The difference in pressure between alveolus and surroundings is established and the air flow is followed. The main focus, however, is on normal and pathological inhomogeneities which are implemented into the model as decreased tissue elasticity, surfactant deficiency, and constriction of airways. Under such conditions, particle transport and deposition patterns distinguish from those observed in normal healthy lungs.

Keywords: aerosol deposition, surfactant, lung elasticity, pulsating gas flow.

1 Introduction

The air quality presents a serious problem even in places located far from metropolitan and industrial centers. Tests taken of air near certain industrial complexes reveal that more than a tonne of pollutants are descended daily on each square kilometer. Exposure to air pollution may be few orders of magnitude grater at places of employment; miners, factory workers, and construction workers are most at risk. Various diseases of the respiratory system are triggered by exposure to air pollutants. People vary widely in their susceptibility to air pollution. Certain weather conditions, various health disorders, normal and pathological inhomogeneities of the lungs can enhance potential medical problems associated with inhaling of pollutants. Many

*To whom correspondence should be addressed

J. C. M. Marijnissen and L. Gradoń (eds.), Aerosol Inhalation: Recent Research Frontiers, 195–204.

studies of particle deposition in human respiratory tracts have been conducted in an attempt to better understand underlying phenomena and to assess the health risk of inhaling aerosol particles.

Although a substantial efford has been put to develop mathematical models capable to simulate gas flow and aerosol deposition in human respiratory tracts there is still a discrepancy between theoretical predictions and experimental results. The problem is attributed to the inherent geometrical complexity of respiratory tracts and the multitude of physical phenomena involved. The intricate geometry of lungs is commonly simplified by the system of tubes, which are straight or bent. The most recent models include symmetric [1] or asymmetric bifurcations [2]. A simulation of breathing takes the form of discretizing a predetermined two- or three-dimensional domain into finite elements and solving the conservation equations for mass and momentum in this domain, subject to specific boundary conditions. The presence of particles in a gas phase introduces an additional complication to theoretical modeling. If no interaction between gas carrier and aerosol phase is assumed the description of particles dynamics may be decoupled. The gas phase solution is determined first. Thereafter, the motion of particles in this gas flow field can be found.

The purpose of this work is to extend the previously developed analysis, [1], to simulate the effects of normal and pathological inhomogeneities of the lungs tissue on aerosol transport and deposition inside respiratory tracts. Those inhomogeneities represent in the model locally decreased tissue elasticity, surfactant deficiency, and increased resistance to gas flow. Nonuniform deformation of alveolus wall under periodically changing pressure in thorax or obstruction to the air flow by different parts of respiratory tracts significantly change the deposition pattern in the lungs.

2 Theoretical Analysis

The present study is an attempt to simulate the ventilation and lung mechanics. The air flow field, aerosol dynamics and deposition are deduced from the deformation of lung elastic structures during the breathing cycle.

The lung, which cardinal function is to exchange gas, comprises the series of tubes. Within the lung, there are more than 20 generations of branchings, each resulting in narrower, shorter, and more numerous tubes. Finally, the smallest branches terminate in air sacs called alveoli. The geometrical model was formulated [3] to mathematically define the lung structure as a space domain for numerical simulations. Basic information about number of tubes in a branch, the characteristic diameter and length of tubes in every generation was adopted from Weibel [4].

2.1 Mechanics of Ventilation

The lungs lie within the chest cavity in which the ribs form the sides and the diaphragm forms the floor. When the thoracic dimensions are changed due to the action of different muscles, the lungs volume is modified virtually to the same degree because of the change in intrapleural pressure.

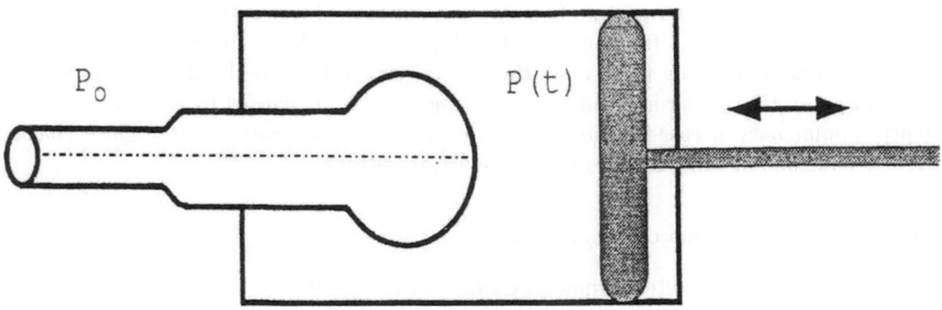

Figure 1: **Model of lungs ventilation**

A model to simulate the ventilation process consists of a tube (a respiratory bronchiole) closed at one end by an elastic spherical balloon (an alveolus) and the other end connects with surroundings through an extra resistance to flow (it represents the resistance to air flow through the nose, the pharynx, the larynx, the trachea, and all generations of airways preceding respiratory bronchiole). This structure is placed inside a cylinder in such a way that the tubular part protrudes through the cylinder wall and it opens to the environment Fig. 1. The cylinder (a thorax) is closed by moving piston (a diaphragm) which regulates the inside pressure (intrapleural pressure). When the piston is pulled out by an external force the pressure inside the cylinder decreases, the elastic walls of the balloon stretch, the pressure inside the balloon drops below atmospheric pressure and the air is drawn through the extra resistance. This sequence of events corresponds to the inspiration part of the breathing cycle. When the external force is relaxed the piston moves inside until the pressure inside and outside equalizes. The elastic walls of the balloon recoil and the air flows outside. This sequence of events represents the expiration part of the breathing cycle.

2.2 Resistance to air flow

Air moves into and out of the lungs in response to changes in lung volume. The respiratory system imposes resistance to the gas flow. Resistance is directly proportional to the volumetric rate flow, the viscosity of the air and the length of the airway, and inversely proportional to the fourth power of the airway radius.

$$\Delta P \propto \frac{L\mu}{D^4} Q \tag{1}$$

The total resistance to the movement of air into and out of the lungs is the sum of the upper and lower airways resistances. A functional relationship $\Delta P = f(Q)$ can be constructed, [5]. The resistance to flow is normally very small, the average atmosphere to alveoli pressure difference is less than 1 mm Hg. Under abnormal circumstances which are caused by asthma and chronic obstructive pulmonary disease functional characteristic of airways resistance is seriously changed. Some of the factors affecting airways resistance are as follows:

- mucosal congestion, edema or inflammation,
- plugging, partial or complete, by mucus, exudate or foreign bodies,
- cohesion of mucosal surfaces by surface tension forces,
- infiltration, compression or fibrosis,
- compression or closure caused by increased intrapleural and decreased transpulmonary pressure,
- collapse or kinking of airways walls due to decrease in elasticity

2.3 Wall Mechanics

The airways can be classified into three sequential groups. The trachea and bronchi are cartilaginous conducting airways, the bronchioles are membranous airways, and the respiratory bronchioles and alveolar ducts are transitional gas exchange structures that distribute air into the alveoli. The alveoli walls contain a dense network of capillaries and reticular, collagenous, and elastic fibers. The non-linear response of an airway wall tissue can be represented by a two-phase continuum model based on the theory of mixtures [6]. The tissue (Fig. 2) consists of an incompressible, elastic porous solid skeleton (β - phase) saturated with viscous fluid (α – phase).

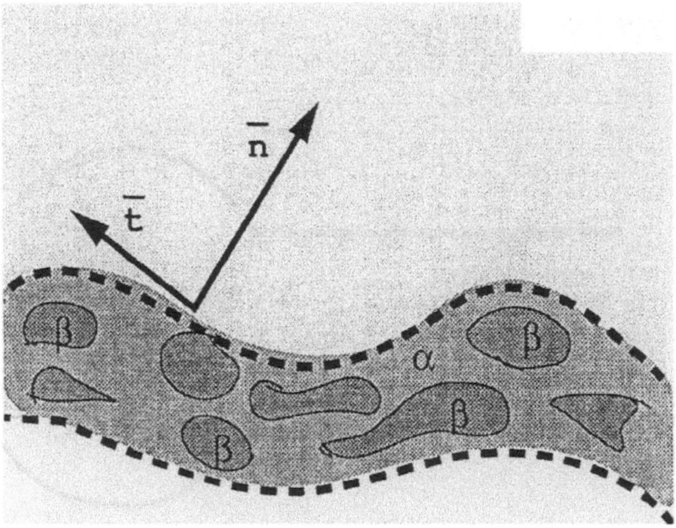

Figure 2: **Two-phase continuum model of lung wall tissue**

3 The Model

A model that describes the dynamic behavior of an alveolus during the breathing cycle consists of the conservation equations for mass, and momentum as well as the kinematic equations for wall deformation. The air flow and aerosol transport and deposition are decoupled from the description of wall deformation with significant saving in computing time.

3.1 Wall Deformation Model

Fig. 3 shows geometry of alveolus model used in this study. A local normal force balance between viscous and elastic stress, surface tension, and gas pressure is used to evaluate walls deformation. Conservation equations for mass and momentum are derived. Continuity equation,

$$\vec{\nabla} \cdot (\phi_\alpha \vec{v}_\alpha + \phi_\beta \vec{v}_\beta) = 0 \tag{2}$$

Momentum equations, neglecting body forces,

$$\rho_\alpha \frac{D\vec{v}_\alpha}{D\,t} + \vec{\nabla} \cdot \bar{\bar{\sigma}}_\alpha - \vec{\pi}_\alpha = 0 \tag{3}$$

$$\vec{\nabla} \cdot \bar{\bar{\sigma}}_\beta + \vec{\pi}_\beta = 0 \tag{4}$$

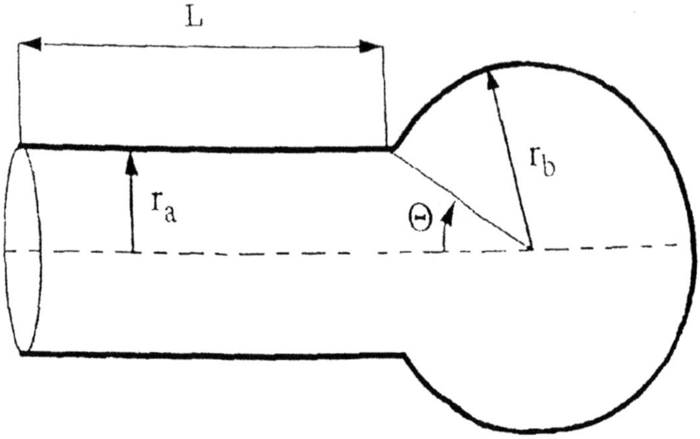

Figure 3: **Geometrical configuration of alveolus model**

Constitutive equations,

$$\bar{\bar{\sigma}}_\alpha = -\phi_\alpha P I \quad (5)$$
$$\bar{\bar{\sigma}}_\beta = -\phi_\beta P I + \lambda_\beta \mathbf{e}_\beta I + 2\mu_\beta \varepsilon_\beta \quad (6)$$
$$\vec{\pi}_\beta = -\vec{\pi}_\alpha = \kappa(\vec{v}_\alpha - \vec{v}_\beta) \quad (7)$$

Here, ∇ is the gradient operator, ϕ is a volume fraction and $\sum_{i=\alpha,\beta}\phi_i = 1$, ρ is a density, σ is a stress tensor, λ and μ stand for Lamé constant and shear modulus of elasticity respectively, and κ is the drag coefficient. Appropriate boundary conditions are required,

$$\vec{n}_I \cdot \bar{\bar{\sigma}}|_{r=R_I} = \vec{n}_I P_I + \vec{n}\aleph_I \quad for\ inner\ surface \quad (8)$$
$$\vec{n}_o \cdot \bar{\bar{\sigma}}|_{r=R_O} = \vec{n}_I P_o(t) \quad for\ outer\ surface \quad (9)$$

Kinematic equations are used to find alveolus shape,

$$\vec{n} \cdot \frac{d\,\vec{r}}{d\,t} = \vec{n} \cdot \vec{v}_\alpha \quad (10)$$

3.2 Air Flow Field

The deformation of alveolus walls creates pressure gradients inside the alveolus and the air flow follows. To determine the fluid flow field the continuity and momentum balance equations are

solved.
Continuity equation,

$$\vec{\nabla} \cdot \vec{v}_{air} = 0 \tag{11}$$

Momentum balance equation,

$$\rho_{air} \frac{D\,\vec{v}_{air}}{D\,t} = -\vec{\nabla} \cdot \bar{\bar{\sigma}}_{air} \tag{12}$$

The boundary conditions used in the calculations are as follows:

$$\vec{v}_{air} = \vec{v}_{wall} \tag{13}$$

$$\vec{n} \cdot \bar{\bar{\sigma}}_{air} = \vec{n}(P_{atm} - \Delta P(\|\vec{v}\|)) \quad alveolus \ \ entrance \tag{14}$$

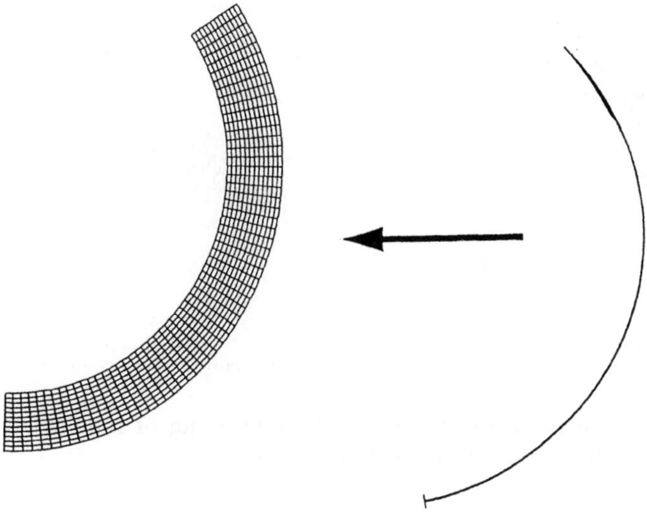

Figure 4: **Transformation of physical domain**

3.3 Transport of Particles

The volume fraction of particles in the inhaled air is usually very low so the interaction between the gas carrier and aerosol particles can be neglected and the transport of aerosol phase is solved independently. A dynamic behavior of aerosol particles is deduced from the balance of forces acting on a single particle:

$$m \frac{\mathbf{d}\vec{\mathbf{v}}}{dt} = \sum_i \vec{F}_i \quad i = D,\ G \tag{15}$$

where individual forces are expressed as following

$$\textit{viscous drag force} \quad \vec{F}_D = 3\pi\phi_D\mu d_p(\vec{\mathbf{v}}_{\mathbf{air}} - \vec{\mathbf{v}}_{\mathbf{p}}) \tag{16}$$

$$\textit{gravitational force} \quad \vec{F}_G = \frac{\pi d_p^3}{6}\vec{g}(\rho_p - \rho_{air}) \tag{17}$$

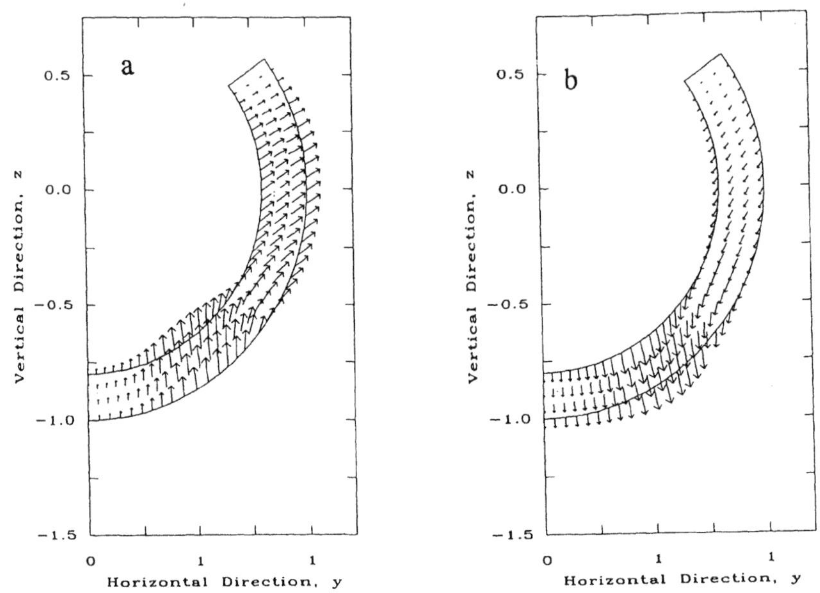

Figure 5: **Flow field of an interstitial fluid at the beginning of inhalation; (a) local surfactant deficiency, (b) locally decreased elasticity**

4 Numerical Solution Techniques

The problem is solved by employing the finite element method where the functional is defined by the Galerkin method. The domain is discretized into isoparametric rectangular elements and Lagrangian polynomials are used as basis functions. For time integration, the explicit Adams-Bashfort predictor method with implicit trapezoid corrector is used. As a result of these proceedings the non-linear set of equations is obtained, which is solved by the modified Newton-Raphson iterative method.

In order to reduce the computational effort the physical domain, which undergoes changes of shape, area and volume, is mapped onto the unchangeable computational domain, Fig. 4 . Such a procedure makes it possible to model the deformation of bodies with complex geometry.

The computer simulation of a particle motion under these conditions is performed using a leapfrog method (Birdsall and Langdon 1985).

4.1 Results and Conclusions

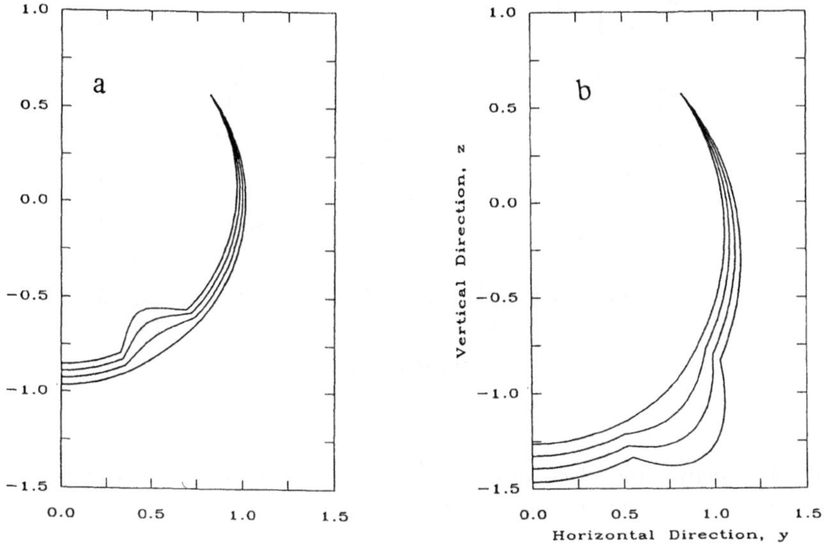

Figure 6: **Instantaneous shapes of alveolus walls; (a) local surfactant deficiency and inhalation process, (b) locally decreased elasticity and exhalation process**

A gas flow field is created inside the alveolus as a result of changes in volume and generation of pressure gradients during elastic deformation of the wall structure. The more nonuniform deformation of walls the more complex flow field appears. This flow is also characterized by very low values of Reynolds numbers for all realistic breathing patterns. Numerous experiments indicate that only very small aerosol particles can penetrate to alveoli. Such particles have low inertia and follow closely carrier gas streamlines. Two major deposition mechanisms are considered; gravitational settling, and due to diffusion. Our preliminary results suggest that there are topological reasons for nonuniform deposition patterns. Combination of concave and convex features on a wall can create a shadowing effect. In addition, it changes the surface area available for deposition.

Figure 5 shows the flow field of an interstitial fluid during wall expansion; a) corresponds to local deficiency of surfactant, b) shows local decrease in wall elasticity. Figure 6 illustrates the instantaneous shapes of alveolus walls. The local deposition rate is shown in Figure 7.

Numerical methodology has been developed to simulate the ventilation process. This unique

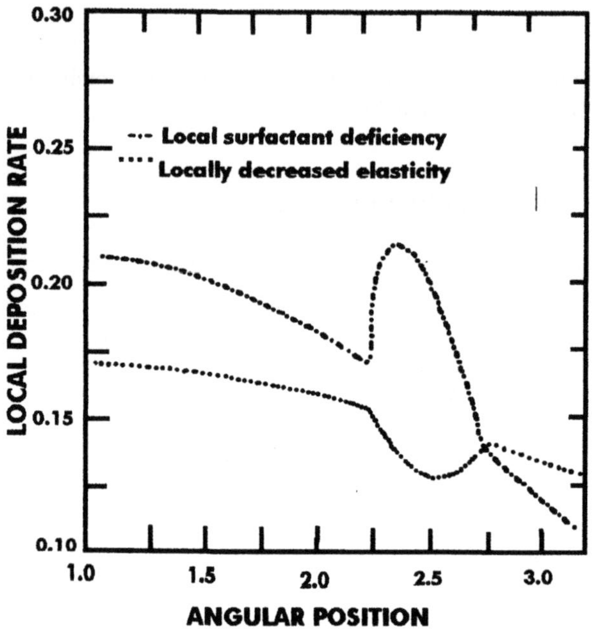

Figure 7: **Local deposition rate**

model allows to analyze the effects of pathological changes in human lungs on transport processes. Currently, we are working on a model of airways bifurcation having elastic walls.

References

[1] Gradon L.,Orlicki D., *Deposition of inhaled Aerosol Particles in a Generation of the Tracheobronchial Tree,* J. Aerosol Sci. 22, 15-30, 1990,

[2] Balashazy I., Hofmann W., *Deposition of Aerosols in Asymmetric Airway Bifurcations*, J. Aerosol Sci. 26, 273-292, 1995,

[3] Orlicki D., *Mathematical Modeling of Aerosol Deposition in Human Respiratory Tracts*, Ph.D. Thesis, Warsaw University of Technology, Warsaw, Poland, 1988,

[4] Weibel E. R., *Morphometry of the Human Lung*, Academic Press, New York, 1963,

[5] Olson D. E., Dart G. A., and Filley G. F., *Pressure drop and fluid flow regime of air inspired into the human lungs*, J. Appl. Physiol. 28: 482-484, 1970,

[6] Gallagher R. H., Simon B. R., Johnson P. C., and Gross J. F., *Finite Elements in Biomechanics*, Wiley, New York, 1982.

Methods For Predicting Total Size Dependent Particle Collection Efficiencies With More Than One Mechanism Acting

James W. Gentry , N. C. Liu

Department of Chemical Engineering
University of Maryland, College Park , Md.

Key Words Lung Deposition, Collection Mechanisms, Diffusion , Sedimentation

Abstract

A protocol for the estimating the efficiencies where more than one collection mechanism is present are examined. Theoretical expressions are obtained for collection with more than one mechanism present simultaneously. A protocol is introduced in which the collection mechanisms are classified into four different categories. This protocol is used to examine two cases: the interaction of locally and globally attraction as in sedimentation and electrostatic attraction and in the case where there is global attraction and a stochastic removal process as is the case for diffusion and sedimentation.

Introduction :

In a typical application for particle deposition in the lung or in a granulated bed, the collection efficiency is experimentally measured for particles of different sizes. In the data analysis , collection mechanisms were assumed to be due to two collection mechanisms- gravitational settling or diffusion. The steps in the analysis were the following:

(1) For sufficiently small particles , the collection mechanism was considered to be only diffusion. An empirical expression was obtained relating the experimental efficiency η_d to the diffusion parameter χ_d.

(2) For sufficiently large particles, only sedimentation was considered as a collection mechanism. An empirical expression was obtained relating the efficiency η_s to the sedimentation parameter χ_s.

(3) The overall efficiency η_E was measured in the intermediate region where both sedimentation and diffusion are important. Using this efficiency values of a critical angle one obtains a critical angle θ_E defined below. In addition two calculated efficiencies η_{30} and η_3 are calculated. These efficiencies , in many cases, represent asymptotic limiting curves to the experimental efficiencies. These expressions do not depend on the theoretical expressions for either diffusion or sedimentation. It was found that the experimental efficiencies for sedimentation in a packed bed or in the lung where both diffusion and sedimentation were important lay between the curves η_{30} and η_3.

MODEL DEVELOPMENT

A vector model was found to be efficient in establishing bounding solutions for experimental measurements of particle collection in granulated bead filters and in models of the lung. The assumption was that the overall efficiency can be represented as the vector product of the two

J. C. M. Marijnissen and L. Gradoń (eds.), Aerosol Inhalation: Recent Research Frontiers, 205–213.

component efficiencies η_1 and η_2 . In the original development these efficiencies were the constituent efficiencies for settling and η_s for diffusion η_d. A pictorial representation of the models is given in Figure 1.

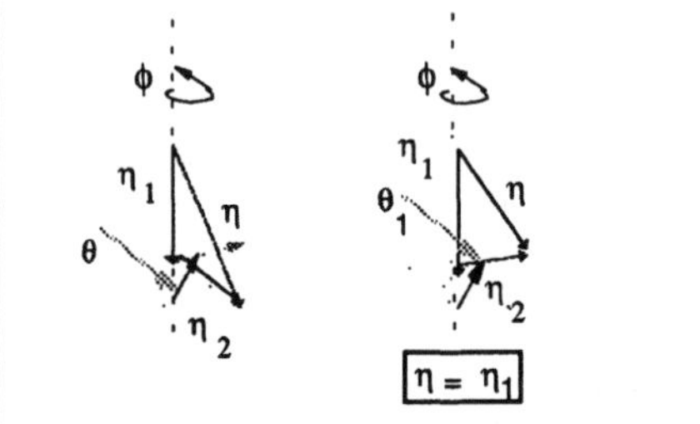

FIGURE 1: SCHEMATIC OF VECTOR MODEL

The two constituent efficiencies are η_1 and η_2. The azimuthal angle between the smaller of the two efficiencies and the larger (taken as η_1 in the schematic) is θ. The combined efficiency or resultant as a function of the angle θ is given by

$\eta^2 = \eta_2^2 + \eta_1^2 + 2\,\eta_2\,\eta_1 \cos\theta$ (1) .

A mean efficiency can then be obtained by averaging the efficiency over all polar angles φ from 0 to 2π and over all azimuthal angles θ (the angle between the major axes of the two efficiencies) from 0 to θ_A . The value of θ_A indicates which of the models is being used. In the first model pictured, the value of θ_A is π. This model assumes that the all solid angles are considered and that the relative alignment is random. In the second model the maximum value of the angle is chosen so that the overall efficiency is always greater than the larger of the two constituents. This angle θ_1 defined by :

$$\theta_1 = \cos^{-1}(-\eta_2 / 2\,\eta_1). \qquad (2)$$

For the second model it is the maximum value of the azimuthal angle . This restricts the averaging process to conditions where only deposition efficiencies which satisfy $\eta \geq \eta_1$ (the larger of the two efficiencies) are allowed. This assumption may be important for otherwise , one can have cases where the efficiency for removal by two mechanisms is lower than when only one mechanism is included.

A third correlation variable θ_E is chosen to satisfy the equation

$$\eta = \frac{\int_0^{2\pi} d\phi \int_0^{\theta_E} \eta(\theta)\, sin\,(\theta)\, d\theta}{\int_0^{2\pi} d\phi \int_0^{\theta_E} \sin(\theta)\, d\theta} \qquad (3)$$

where $\eta(\theta)$ is given by equation 1. For most of the comparisons with experimental measurements the value of θ_E was between θ_1 and π. .

We believe that the explanation as to why the vector relationship is applicable to efficiencies , is that the deposition within the lung (or packed bed) can be related to a distance. For example the sedimentation efficiency for uniform flow between two parallel plates is given by the terminal settling velocity divided by the separation distance times the residence time. Thus this efficiency represents the distance settled divided by the distance between the planes. Moreover if the two plates are inclined at angle ζ then the efficiency is multiplied by cos (ζ). The argument for the approach outlined in this paper is that there is a relationship between distance (or velocity) and efficiency. Since the distance is a vector then the efficiency can be treated as a vector as well. Secondly in the case of the lung or packed bed , the two dominant mechanisms were diffusion which is stochastic without a preferential direction and sedimentation which has a preferred direction parallel to the gravitational vector.

For the first of the models one obtains the result that :

$$\eta = \frac{1}{2}\int_0^{\pi} \sqrt{\eta_1^2+\eta_2^2+2\eta_2\eta_1\cos(\theta)}\ \sin(\theta)d\theta \tag{4}$$

$$\text{or} \quad \eta = \left[\frac{|\eta_1+\eta_2|^3 - |\eta_1-\eta_2|^3}{6\eta_1\eta_2}\right] \tag{5}$$

The total efficiency is constrained to be always less than one. To account for this constraint the angle θ_0 is defined by the relation

$$\theta_0 = \cos^{-1}[\{1-(\eta_2^2+\eta_1^2)\}/(2\eta_2\eta_1)] \tag{6}$$

This insures that the total efficiency is always less than one. The resulting expression for the average efficiency then becomes

$$\eta = \frac{1}{2}\int_{\theta_0}^{\pi} \sqrt{\eta_1^2+\eta_2^2+2\eta_2\eta_1\cos(\theta)}\ \sin(\theta)d\theta + \int_0^{\theta_0}\sin(\theta)\,d\theta \tag{7}$$

$$\text{or} \quad \eta = \left[\frac{1-|\eta_1-\eta_2|^3}{6\eta_1\eta_2}\right] + \frac{\left[\eta_1^2+\eta_2^2\right]-1}{4\eta_1\eta_2} \tag{8}$$

In the second model the resultant efficiency is constrained so that the efficiency is always as great as the larger of the two constituent efficiencies. In this case the appropriate expressions are

$$\eta = \frac{\int_0^{\theta_1}\sqrt{\eta_1^2+\eta_2^2+2\eta_2\eta_1\cos(\theta)}\ \sin(\theta)d\theta}{\int_0^{\theta_1}\sin(\theta)d\theta} = \frac{\left[|\eta_1+\eta_2|^3-\eta_1^3\right]}{3\eta_2\eta_1\left[1+(\eta_2/2\eta_1)\right]} \tag{9}$$

When the constraint that $\eta_1 + \eta_2 < 1$ is violated then one has the equation

$$\eta = \frac{\left[1-\eta_1^3\right]}{3\eta_2\eta_1\left[1+(\eta_2/2\eta_1)\right]} + \frac{\left[|\eta_1+\eta_2|^2-1\right]}{2\eta_2\eta_1\left[1+(\eta_2/2\eta_1)\right]} \tag{10}$$

In summary for the first model the angle θ varies from 0 to π and for the second model the angle varies from 0 to θ_1 . It is important to note that these equations assume that there is no preferential direction for the attraction. One would expect that this would not be unreasonable for the case where one has sedimentation and diffusion, but might fail when the two mechanisms are both directional as is the case when one has sedimentation and electrostatic attraction.

Classification:

In considering the modeling of interacting forces, two categories were considered . First is the mode of attraction - is stochastic as in the case of diffusion or is it globally directional as is the case of sedimentation. It is obvious that there can be at least four different types of collection mechanisms. The second category which one can assign is to the geometry of the system, specifically flow through a cylindrical tube or between flat plates as well as uniform or parabolic flow. The approach which we have selected is to begin with the simplest geometry and flow and attempt to find analytical solutions for the simultaneous interaction of mechanisms.

Interaction between collection mechanisms can be characterized by four different modes - globally directional, locally directional , stochastic , and inertial.

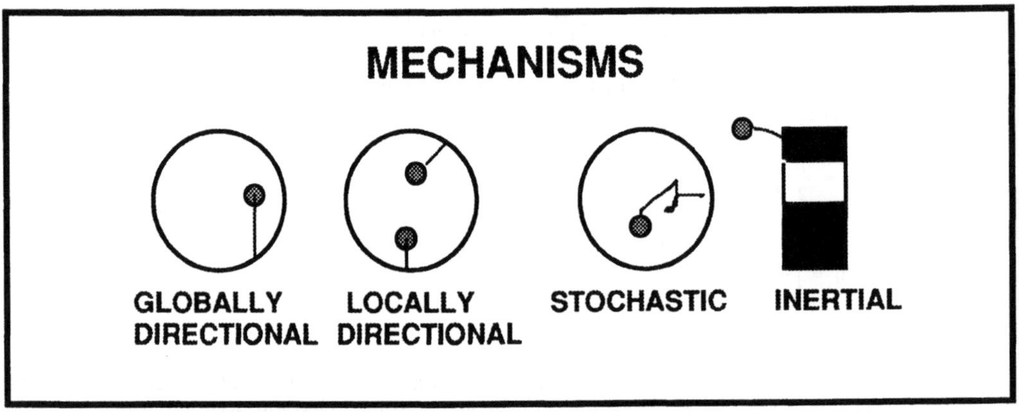

FIGURE 2 COLLECTION MECHANISMS

The best example of a globally directional mechanism is sedimentation in which the deposition force is directed parallel to the gravitational force. In the case of a locally directional collection mechanism the force on the particle acts in the direction where the distance between the particle and the wall is a minimum. Examples of such mechanisms are electrostatic and thermophoretic forces. Analytical expressions for globally and locally directional forces can often be obtained, even when the collection mechanisms are operating simultaneously.

The next step in the analysis was a geometrical hierarchy. This hierarchy were to progress from uniform flow to parabolic or laminar flow and from Cartesian to cylindrical geometries. Exact Fourier or Fourier-Bessel series solutions exist for the diffusion (i.e. stochastic case) when the flow is uniform. Similarly one has an asymptotic expansion for the case of laminar flow in both Cartesian and cylindrical geometry. Using the method of characteristics the analytical solution for sedimentation can be obtained for both geometries and velocity profiles. The solutions are especially simple when the geometry is Cartesian. However the case for two or more mechanisms operating independently is much more difficult, and thus far the solutions are limited to Cartesian geometry and the interaction of mechanisms which are locally or globally directed. In this paper we consider the case where the locally directed collection mechanism is described by a power law.

Classical Problems:

The classical problems are based on the studies of Heyder and Gebhart ((Gebhart et. al. (1984,1985), Heyder et.al. (1986) at Frankfurt . The collection efficiency for butyl sebacate particles (ρ =0.91) of different sizes are measured in a granulated bed and in the lungs. The collection efficiency data for lung deposition (Heyder (1985 et. al.) is shown in Figure 3. In obtaining these data it is assumed that for the smallest particle size, only the diffusion mechanism contributes and for large particles only sedimentation contributes. An empirical expression is then obtained which describes the measured efficiency in terms of the diffusion parameter and sedimentation parameter respectively.

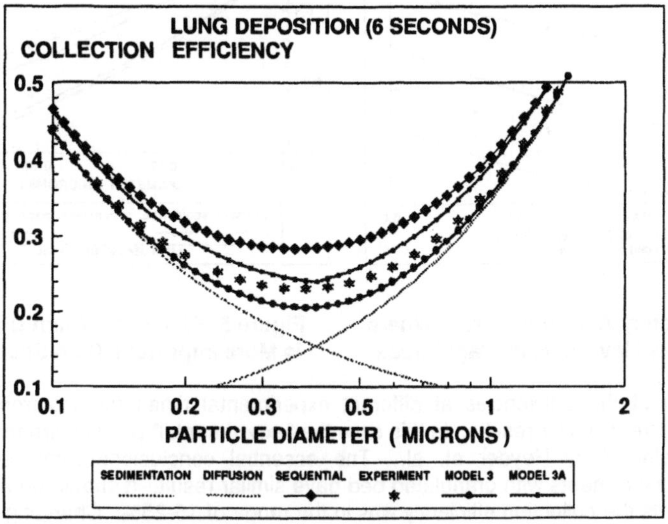

FIGURE 3 EXPERIMENTAL MEASUREMENTS OF LUNG DEPOSITION

The results show first that the experimental efficiency (stars) as a function of particle diameter is bounded by an envelope formed from the two limiting cases of η_{30} and η_3. The two monotonic curves are diffusion (descending) and sedimentation (ascending). The minimum in the efficiency is very near where the component efficiencies are the same. The efficiency calculated assuming that the two mechanisms operate sequentially is substantially above the envelope indicating that this model predicts too large a collection mechanism. This is consistent with the results observed by Gebhart. In previous papers (Gentry et. al. 1993,1994), comparisons were made with three sets of experiments with granulated beds as well as the lung data. In all cases the experimental data were bounded by η_{30} and η_3. In the earlier papers a second model based on only two dimensions and not accounting for rotation about the φ axis were developed. The envelope of these two values η_{20} and η_2 enclosed the bounds. Several widely used empirical expressions can be derived as limiting cases of the expression for η_2.

This simulation is shown in greater detail in Figures 4 and 5 where the efficiency is shown for the narrower size ranges from 0.3 to 0.6 and from 0.1 to 0.16 respectively. The light dashed line in Figure 4 , is the empirical formula of Rudolf (Rudolf et.al.(1988) . This model had been developed as a correlation for the studies with granulated beds. The most significant point in the figures is that the experimental values are near the middle when the efficiencies are of comparable magnitude, but is very close to the diffusion value and the lower value of η_3 when the particles are very small.

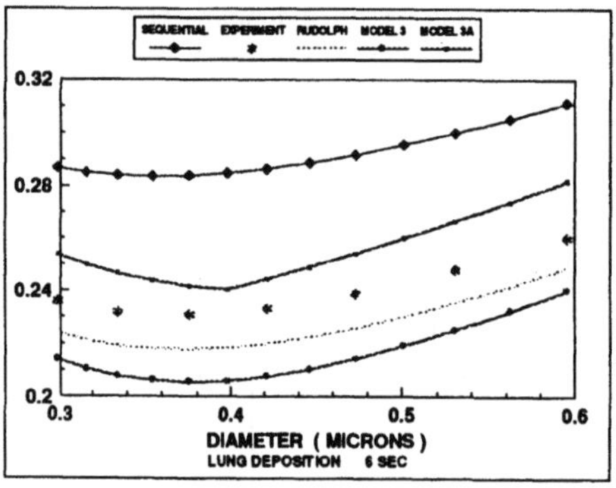

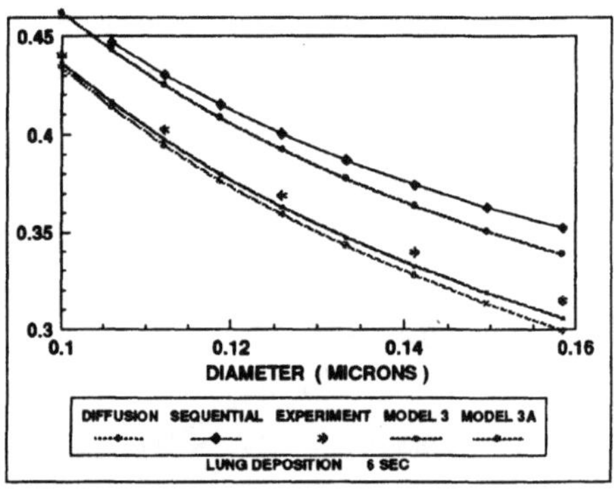

Figure 4 Efficiency At Particle Sizes Where The Mechanisms Have Similar Magnitudes

Figure 5 Efficiency When Diffusion Is More Important Than Sedimentation

An investigation of the efficiencies at different experimental times ranging from 0.01 to 100 seconds showed very similar results. In this case the "experimental" points were the values of the correlation developed by Heyder et. al. The essential conclusions remained unchanged. Comparable measurements with granulated bed gave similar result. It should be noted that in all of the experiments the collection efficiency was in the range of 15-25%. Changing the residence time or particle size and extrapolating the empirical expressions allowed the examination of bead size and the relationship between beads and particle diameter. For these extrapolated conditions there were cases where the "experimental" values crossed the bounding envelope. This raises the using models for which analytical or numerical solutions exist.

The simplest models which permit analytical solution for the combined mechanisms of globally and directional collection forces is the case of flow between two parallel plates assuming parabolic or uniform flow. It is probable that a good approximation can be developed for the case

the geometry is Cartesian for the case where there are stochastic and directional deposition. It is interesting to note that even in the simplest case of two locally attractive mechanisms, the description is complicated.

Uniform Flow , Cartesian Geometry :

We begin by defining the dimensionless variables parallel (Z) and perpendicular (Y) to the flow. The dimensionless direction Y = y/ h takes the values 0 and 1 at the walls. The dimensionless variable is defined as Z = z/L where h is the separation distance between the two plates and L is the length of the section. The trajectory equation in terms of the dimensionless distances Y and Z is given as :

$$\frac{dY}{dZ} = -\chi_S - \frac{\chi_E}{Y^n} \quad \text{where} \quad \chi_S = \frac{V_s L}{U_0 h} \quad \text{and} \quad \chi_E = \frac{\alpha L}{U_0 h} \tag{11}$$

There is an attractive force to the wall = α / y^n which is proportional to the **n**th power with distance. In the case of electrostatic attraction **n** would be equal to 2. This means that in some cases the net force would be either directly complimentary or directly opposed to each other. The efficiencies for globally directional , locally directional , and stochastic forces will be labeled as η_s, η_{el}, and η_d for sedimentation, electrostatic, and diffusion forces respectively. These are the best known of the different mechanisms.

Equation (11) was integrated and the values of the overall efficiency η , the bounding efficiencies η_3 and η_{30} , and the bounding angle θ_E computed . In Figure 6 , the angle is plotted as a function of the sedimentation efficiency for the case where $\eta_{el} = \eta_s$. The second parameter is the power. It was found that the efficiency is not necessarily bounded by the asymptotic values. The areas where the bound is exceeded in a complex function of power. Between power exponents of 0.925 and 1.994 the efficiency is always bounded. For smaller values of the power exponent there are certain values of η_s where the constraint that $\eta_E > \eta_3$ is violated. For exponents above 2 the constraint that $\eta_E < \eta_{30}$ does not apply. Note that the curves are flat and θ_E almost constant for $\eta_s < 0.4$. In Figure 7, the zones where the calculated overall efficiency lies outside the bounds is indicated by the shaded area where η_{el} and η_s are the parameters. The isolated point at $\eta_{el} = \eta_s = 0.8$ is the point of discontinuity in Figure 6. There are three such regions of discontinuity. The region nearest the plane where $\eta_S = 0.0$ and the isolated point correspond to regions where $\eta_E > \eta_{30}$. The intermediate region where η_s is comparatively small , and where η_{el} is large is the region where $\eta_E > \eta_{30}$. This figure clearly illustrates that the bound is not general, although there are wide regions where it is applicable. Also the effect of different efficiencies for local and global attraction. Plots such as those shown in Figure 7 have the potential for elucidating fundamental points regarding the interaction of mechanisms.

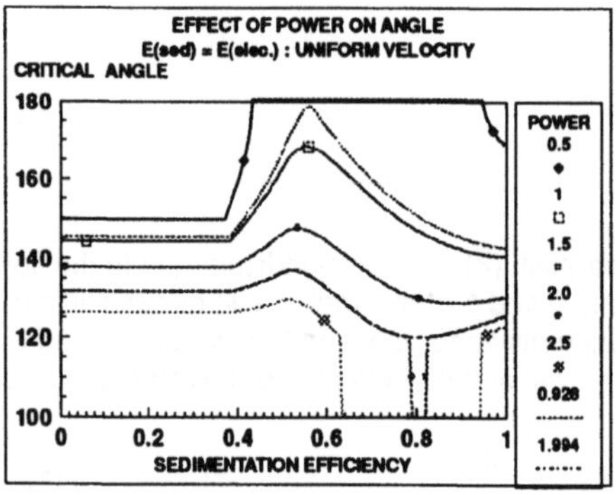

FIGURE 6 EFFECT OF POWER ON ANGLE WITH UNIFORM FLOW

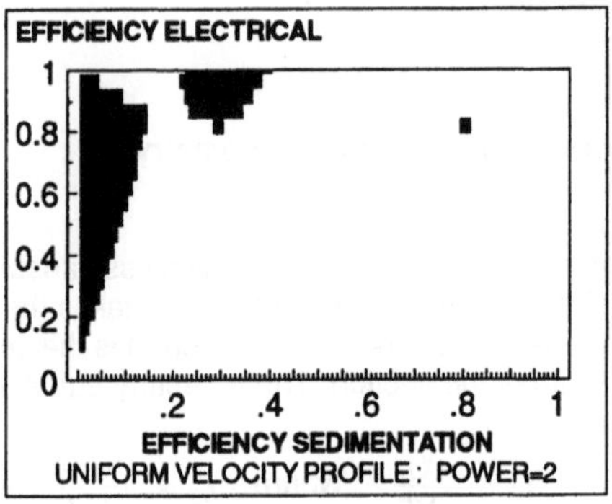

FIGURE MAP OF EFFICIENCIES FOR n = 2 and UNIFORM FLOW

Parabolic Flow , Cartesian Geometry :

For parabolic flow the velocity is given by 6 U_0 Y (1-Y). For parabolic flow as well as uniform flow the collection efficiency for sedimentation alone is equal to χ_S. The expression for the collection efficiency to electrical effects is given by

$$\frac{dY}{dZ} = \frac{\left[-\chi_S - \frac{\chi_E}{Y^n}\right]}{[6Y(1-Y)]} \qquad (12)$$

In Figures 8 and 9 the analogous results for parabolic flow to that of uniform flow are shown. The regions where the bounds are violated are larger than was the case for uniform flow. Note that the discontinuity point at $\eta_{el} = \eta_S$ does not exist. (In the graph the nomenclature E(Sed.) and E(Elec.) are used for $\eta_{el} = \eta_S$. Again the patch at the right of Figure 9 corresponds to the region where $\eta_E > \eta_{30}$ and the central patch at the top corresponds to the case where $\eta < \eta_3$. The nature of the graph for the limiting angle when $\eta_{el} = \eta_S$ as a function of power is significantly different for laminar flow than for uniform flow. There appears to be no case where $\eta_E > \eta_{30}$. This would be indicated by a discontinuity at the lower angle values as was the case for powers >2 for uniform flow. At values of n < 1.65 there are a number of conditions where $\eta < \eta_3$ indicated by the angle exceeding 180 degrees.

The increase in the size of the second patch can be explained qualitatively. For laminar flow the magnitude of the velocity near the wall is much lower than the bulk flow, so that to have the same collection efficiency the relative magnitude of the electrical force must be stronger. This means that it overwhelms the moderate sedimentation forces and the net affect is to have a greater collection efficiency. Although not shown here, it was found that the angle was highly asymmetric in that the angles for $\eta_{el} = \kappa \eta_S$ were not the same as for $\eta_{el} = \eta_S / \kappa$.

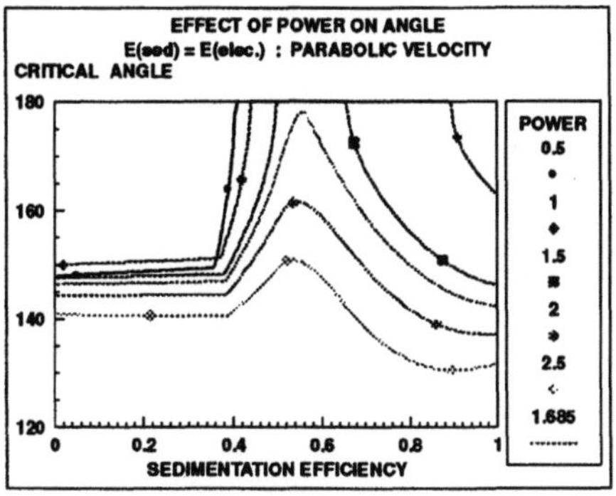

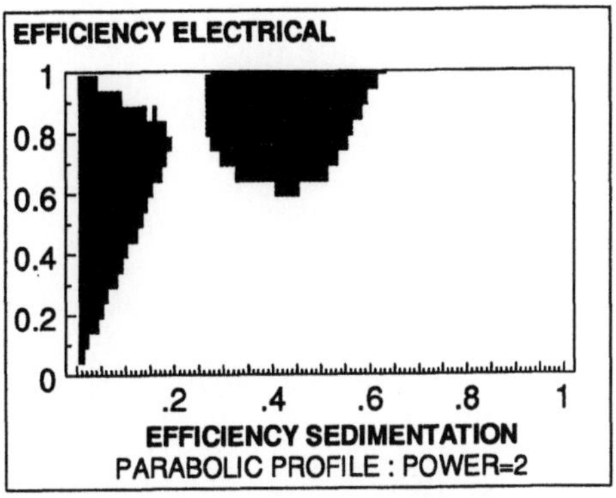

FIGURE 8 EFFECT OF POWER ON ANGLE WITH PARABOLIC FLOW

FIGURE 9 MAP OF EFFICIENCIES FOR n = 2 and PARABOLIC FLOW

SUMMARY:

The description of experimental collection efficiencies when there is simultaneous interaction of more than one mechanism has been successfully treated using a vector addition of collection efficiencies. The caveats are that agreement with experiment was obtained only when the total efficiency was comparatively small and the two collection mechanisms were sedimentation and diffusion - a stochastic and a globally directed collection mechanism. Mathematical expressions were obtained establishing a bounding envelope for the interaction of the two mechanisms. Four classification categories-stochastic, sequential, globally directional, and locally directional- were developed to describe different modes of particle removal.

REFERENCES

Gebhart , J. , J. Heyder , W. Stahlhofen , (1984) in **Aerosols** , ed. B. Liu , D. Pui , H. Fissan , Elesvier Science , 1023-1026

Gebhart , J. , J. Heyder , (1985) ,, **J. Aerosol Sci. , 16** , 175 - 187

Gentry , J.W. , J. Gebhart , G. Scheuch ,(1994) , **J. Aerosol Sci. , 25** , In Press

Gentry , J.W. , J. Gebhart , G. Scheuch ,(1993) , **J. Aerosol Sci. , 24 S** , p S507-S508

Heyder , J. , J. Gebhart , G. Rudolf , C. Schiller , W. Stahlhofen , (1986) , , **J. Aerosol Sci. 17** , 811- 25

Heyder , J. , J. Gebhart , G. Scheuch , (1985) , **J. Aerosol Sci. & Tech.** , **4** , 315- 326

Rudolf , G , J. Gebhart, J. Heyder , G. Scheuch , W. Stahlhofen , (1988) , **Ann. Occup. Hyg. Inhaled Particles VI , 32** , 919-938

The Retention of Inhaled Particles in the Human Respiratory System. Cytotoxic Effects.

Leon Gradoń and Albert Podgórski

Department of Chemical Engineering, Warsaw University of Technology, Warsaw, Poland

Keywords: local deposition, pulmonary clearance, mucociliary clearance, alveolar macrophages, surfactant, membrane

INTRODUCTION

Solid and liquid aerosols in the atmosphere originate from both natural and industrial sources. Inhalation of these aerosols from the ambient air results in the introduction and deposition of various particulate substances in the human respiratory tract, creating a potential health hazard. The assessment of this hazard and appropriate countermeasures to be taken require understanding and estimation of the local and regional deposition of inhaled particles in the respiratory tract.

The human organism is supplied with a system of natural defense mechanisms which accelerate the removal and clearance of foreign bodies deposited in the lungs. Knowledge of the residence time of inhaled particles in the lungs can provide valuable insight into the etiology of many diseases such as pneumoconiosis, emphysema and carcinogenesis. Perhaps the most valuable tool that describes the variation of deposited mass in the lungs with time is the retention curve, whose characteristics are governed by the kinetics of the deposition and clearance processes. Recent years have witnessed not only increased interest in the experimental investigation of the retention curve, but also the development of theoretical models which attempt to duplicate and generalize experimental results. In addition, retention curves can be used for a variety of medical purposes such as optimizing the administration of drugs through inhalation.

Assessment of the toxicity of airborne pollutants can be accomplished through epidemiological studies of morbidity and mortality patterns in humans. Pollutant toxicity is also determined by studies in which animal subjects are exposed to a toxic aerosol in order to predict possible toxic effects on humans.

J. C. M. Marijnissen and L. Gradoń (eds.), Aerosol Inhalation: Recent Research Frontiers, 215–224.

Mammalian lungs are very similar in structure; all evolve to fill the chest cavity in order to supply oxygen and remove metabolites in the most efficiency way. Maldenbrot (1983) has suggested a theoretical model of the bronchial tree which consists of continually branching cylindrical pathways of diminishing length and cross section. A trunk with barren sides has a fertile bud at the end, which given birth to two branches of shorter length and diameter. These two branches also bud at their ends to form four additional branches which are smaller in size than their parents.

The aim of the present work is to provide a quantitative description of aerosol retention in the model of human lung for suggested cytotoxic effects which can change the structure of the lung membranes and their transport properties.

MATHEMATICAL MODEL OF PARTICLE RETENTION

The time dependent mass of particulates in the alveoli and airways in human lungs are described in the present model. An accurate quantitative description of aerosol retention requires a clearly defined region of observation, such as an element of the respiratory system or a differential element in the generation of the bronchial tree, alveoli, or nasopharynx, It is assumed that material deposited in the nasopharynx is removed much more rapidly than particulates in the remaining regions, and thus plays no role in the retention phenomena (Dahlback et al., 1989).

In the differential equations on which this model is based, the smallest region for the mass balance of the particulate matter is a single generation in the dichotomy model of the lungs. Therefore, the mass balance equation for particle retention in the lung is:

$$\frac{dm_i}{dt} = Q_{i+1} - Q_i + R_i - (A_i)_{pulm} , \qquad (1)$$

where m_i denotes the mass of material contained in the *i*-th generation, Q_{i+1} is the mass flow rate into the i-th generation from the i+1 generation through mucociliary action, Q_i is the mass flow rate leaving the i-th generation, R_i is the mass deposition rate in the i-th generation originating from the particulate rich air flowing through that generation, and $(A_i)_{pulm}$ is the rate of particle removal through pulmonary clearance. In addition to the generalized mass balance, a complete mathematical model of particle retention in rat lungs includes the morphology and geometry of the respiratory system, the rate of particle deposition, and the rate of clearance at each particular point of the respiratory tract.

Geometry of the lung and rate of particle deposition

The simplest and most widely used geometrical configuration for the lung is the symmetrical lung model proposed by Weibel (1963). The bronchial tree is assumed to consist of both an expandable pulmonary region where respiratory gas exchange takes place and a bifurcating series of parallel and semi-rigid cylindrical tubes.

A through description of the branch couplings is provided by three classes of second order functions (Gradoń and Orlicki, 1990). Combining the above theoretical models, a detailed framework for the rat lung structure is obtained.

The air velocity profile within each bifurcation is found from the solution of the quasi steady state Navier-Stokes equation. The average velocity at the entrance to any generation is estimated from the volumetric flow rate, which in turn is governed by both the lung tidal volume *TV* and breathing frequency f_b. The flux of particles to the bifurcation walls is found through the solution of the non-steady state convective/diffusion equation coupled with the Stokes equation of particle motion. The particle concentration, C(x,t), at point x within the i-th generations gives us the particle flux, **r**(x,t):

$$\frac{d\mathbf{R}}{ds} = \boldsymbol{r}(x,t) = \boldsymbol{u}_n \cdot C(x,t) - D\frac{\partial C(x,t)}{\partial \boldsymbol{n}}, \tag{2}$$

where $\boldsymbol{n}$ is a vector normal to the wall of the airway, $\boldsymbol{u}_n$ is the particle velocity vector normal to the airway, and D is the diffusion coefficient of the particles.

The alveolus is assumed to consist of a sphere with radius R_a attached to a stiff cylinder with a radius of intersection R_c. The radius of the alveolus, $R_a(t)$, changes with time symmetrically and uniformly according to the volume balance of inhaled air. The class of equations applicable to single bifurcations are used for the calculation of particle flux and deposition. For particles of higher mass, additional forms of particle settling by gravity are included. The above model provides the local flux $\boldsymbol{r}$(x,t) to the wall of the respiratory system. The calculated fluxes are averaged over the space and time of inhalation an expiration for each generation of the respiratory system.

Mucociliary clearance

The analysis given below applies only to inhaled insoluble particles. These, when deposited in the lungs, are gradually eliminated by means of several mechanisms. In the tracheobronchial (TB) or conducting airways, the most important mechanism is mucociliary clearance. Particles deposited on the surface of the mucus are transported into the pharangeal region as the mucus is propelled by the rhythmic beating of cilia. Numerous papers describe theoretical studies of mucociliary clearance (see for example Ross and Corsin, 1974; Blake and Winet, 1980; Blake and Fulford, 1984; and others). In the model recently of cilia activity and the geometry of the lung on which it strongly depends. Depending on the particular data of the model, which are measured elsewhere, different linear velocities can be reached. For example, the mucus velocity for the trachea can have values ranging from 1 to 50 mm/min. Such values are observed in experimental works.

As an example of the application of our calculations to the model of retention we choose the mucus velocity of 10.48 mm/min in the trachea. The velocities for other generations of the tracheobronchial tree were calculated on the assumption of the continuity of mucus flow (Lee et al., 1979; Yu et al., 1983).

Pulmonary clearance

Relatively low doses of insoluble and non-cytotoxic particles can be ingested by alveolar macrophage (AM) within hours after deposition (Lehnert and Morrow, 1985). The particles deposited in the pulmonary region of the respiratory system (generations 17-23) exist as "free" and/or phagocytosed particles at any given instant. Both forms of deposited particles are withdrawn by a mucociliary escalator and subsequently propelled into the lung's interstitial, the lymph nodes and to the blood system.

The estimation of residence time for both forms of particles in the pulmonary region of the lung is uncertain. Recent publications on lung clearance include relatively insufficient information on macrophage mobility and its translocation velocity from the alveoli to the mucociliary escalator. Morrow (1989) suggests two general mechanisms of AM translocation. The first is that of directed migration through airways to the mucociliary escalator. In this case, AM migration is presumed to be due to either directed motion of the alveolar lining layer on which the AM, as well as "free" deposited particles are passively transported by hydrodynamic effects, or due to the surface tension gradient in the surfactant monolayer adsorbed on the hypophase (Gradoń and Podgórski, 1989; Podgórski and Gradoń, 1990, 1991, 1993). Motion may also arise from the chemoattractant gradient which directly polarizes and facilitates AM migration along the alveolar epithelial surface. Macklin (1955) proposes another concept of directional movement based on the assumption of a continuous capillary effusive into the alveoli, but this theory is not elaborated in detail. The second mechanism for translocation of the macrophages assumes the motion not to be directional, but a random walk of the AM.

A mathematical analysis of pulmonary clearance has been presented by Gradoń and Podgórski (1993, 1994). According to that study, the "free" particles and chemotactically inactive AM are transported in response to hydrodynamic forces. This movement is directional by ordering the surface tension gradient in the breathing cycle. The particle trajectory, $\mathbf{x}(t)$, is defined by the deterministic equation:

$$d\mathbf{x} = \mathbf{v}\, dt \quad (3)$$

where $\mathbf{v}$ is the directional velocity vector of the hypophase.

In the case of simultaneous random and directional chemotactical displacement of the AM, a stochastic element of movement of the alveolar macrophage is included in the trajectory of the ingested particles. For this case, the stochastic equation describing AM displacement has the form:

$$d\mathbf{x}_\epsilon = \mathbf{v}dt + \epsilon\delta\, dW(t) \quad (4)$$

The deterministic term $\mathbf{v}dt$ affected by stochastic noise produced by the chemotaxis of the AMs. The stochastic part of the trajectory is determined by the Wiener process, W, which is defined by the coefficient of chemotaxis D_{ch} (or δ) and is the mathematical analog of Brownian displacement. The variance of the Wiener Process is δ^2. The parameter ϵ in Eq. (4) describes the contribution of the stochastic element to the trajectory, and is a function of the concentration of deposits on the surface of the alveoli. Physically, ϵ estimates the contribution of the side displacement of the AM from the directional movement of the hypophase which carries both macrophage loaded particles and "free" particles.

The integration of Eq. (3) or (4) produces an estimate for the residence time of particles in the generation comprising the alveolar region of the lung. The average residence time for balanced deposits is:

$$\tau_a = \phi\tau_d + (1 - \phi)\tau_{ch} \quad (5)$$

where ϕ is the mass fraction of the deposits which exists as a free particles, τ_d is the average particle residence time due to the directional displacement of the hypophase as calculated from Eq. (3), and τ_{ch} is the average residence time of the AMs loaded with particles calculated from Eq. (4).

There is a second proposed pathway for deposits to enter the pulmonary region. It is the slow phase of displacement of the particles into the lung parenchyma, followed by entrance into the blood system. In our model, we assume that this slow phase is composed of two stages. In the first stage, part of the deposited material contained in the alveoli and terminal bronchiole of a given generation is transferred to the wall of the tissue at a constant rate K. The material is then transported to the lymphatic and blood system at a rate Λ. The value of the transfer constant, K, is determined by the permeability of the lung parenchyma membrane subjected to "free" and phagocytozed particles. The value of the coefficient Λ is a function of the cell metabolism.

Toxic effects

Toxic effects increase the contribution of the slow phase of clearance in the process of particle retention in the human lung. The influence of toxic effects on the clearance process is intricate and involves many phenomenological aspects. With respect to the proposed model and transport processes described previously, the toxic effects can change the value of the transfer constant K and/or increase the mass fraction, ϕ, of deposits in the form of "free" particles. Compared with chemotaxis, free particles withdraw slowly from the pulmonary region due to hydrodynamic effects. This sluggish movement increases the fraction ϕ, thereby increasing the residence time of the material and increasing the probability of ingestion of the particles into the parenchyma.

The cytotoxicity of the particulate material also increases the value of ϕ. After the toxin-induced autolysis of the cell, particles are released and exist (until re-phagocytosis) in the form of free particles. Another effect of overloading is the moderation of chemotaxis of the macrophages. For high deposition rates and/or high exposure times, the loading of the AM with particles decreases cell mobility.

For a purpose of this paper we will consider particular toxic effect which can affect the retention of inhaled insoluble particles, related to the simultaneous exposure of the human lung to the radioclides presented in an inhaled air. Taking into account proposed mechanistic elements of the clearance process, the influence of irradiation on the transport rate should be analyzed.

The early reaction of the alveolar structure on the irradiation is a reduction of the Type 1 and Type 2 pneumocytes and, as a result of that, reduction of surfactant production (Kapaci et al., 1969; Yuen et al., 1971) changes in surface concentration of surfactant reduce the transport rate of deposits in the air space of the lung. High dose of irradiation and long-term effects change the structure of epithelium and endothelium (Penney et al., 1977; Phillips, 1966), which change diffusional and convectional transport in the lung parenchyma (K, Λ). Because of a lack of quantitative data, and the very complicated picture of the structural changes in the cells under irradiation, we will only estimate the values of the transport rate constant to observe how they change the retention curve of inhaled particles.

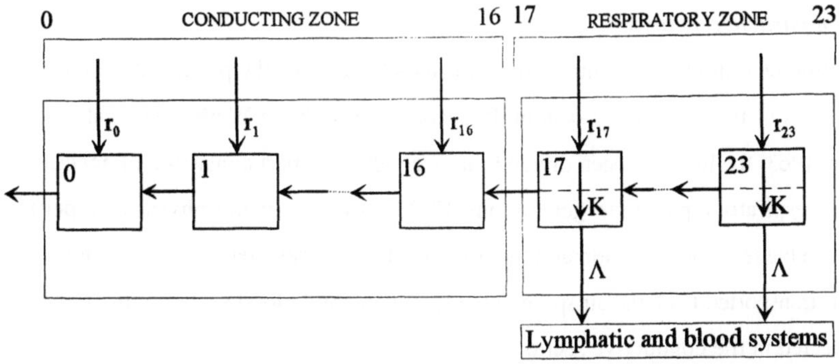

Fig. 1. Schematic representation of the retention model based on Weibel's 'model of the lung.

Table 1. Total areas, S_i, mean residence times, τ_i, and deposition fluxes, r_i, for generations of the human respiratory system

generation	0	1	2	3	4	5	6	7	8	9	10	11
$S_i \times 10^{-3}$, mm^2	6.786	3.649	1.982	1.070	2.873	3.765	5.067	7.092	9.574	13.38	19.24	27.35
τ, min	4.90	4.54	2.33	1.11	3.31	3.70	4.66	6.03	7.44	9.64	12.7	16.3
$r_i \times 10^6$, m^{-2}s^{-1}	4	4	5	8	9	9	10	20	20	20	30	30

generation	12	13	14	15	16	17	18	19	20	21	22	23
$S_i \times 10^{-3}$, mm^2	40.34	56.98	87.61	135.9	203.8	442.2	893.6	1900	4118	8236	16470	28000
τ, min	22.0	27.0	38.3	50.0	55.0	100	250	400	600	700	800	1000
$r_i \times 10^6$, m^{-2}s^{-1}	30	30	30	30	30	28.4	27.0	33.2	26.9	19.3	11.6	7.72

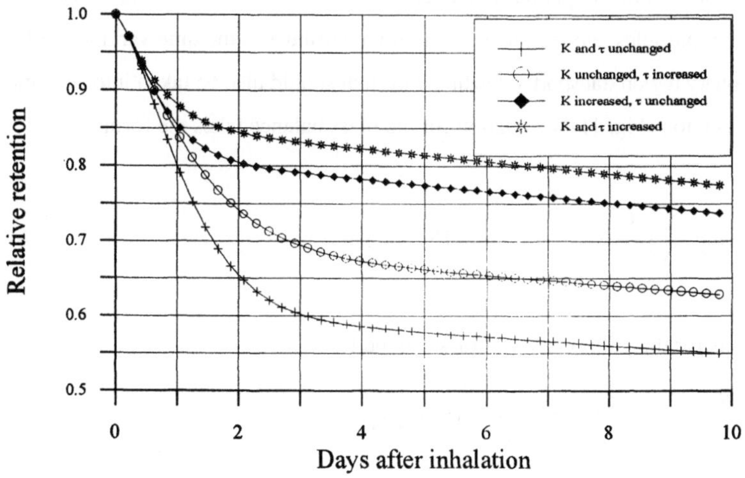

Fig. 2. Retention curves calculated (explanatation in text).

Model of particle retention

Figure 1 shows the control areas for mass balance of the particulate matter retained in the airways of the respiratory system. In accordance with the Weibel schematic presentation (Weibel, 1963) the lung has been divided into a conducting part consisting of generations 0-16, and the respiratory part with generations 17-23. Because of the physical properties of the surfaces in the respiratory region and the long residence times there, a fraction of the deposited matter is transported from the air space to the lymphatic and blood systems, an effect which can be neglected in the conductive section.

According to the proposed model, in the first stage part of deposited material contained in the alveoli of a given generation (the actual proportion depending on concentration) is transferred across the epithelium barrier at a constant rate K, and is then transported to the lymphatic and blood systems at rate Λ. The value of the transfer constant, K, determines the amount of deposited material remaining in the alveoli in a free state and subject to fast transport. At higher values of K an increasing fraction of the deposited matter undergoes the slow transportation.

The kinetic equation of retention describes the time variation of the mass of material in the region under consideration. The amount of deposited particles can be expressed in different ways, the most suitable being to express than as a surface concentration of particle, C_i (number of particles per unit surface of the airways of the respiratory system at a given i-th generation).

In the balance equation for a selected generation i, an average flux r_i of the particles deposited in the generation, a flux of particles transferred from the generation i+1 in a cleansing transport, and a flux of particles leaving the generation i, contribute to the time variation of C_i.

In the pulmonary region transport to the blood system should also be taken into account. For the conductive section (Fig. 1) the equation for the mass balance takes the form:

$$\frac{dC_i^a}{dt} = r_i + \frac{S_{i+1}}{S_i}\frac{C_{i+1}^a}{\tau_{i+1}} - \frac{C_i^a}{\tau_i}, \qquad i \leq 16 . \tag{6}$$

The equations for the pulmonary section have the form:

$$\frac{dC_i^b}{dt} = r_i + \frac{S_{i+1}}{S_i}\frac{C_{i+1}^b}{\tau_{i+1}} - \frac{C_i^b}{\tau_i} - KC_i^b, \qquad 17 \leq i \leq 23, \tag{7}$$

$$\frac{dC_i^c}{dt} = KC_i^b - \Lambda C_i^c . \tag{8}$$

Retention of particles, R(t), determined as the amount of particles retained at a given moment in the respiratory system, is calculated from:

$$R(t) = \sum_{i=0}^{23} \left[C_i^a(t) + C_i^b(t) + C_i^c(t) \right] S_i . \tag{9}$$

RESULTS AND DISCUSSION

Calculations of relative retention were performed for four sets of data. Set number 1 corresponds to the normal conditions with no radionuclides effects and is characterized by the values of the transport rates $K = 5 \times 10^{-4}\text{min}^{-1}$, $\Lambda = 7.1\text{x}10^{-6}\text{min}^{-1}$, and by the residence times in the subsequent generations given in Table 1. Set number 2 differs from the previous one only in the value of K increased to 1.2×10^{-3} min^{-1}. The third set has again unchanged value of $K = 5 \times 10^{-4}\text{min}^{-1}$, but increased residence times in the respiratory zone as effect of decreased activity of type II pneumocytes and pulmonary surfactant. For this case values of τ_i, $i \geq 17$, were estimated from the mathematical model of pulmonary surfactant hydrodynamics, (Podgórski, 1991), assuming drop of surfactant concentration by 12% compared to the optimal conditions (when the self-cleansing is the fastest). This yields the following values of the residence times in the respiratory zone (subscript denotes generation number): τ_{17} =100min, τ_{18}=283.75min, τ_{19}=508min, τ_{20}=843min, τ_{21} =1078min, τ_{22}=1340min, τ_{23}=1810min. The fourth set combines effects of increased values of $K = 1.2 \times 10^{-3}\text{min}^{-1}$ and τ_i given above.

Results of computations for these data are shown in Fig. 2, where prolonged retention for increasing values of K and τ_i can be observed. The half-times of clearance (when relative retention falls to 0.5) are following: set #1: $\tau_{1/2}$ = 19 days, set #2: $\tau_{1/2}$ = 47.9 days, set #3: $\tau_{1/2}$ = 32.2 days, set #4: $\tau_{1/2}$ = 52.6 days.

The retention curve for set #1 corresponds to the reference state which assumed unchanged structure of parenchyma and defence mechanisms of clearance. Three other cases (sets #2, #3 and #4) correspond to the situation when clearance mechanisms are reduced due to structural changes of the lung tissue and activity of surfactant due to decreasing of its concentration. A slow phase mechanism of a transport through the lung parenchyma (higher K) has stronger influence on the retention then passive air-space clearance related to the surfactant activity. The model proposed above can be used for quantitative analysis of the influence of particular mechanisms of clearance on the retention process.

REFERENCES

Blake, J.R. and Winet, M., (1980), *Biorheology*, **17**, 125-134

Blake, J.R. and Fulford, C.R., (1984), *Phys.-Chem.Hydrodynamics*, **5**, 401-411

Dahlback, M., Eirefelt, S., Korlberg, J., Nebrink, O., (1989), *J.Aerosol Sci.*, **28**, 1325-1327

Gradoń, L. and Podgórski, A., (1989), *Chem.Eng.Sci.*, **44**, 741-749

Gradoń, L. and Orlicki, D., (1990), *J.Aerosol Sci.*, **21**, 3-19

Gradoń, L. and Podgórski, A., (1993), Abstracts of 12-th Annul Meeting of AAAR, p.153, Chicago, Illinois

Gradoń, L. and Podgórski, A., (1995), *Medical and Biomedical Eng.Computing*, **33**

Kapanci, Y., Weibel, E.R., Kaplan, H.P., Robinson, F.R., (1969), *Lab.Ivest.*, **20**, 101-118

Lee, P.S., Gerrity, T.S., Hass, F.J., Lourenco, R.V., (1979), *IEEE Trans.Biomed.Eng.*, **26**, 624-630

Lehnert, B.E. and Morrow, P.E., (1985), *Exp.Lung.Res.*, **9**, 1-16

Macklin, C.C., (1955), *Acta Anat.*, **23**, 1-23

Maldenbrot, B.B.,(1983), The fractal geometry of nature, W.H.Freeman and Co, New York

Morrow, P.E., (1989), *Fund.Appl.Toxicol.*, **10**, 369-379

Penney, D.P. and Rubin, F., (1977), *Int.J.Radiation Oncol.Biol.Phys.*, **2**, 1123-1132

Podgórski, A. and Gradoń, L., (1990), *Ann.Occup.Hyg.*, **34**, 134-147

Podgórski, A. and Gradoń, L., (1991), *Chem.Eng.Comm.*, **110**, 143-162

Podgórski, A., (1991), PhD Dissertation, Warsaw Univ. of Technology

Podgórski, A. and Gradoń, L., (1993), *Ann.Occup.Hyg.*, **37**, 347-365

Ross, S.M. and Corrisin, S., (1974), *J.Appl.Physiol.*, **37**, 333-340

Weibel, E.R., (1963), Morphometry of the human lung, Springer, Berlin

Yu, C.P., Hu, J.P., Leikauf, G., Spector, D. and Lippman, M., (1983), in Aerosols in the mining and industrial work env., pp177-184, Ann.Arbor.Sci.Lab.

Yuen, T.G.H. and Sherwin, P., (1971), *Arch.Environ.Health*, **22**, 178-188

Quantitative Fit Testing Techniques for Respirators Protecting the Human Respiratory System against Environmental Aerosols

Klaus Willeke, Don-Hee Han *

Aerosol Research and Respiratory Protection Laboratory, Department of Environmental Health, University of Cincinnati, Cincinnati, OH 45267-0056, USA
*On leave from Department of Industrial Health, Inje University, Kyongnam-do, Kimhae, Republic of Korea

Keywords

Cartridge; Condensation Nuclei Count; Face Seal Leak; Filtering Facepiece; Leak Flow Measurement; Mask; Photometric Aerosol Measurement; Quantitative Fit Testing; Respirator; Respiratory Protection.

Introduction

Occupational diseases caused by breathing contaminated air is avoided primarily by engineering control measures that prevent atmospheric contamination. When effective engineering controls are not feasible or while they are being instituted, an appropriate respirator must be used. Respirators can be classified into two types: air-purifying respirators that remove contaminants from the ambient air and atmosphere-supplying respirators that provide air from a source other than the surrounding atmosphere (National Institute for Occupational Safety and Health, 1987a). Except for special cases, air-purifying respirators are used most frequently in routine workpractices because they are smaller, more easily maintained, and restrict the wearer's movements the least. Thus, the focus of this article is on air-purifying respirators.

Even the best respirator may not protect the wearer sufficiently if there is an improper match between the edge of the respirator and the facial skin. Therefore, respirator fit testing is required before entering specific work environments to ensure that the respirator worn satisfies a minimum of fit and that the user knows when the respirator fits properly (American Industrial Hygiene Association, 1991). The fit of a respirator can be determined by qualitative or quantitative methods. The qualitative fit test (QLFT) methods rely on the subject's voluntary or involuntary response, i.e., taste, smell, or irritation to a test agent. The quantitative fit test (QNFT) methods provide an objective and numerical basis by measuring the fit factor (FF) (National Institute for Occupational Safety and Health, 1987b). These factors do not depend on the subject's voluntary or involuntary response. QLFT methods have not been universally accepted because of their dependence on the wearer's subjective response.

Until a few years ago, only one QNFT method was available and accepted by U.S. regulations. When using this method, an aerosol of about 0.5 μm mass median aerodynamic diameter (MMAD) is generated and dispersed in a test enclosure in which the respirator fit is tested on an individual by photometrically measuring the aerosol concentration inside and outside the respirator. In the 1980's and 1990's several new and fundamentally different QNFT methods have been developed. Two of the newer methods are also commercially available and are accepted by the U.S. Occupational Safety and Health Administration (OSHA) as suitable alternatives: the continuous-flow condensation nuclei counting (CNC) method, commercially known as

J. C. M. Marijnissen and L. Gradoń (eds.), Aerosol Inhalation: Recent Research Frontiers, 225–234.

"PortaCount™", and the controlled negative pressure (CNP) method, commercially known as "FitTester 3000™". Other methods, such as the dichotomous-flow QNFT (Dichot) method, offer new ways to quantitatively assess respirator fit and may become commercially available in the future. Since these various methods have different advantages and disadvantages, and use different techniques to yield a quantitative measure of fit, the numerical values resulting from their use may differ from each other (Biermann and Bergman, 1988; Crutchfield, et. al, 1995). In this article, the principle of operation of each QNFT technique is explained with special focus on the aerosol methods.

QUALITATIVE VERSUS QUANTITATIVE FIT TEST

When a qualitative fit test is performed, the QLFT method requires the introduction of a gas, vapor or aerosol challenge agent into the area around the respirator while it is worn. A determination is then made as to whether the wearer can detect the presence of the challenge agent through subjective means such as odor, taste or nasal irritation. If the presence is detected, the respirator fit is considered to be inadequate. Three types of QLFT agents are currently accepted by OSHA: isoamyl acetate, sodium saccharin, and irritant fume (Asbestos, 1994). The QLFT methods have the advantages of being inexpensive, fast, relatively simple, and easy to perform in the workplace. The disadvantage of the QLFT methods is that their output relies on the subjective response of the wearer as to whether or not the challenge agent has penetrated into the respirator. It is also impossible to rank the degree of fit in order to select the best respirator for an individual. If the wearer has a disability for detecting the challenge agent, the test result is invalid.

When a quantitative fit test is performed, the respirator is worn in a stable test atmosphere containing aerosol particles. The adequacy of the QNFT is determined by measuring the aerosol concentrations outside and inside the facepiece of the respirator. Since talking and head movement may distort the face where the respirator seals against it, the QNFT is performed in several head and mouth positions reflecting work situations. During the test the cartridges of the test respirator are of the high efficiency particulate air (HEPA) filter type so that the aerosol particles measured inside the facepiece can be assumed to have penetrated there through the respirator faceseal interface, not through the filter material of the cartridges. The ratio of the aerosol concentration outside the respirator (C_o) to the one inside (C_I) is referred to as the respirator's fit factor (FF). The FF measures the respirator's fit to the wearer's face under the conditions of the test. As such, the FF does not predict the degree of protection during actual wear in an occupational exposure. During actual wear, the amount of aerosol particles reaching the wearer's breathing zone from the outside environment depends on the concentration and size distribution of the aerosol particles surrounding the wearer, the type of cartridges used, the amount of sweat and dirt on the wearer's face, the degree of facial movements, the wearer's care in keeping the respirator well placed on the face, and other factors. In the case of gaseous contaminants, the degree of penetration to the wearer's breathing zone depends also on the adsorption characteristics of the cartridge(s) and the surfaces of the face seal leak. The ratio of contaminant concentration outside the respirator to inside the respirator during actual wear is referred to as the workplace protection factor (WPF).

The major advantage of the QNFT is that this test yields a numerical and objective FF that is compared with the minimum FF required by the standards. Thus, a decision can be made as to whether the respirator has an acceptable fit. The QNFT is therefore recommended when respirator leakage must be minimized for work in a highly toxic workplace. However, the fit of the same respirator on an individual may vary from one wearing to the next, because the human facial features are complex and the respirator may not attach to the face in the same way every time it is worn. The fit factor determination may also be affected by a number of other factors, particularly by the type of fit test technique used (Rose, et al., 1990; Krishnan, et al., 1994). When the QNFT is performed, aerosol particles from outside the respirator may streak into the interior space of the respirator through one or more leak sites. Since the wearer's breathing cycle is too short for complete mixing of the aerosol inside the respirator, the measured FF depends on the location of the leak site(s) relative to the sampling port leading to the aerosol sensor. Thus, there may be considerable sampling bias in a QNFT utilizing an aerosol sensor (Holton, et al.,1987; Myers, et al., 1988; Willeke and Krishnan, 1990) .

ELASTOMERIC VERSUS FILTERING FACEPIECE RESPIRATOR

The only respirator that is legally permitted at this time to be worn in U.S. occupational environments where fit testing is required before its use is the elastomeric kind with one or two cartridges. Due to the flexible material that is mostly used for the respirator body in recent designs, such a respirator generally follows the contours of the wearer's face quite well. When fit testing an elastomeric respirator with aerosol particles, face seal leak penetration is distinguished from cartridge penetration by using HEPA filters in the cartridge(s). The HEPA cartridges are assumed to be 100% efficient so that only face seal leakage causes aerosol transport from the outside to the respirator interior. It is also assumed that there is no leakage through the exhalation valve during fit testing with a clean respirator.

There is currently no fit testing regulation for filtering facepiece respirators. These devices are also referred to as low-maintenance or disposable respirators. In such a device, the filter material and its support structure make up most of the respirator body. Thus, only in a filtering facepiece made of HEPA filter material can environmental aerosols be assumed to penetrate to the respirator interior through the face seal leak(s) alone. With filtering facepiece respirators made of less efficient filter material, aerosol penetration through the face seal(s) is not easily distinguished from aerosol penetration through the filter material itself. The construction of such devices frequently includes now soft material at the facial interface so that the filtering facepiece respirator follows the facial contours better. Therefore, a need exists for the development and commercialization of fit testing methods suitable for fit testing such devices so that fit test regulations may be considered for them in the future. Figure 1 categorizes the principal QNFT techniques that have been developed for both types of respirators.

QNFT METHODS FOR ELASTOMERIC RESPIRATORS WITH CARTRIDGE(S)

Fit testing is performed to minimize face seal leakage. When a leak occurs, air flows through the leak site(s) from the exterior to the respirator interior during inhalation when the pressure in the interior is less than that outside. Therefore, one type of QNFT is the measurement of the air flow into the respirator interior. Environmental contaminants, be they particulate or gaseous, may be carried inward with this air flow. Measurements of the aerosol concentration inside the respirator relative to that outside constitute the other type of QNFT in use today. Methods of the latter type will be discussed first in the order shown in Figure 1.

Aerosol Measurement Methods

The first aerosol QNFT method was developed about 35 years ago (Burgess et al., 1961). The first practical aerosol system for routine use utilized dioctylphthalate (DOP) aerosol as the challenge agent and forward light-scattering photometry as the aerosol detection technique. Since then other aerosol instruments have been used for detection and a variety of other aerosol particles have been utilized as test agents, e.g., sodium chloride, corn oil, and ambient air aerosol particles. During fit

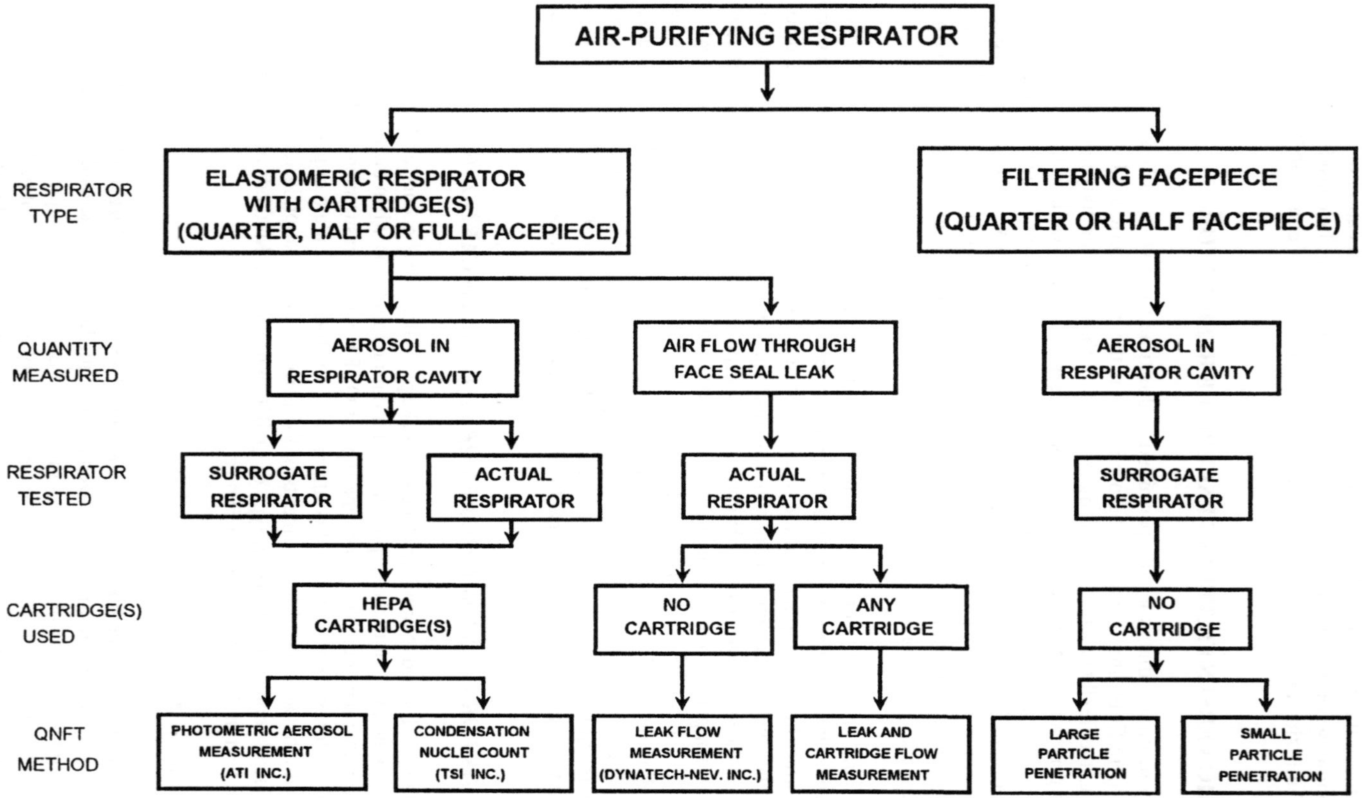

FIGURE 1. Fit testing methods for air-purifying respirators

testing with an aerosol detector, the QNFT operator selects one or more respirator brands manufactured by one or more companies and chooses an available respirator size for each brand that will most likely fit the respirator wearer. Since the HEPA filters used in the cartridges are at least 99.97 % efficient measuring the most penetrating particles of about 0.3 μm aerodynamic diameter, 0.03 % penetration of 0.3 μm particles yields a FF of 3,333, if no face seal leak exists. Thus, FFs above 1,000 may have limited meaning, if the aerosol instrument efficiently detects particles penetrated through the HEPA cartridges(s).

During actual wear in an occupational work environment the cartridges may not contain HEPA filters, but filters of lower efficiency which produce a lower pressure drop across them during inhalation. The respirator's fit to the face may be less efficient during inhalation, if the pressure drop to the respirator interior is less. Thus, the aerosol test is useful for selecting a suitable respirator, but may not predict the face fit during actual wear. Another problem in this technique is that the aerosol instrument measures the aerosol concentration continuously inside the respirator during inhalation and exhalation. During exhalation the aerosol concentration in the respirator is reduced by the amount of aerosol particle deposition in the wearer's respiratory tract. The aerosol measurement method thus measures primarily face seal leakage, but the data may be affected by lung deposition and cartridge penetrations.

While the QNFT is performed with an aerosol detector, the respirator wearer undertakes several exercises, such as up and down movement or side-to-side movement of the head, or talking while keeping the head in the forward position. These movements simulate actual wear practices, that may affect the respirator's seal to the face. However, these movements are not defined as to the angle of exertion, length of time and breathing rate. As a result, wearers with the same fit are likely to have different fit factors resulting from this activity.

Photometric Aerosol Measurement: The first practical QNFT systems used an aerosol photometer to detect the aerosol inside and outside the respirator, as schematically represented in Fig. 2A. Photometers are most sensitive to aerosol particles of about 0.5 μm in MMAD. As the aerosol concentration inside a well-fitting respirator is much less than that outside, and there are not that many 0.5 μm particles present in normal indoor or outdoor air environments, it is necessary to generate an aerosol in that size range when photometer aerosol detection is used. This leads to the need for an enclosure, either a fixed booth or a portable hood that covers at least the head. The advantage of this system is that the aerosol concentration can be rigidly kept constant. Also, the size of the test particles is in the respirable size range. Therefore, this system is frequently used as the benchmark for evaluating other systems. The disadvantages are the size and expense of the entire system and the aforementioned inherent difficulties with particle deposition in the lung and inhomogeneity in aerosol mixing inside the respirator.

Condensation Nuclei Count: In a natural outdoor environment there are generally 10^3 to 10^4 more 0.05 μm particles present than 0.5 μm ones. In indoor environments that ratio is similar or less. This led to a QNFT with an aerosol measurement device that detects particles smaller than 0.5 μm (Willeke, et al., 1981). A continuous-flow condensation nuclei counter (CNC) was used instead of the photometer and was found to perform the QNFT quite well. TSI Inc. (St. Paul, Minnesota) developed this technique into a commercial product, termed "PortaCount™", which is currently in wide use. As shown in Fig. 2B, it requires neither an aerosol generator nor an enclosure. However, the ambient air concentration must be constant, unperturbed by smoking and other sources, that might vary the external aerosol concentration. A highly filtered room is also not desirable, as the aerosol concentration in such as a room may be too low.

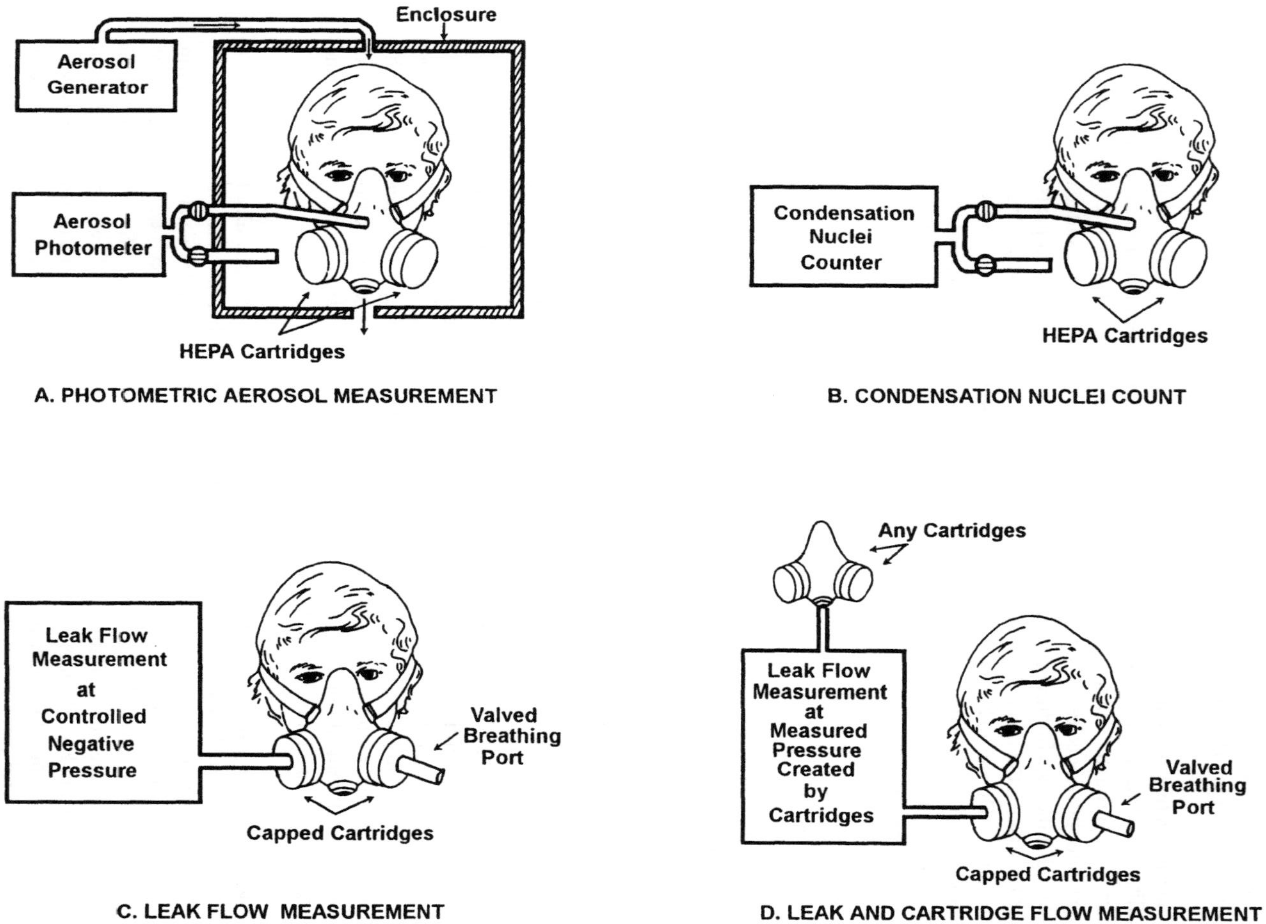

Figure 2. Schematic representation of QNFTs for elastomeric respirators with cartridges

The major advantage of this system is its smaller size. The QNFT is relatively simple and economical to administer. It has the same disadvantages of lung deposition and imperfect aerosol mixing in the respirator as the photometric QNFT system. In addition, if small aerosol droplets are generated by the test subject's respiratory system during exhalation, these exhaled particles will be added to the total count attributed to face-seal leakage. If the test subject has smoked within 30 to 45 minutes before the QNFT, smoke particles retained in the respiratory tract may also increase the particle count from inside the respirator. Although the present OSHA regulations are written for QNFTs with photometric measurement, the CNC technique is also accepted as a "de minimus violation".

Air Flow Measurement Methods

Fit testing is performed to select a respirator that minimizes the leakage of external contaminants into the respirator and thus the wearer's breathing zone. The contaminants may be particulate or gaseous. The QNFTs discussed above measure the amount of aerosol present in the respirator wearer's breathing zone. Particulate and gaseous contaminants can be present in the respirator interior only, if they are carried there by air leakage through face seal leakage. This led to QNFT methods that measure the leakage of air into the interior of the respirator. The two major techniques are listed in Fig. 1 and are schematically represented in Figs. 2C and D.

Leak Flow Measurement: The first leak measurement method utilized an indirect method to measure face seal leakage (Carpenter and Willeke, 1988). The cartridges of an elastomeric respirator were replaced with two capped cartridges. One capped cartridge had a large opening in it, so that the wearer could breath normally. The wearer then closed a valve, inflated the chest a little to create a negative pressure in the respirator interior and then held the breath for a few seconds. A pressure sensor attached to the same or to the other capped cartridge monitored the pressure inside. If the pressure remained constant, there was no leakage. If the pressure difference to the outside decayed, the decay rate was an indication of the leakage rate. This method, commercialized for a while by Dynatech-Frontier (Albuquerque, New Mexico) in the late 1980's and early 1990's, worked well with rigid respirator bodies (McGlone, 1988). However, as the respirator bodies become more pliable in the early 1990's, the volume of the respirator interior could no longer be assumed to remain constant, independent of the pressure difference; i.e., with the more pliable respirator material the pressure decay rate become a function of air leakage and change in respirator cavity volume. Although calibration of the volume change could have been entered into the leak flow calculation, use of this method was discontinued in favor of a more direct leak flow measurement.

The face seal leak flow is measured more directly in the commercially available FitTester 3000™ (Dynatech Nevada, Carson City, Nevada) (Crutchfield, et al., 1995). When using this device, the cartridges are also capped, see Fig. 2C, and the wearer holds the breath for a few seconds while the instrument reduces the pressure inside the respirator to a predetermined level that is typical of inhalation. While maintaining this controlled negative pressure (CNP), a mass flow meter measures the air flow rate out of the respirator that is needed to maintain the negative pressure. In the absence of exhalation valve leakage, this flow rate is equal to the face seal leak rate at that pressure. The FF is then determined by dividing the inhalation flow rate, that is typical for that pressure, by the measured leak flow rate. The test is repeated in several head positions, e.g., with the head turned upward or sideways. Such static tests are more repeatable than aerosol measurements during head movements, because aerosol measurements are dependent on the location(s) of the face seal leak(s). When air leaks into a volume at negative pressure, the air molecules virtually instantaneously adjust relative to each other so that the measured pressure is the same anywhere in the volume. In contrast, aerosol particles take time to physically mix throughout the entire volume to the same concentration level.

One disadvantage of the CNP technique is that the leak flow is determined at a predetermined negative pressure. Cartridges, even of the same classification, produce different negative pressures inside the respirator cavity, because different filter materials and areas in the cartridges have different flow resistances at the same flow rates (Xu, et al., 1991; Chen and Willeke, 1992). A major advantage of the FitTester 3000™ is its lower cost. The pressure and flow sensors in a leak flow measurement system are inherently more inexpensive than the sophisticated aerosol measurement technology. Although the FitTester 3000™ is very different from the photometric aerosol measurement system, it has been granted by OSHA the "de minimus violation" status.

Leak and Cartridge Flow Measurement : In order to fit test with the actual respirator to be worn and with the cartridges that are to be attached to the cartridges, be they of the HEPA or any other classification, a new system has been developed (Willeke and Krishnan, 1994) that measures the face seal leakage at the negative pressure created by the wearer's own breathing, see Fig. 2D. In this method, the cartridges are attached to a reference respirator and the pressure is measured inside at a typical average breathing rate. Through a feedback system the same pressure is created inside the respirator worn by the test subject. During breathing holding the leak flow is thus measured at the negative pressure created by the cartridges. The ratio of the cartridge flow to the leak flow determines the degree of fit. This ratio is also the average dilution factor with which contaminants penetrating through the face seal leak(s) are diluted before entering the respiratory tract. This leak and cartridge flow measurement system is not yet commercially available. Since two flows are measured by this system, it is also referred to as the Dichotomous QNFT system or "Dichot" in short. This new system can potentially also be used in field situations for determinations of the actual face fit before and after wear in hazardous environments. Sweat, dirt and strap tension may seriously affect the face seal during actual wear. If the aerosol penetration is experimentally determined for different face seal leak situations, and the aerosol size distribution is measured or known in the work environment, the amount of aerosol inhaled can be calculated for each FF measurement. Thus, an estimate of the workplace protection factor could be made. Another advantage of this and the CNP method is their fast response time. A break of the face seal is detected virtually instantaneously by any of the leak flow measurement methods, making these methods suitable training tools for workers trying to detect how far they can move their head before serious face seal leakage occurs.

QNFT METHODS FOR FILTERING FACEPIECES

When the air purifying filter is not contained in one or two cartridges, but constitutes the respirator body itself, the respirator is refered to as a filtering facepiece, low maintenance respirator or disposable respirator. If the filter is made of HEPA material, the QNFT aerosol methods for elastomeric respirators with cartridges can be applied to filtering facepieces, because the measured aerosol concentration can be assumed to be primarily due to face seal leak penetration. However, the facepiece must be probed, i.e., a surrogate filtering facepiece must be used. Since most filtering facepieces used today are less efficient than the HEPA ones, new methods have been explored specifically for filtering facepiece. For a while, HEPA versions of less efficient filtering facepieces were manufactured for fit testing. However, the QNFT data obtained with the increased rigidity and pressure drop of a HEPA filtering facepiece are not valid indications of the performance of filtering facepieces made of less efficient filter material. Therefore, face seal leakage needs to be detected without making any changes to the filter material.

Large Particle Penetration : Filter penetration is higher in the submicrometer size range. The most penetrating particle size is usually at or near 0.3 μm. A HEPA filter is at least 99.97 % efficient at that size. The latest NIOSH rule, 42 CFR part 84 of 1995, categorizes filters into 3 classes: A is at least 99.97 %, B is at least 99 %, and C is at least 95 % efficient at the mask penetration size (Respiratory Protection Devices, 1995). The previous classifications of dust/mist (DM) and dust/fume/mist(DFM) respirators are generally less efficient than class C respirators. Common to all these respirators is increasingly efficient aerosol particle removal for particle sizes above 1 μm, due to increased impaction and interception of the particles (Chen, et al., 1992).

The almost complete collection of 2.5 μm particles on DM and DFM respirators led the 3M company to the development of the "Large Particle Quantitative Fit Test (LPQNFT)" (Iverson, et al., 1992). The 3M system, consisting primarily of commercially available components, utilizes a monodisperse aerosol generator, an exposure chamber and an aerosol sensing instrument. When using this system, the monodisperse "large" particles are assumed to have penetrated to the respirator interior only through face seal leaks. Thus the measured FF reflects the penetration of the 2.5 μm test particles. In the traditional QNFTs with the photometric measurement the MMAD of the test particles is about 0.5 μm. Thus, the FF values are not comparable between these two types of measurement.

Small Particle Penetration: When a small aerosol particle approaches the filter material of a respirator, it faces a potentially tortuous path through the filter, resulting in removal in most cases. When this particle approaches a leak site, however, that opening may be very large relative to the size of the particle. A 1 μm particle may face an opening of 10, 100 or even 1000 times its own size. Its motion through the leak site and its potential removal by the leak site surface is, therefore, quite different from motion through the filter material. This difference led to the recent development of the "Small Particle Quantitative Fit Test (SPQNFT)" (Myojo, et al., 1994).

In this test the filtering facepiece is probed and the combined aerosol penetration through filter and leak(s) is measured at a low flow rate (e.g., 10 Lpm) and a high flow rate (e.g., 100 Lpm). The face seal leak rate and a corresponding measure of fit is then determined from the penetration ratio for the two flow rates. The test can be performed while breathing normally or during breath holding. The latter is performed, because lung deposition is avoided and the probe samples the entire respirator interior, thus avoiding all of the aerosol sampling problems pointed out before.

ACKNOWLEDGMENTS

This project was partially supported by the Korean Research Foundation and grant No. R01-OH-03244 from the U.S. National Institute for Occupational Safety and Health (NIOSH). This paper's contents are the responsibility of the authors and do not necessarily represent the official views of NIOSH.

REFERENCES

AIHA's Respiratory Protection Committee: Respirator Performance Terminology: Letter to the Editor. *Am. Ind. Hyg. Assoc. J.* 46(5): B22-B24 (1985).

American Industrial Hygiene Association: Respiratory Protection: *A Manual and Guideline*. Akron, Ohio: American Industrial Hygiene Association, 1991. pp. 81-88.

"Asbestos." Code of Federal Regulations Title 29, Part 1910.1001. 1994. pp. 34 - 42.

Biermann, A.. H. and W. Bergman: Filter Penetration Measurements Using a Condensation Nuclei Counter and an Aerosol Photometer. *J. Aerosol Sci.* 19(4): 471-483 (1988).

Burgess, W. A., L. Silverman, and F. Stein: A New Technique for Evaluating Respirator Performance. *Am. Ind. Hyg. Assoc. J.* 22(6): 422-429 (1961).

Carpenter, D. and K. Willeke: Noninvasive, Quantitative Respirator Fit Testing through Dynamic Pressure Measurement. *Am. Ind. Hyg. Assoc. J.* 49(10): 485-491(1988).

Chen, C. C. and K. Willeke: Characteristics of Faceseal Leakage in Filtering Facepieces. *Am. Ind. Hyg. Assoc.* J.53: 533-539(1992).

Chen, C.C., M. Lehtimäti, and K. Willeke: Aerosol Penetration through Filtering Facepieces and Respirator Cartridges. *Am. Ind. Hyg. Assoc. J.* 53(9):566-574 (1992).

Crutchfield, C. D., D. L. Park., J. L. Hensel, M. K. Kvesic, and M. D. Flack: Determination of Known Respirator Leakage Using Controlled Negative Pressure and Ambient Aerosol QNFT Systems. *Am. Ind. Hyg. Assoc. J.* 56(1) :16-23 (1995).

Holton, P. M., D. L. Tackett, and K. Willeke: Particle Size-Dependent Leakage and Losses of Aerosol in Respirators. *Am. Ind. Hyg. Assoc. J.* 48(10):848-854 (1987).

Iverson, S. G., S. G. Danisch, H. E. Mullins, and S. K. Rudolph: Validation of a Quantitative Fit Test for Dust/Fume/Mist Respirators: Part I. *Appl. Occup. Environ. Hyg.* 7(3):161-167 (1992).

Krishnan, U., K. Willeke, A. Juozaitis, M. Lehtimaki, and K. Szewczyk: Development of a Dichotomous-Flow Quantitative Fit Test for Half-Mask and Full-Facepiece Respirators. *Am. Ind. Hyg. Assoc. J.* 55(3):223-229 (1994).

McGlone, J. T.: Quantitative Respirator Fit Testing through Dynamic Pressure Decay Measurement. Appl. Ind. Hyg. 3(11): F27-F29 (1988).

Myers, W. R., J. R. Allender, W. Iskander, and C. Stanley: Causes of In-Facepiece Sampling Bias - I. Half-Facepiece Respirators. *Ann. Occup. Hyg.* 32(3):345-359 (1988).

Myojo, T., K. Willeke, and C. C. Chen: Fit Test for Filtering Facepieces: Search for A Low-Cost, Quantitative Method. *Am. Ind. Hyg. Assoc. J.* 55(9):797-805 (1994).

National Institute for Occupational Safety and Health (USA): *Guide to Industrial Respiratory Protection* (DHHS/NIOSH Pub. No. 87-116). Washington, D.C.: Government Printing Office, 1987a. pp. 3-123.

National Institute for Occupational Safety and Health: *Respirator Decision Logic* (DHHS/NIOSH Pub. No. 87-108). Washington, D.C.: Government Printing Office, 1987b. pp. 50-51.

"Respiratory Protective Devices; Final Rules and Notice." *Federal Register Part II Department of Health and Human Services, Department of Labor.* Vol. 60, No. 110. 1995.

Rose, J. C., R. K. Oestenstad, and V. E. Rose: A Comparison of Respirator Fit Factors Determined by Portable Condensation Nuclei Counting and Forward Light-Scattering Photometric Methods. *Appl.Occup. Environ. Hyg.* 5(11): 792-797 (1990).

Willeke, K., H. E. Ayer, and J. D. Blanchard: New Methods for Quantitative Respirator Fit Testing with Aerosols. *Am. Ind. Hyg. Assoc. J.* 42: 121-125 (1981).

Willeke, K. and U. Krishnan: Present Procedures in Quantitative Respirator Fit Testing: Problems and Potential Solutions. *Appl. Occup. Environ. Hyg.* 5(11): 762-765 (1990).

Xu, M., D. Han, S. Hangal, and K. Willeke: Respirator Fit and Protection through Determination of Air and Particle Leakage. Ann. Occup. Hyg. 35(1): 13-24 (1991).

Aerosol Delivery to the Respiratory tract in Experimental Animals

Magnus Dahlbäck, Stefan Eirefelt, Ola Nerbrink.
Dept. of Pharmacology, Preclinical R&D, Astra Draco AB, Lund, Sweden

Keywords; Dry powder generator, Nebulizer, pMDI, Particle size measurements, Deposition, Lung burden, Body burden, Sephadex induced lung oedema, Kinetics, Inhaled glucocorticosteroids, Budesonide

1. *Introduction*

The inhaled route is a convenient way to deliver drugs to the lung. Inhaled pharmaceutical drugs can either act locally in the lung or get absorbed into the systemic circulation to exert their effect. Since the lung functions as a filter it is important to know the optimal particle size distribution to get an optimal deposition in the lung for the species to be studied. In the search for better drugs for local treatment of inflammatory diseases in the respiratory tract it is important that new candidate drugs are evaluated in inhalation delivery systems that are well characterised. Aerosols can be delivered either nose-only or by intubating rats or dogs. An aerosol generator can be fitted to the inhalation chamber or to the nose mask which allows for the aerosol to be inhaled. Selective aerosol deposition in the lung can be achieved by use of the endotracheal tube. In both the above cases, the inhaled dose can be calculated from a filter sample drawn in the breathing zone of the animal during exposure. Simultaneously the aerosol can be monitored in real-time to measure changes in the particle concentration. This is an advantage as the aerosol concentration can be adjusted during delivery to give a more accurate dose to the animal. In our exposure systems, a tracer aerosol with Evans blue, is used to verify the estimated lung burden with the actual deposition patterns found in these animals. Finally we will also give an example of how the effect / side effects ratio can vary if a glucocorticosteroid (budesonide) is administered by different routes, and how the estimated inhaled dose correlates with the kinetic profile (area under the curve).

2. *Aerosol generation*

Pharmaceutical formulations for inhalation purposes may be manufactured either as a liquid or a dry powder. The technique used when generating these formulations may vary and depends on the type of formulation used. This review will describe some devices frequently used when performing inhalation studies of pharmaceutical formulation on laboratory animals.

2.1 *Generators*

2.1.1 Liquid based generators

The most common generator used for aerosolisation of liquid based formulations is the air jet nebuliser, see Figure 1. The air jet nebuliser can be used for both solutions and suspensions (1). Normally, suspensions will nebulise less efficiently compared to solutions. This is due to the nature of the suspension itself (not all generated droplets contain suspended particles). The air jet nebuliser uses compressed air to generate aerosols. In these nebulisers, the compressed air is also used to transport the aerosol out of the nebuliser. Although the design and output varies between different brands of air jet nebulisers (2) a lot, their generation principles depends on the theory of the Bernoulli equation to operate (3).

J. C. M. Marijnissen and L. Gradoń (eds.), Aerosol Inhalation: Recent Research Frontiers, 235–246.

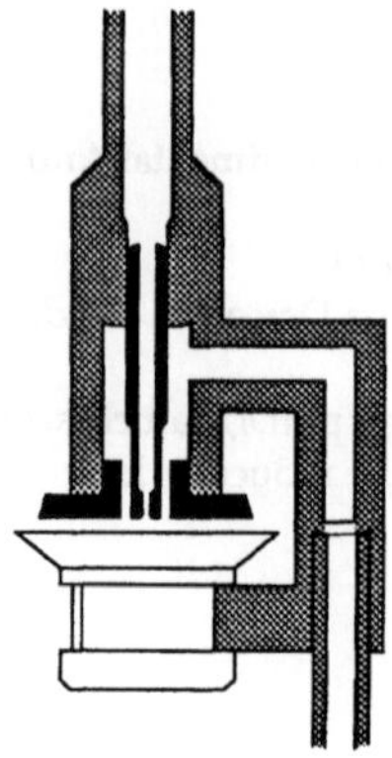

Figure 1, Air jet nebuliser MA2s

The nebuliser concept has widened to also include the ultrasonic nebuliser. An ultrasonic nebuliser generated when the piezo electric element rapidly oscillates. Different solutions are excellently generated from both air jet nebulisers and ultra sonic nebulisers, but suspensions are much less efficiently generated from an ultrasonic nebuliser. This is due to effects caused by the suspended particle inertia and the generated droplets size. As ultrasonic nebulisers do not use compressed air when generating aerosol, they require a fan to transport the aerosol out of the nebuliser unit. The output of both jet- and ultrasonic nebulisers devices are roughly equivalent, but the generation principles are quite different.

Other devices that can be used for generation of aerosols from liquid based pharmaceutical formulations are the concentric jet nebuliser and the liquid spray nozzle. Both devices, though, produce droplets that often are too large and therefore not inhalable. This limits their use in animal exposure set-ups. The produced aerosols also have a large exit velocity which makes them difficult to handle and use together with inhalation equipment. If the wet aerosol is dried, the resulting particles are often small enough to be used for inhalation purposes.

2.1.2 Generation characteristics

All of the above mentioned generators produce aerosols with a polydisperse particle or droplet size distribution. The nebuliser systems often produce droplets in the range of 1 to 10 µm with a geometric standard deviation (GSD) of 1.5 to 2.2 (2). The mass output of liquid from the nebulisers varies from 50 to 100 mg/min up to 1000 mg/min (ultrasonic nebuliser). A jet nebuliser will operate with filling volumes of 1-2 ml and up to more than 10 ml, depending on brand (4, 5, 6). The same applies for ultrasonic nebuliser and also depends on the design of the nebuliser. The jet nebuliser tend to have an excess of liquid output compared to drug output. This is caused by the evaporation of liquid occurring inside the nebuliser compartment during nebulisation. This is not the case for the ultrasonic nebuliser. Typical air flow rates vary from 4 l/min up to 10-15 l/min and driving pressures (jet nebuliser) of 30 to 500 kPa.

2.1.3 Dry powder generators.

On the market today there are several devices used for aerosol generation of dry powder formulations. When performing inhalation studies there are two different devices that are frequently used, the Wright Dust Feed (WDF) mechanism (L Adams Ltd, England) and the RGB 1000 (Palas GmbH, Germany).

The WDF requires a compacted powder in order to operate. Compaction is accomplished by transferring the powder formulation into a substance container or cup and compressing it at high pressure using a plunger. The procedure compacts the powder to a pellet. The cup is fitted on the WDF where a slowly rotating scraper blade removes powder from the pellet. Air is then used to deagglomerate the powder into primary particles and transported out of the device.

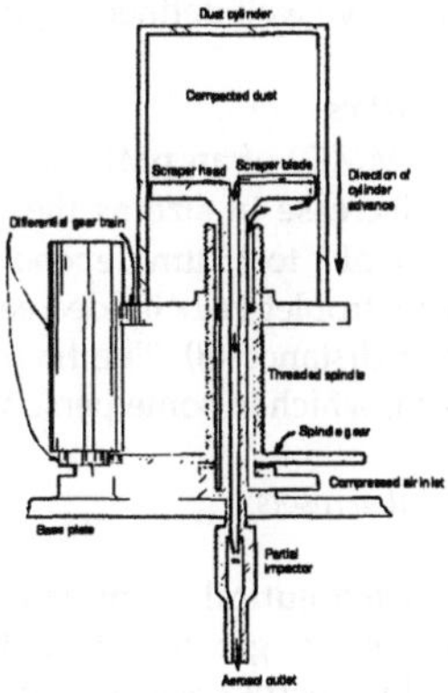

Figure 2, The Wright Dust Feed generator, WDF

The RGB 1000 uses a slightly different technique. A hollow rod with a plunger at the bottom is filled with powder. The rod is fitted vertically, under and perpendicular to a cylindrical brush. When the brush rotates and the plunger moves the powder upwards, and will be scraped off at the top of the rod. Air is fed into the rod/brush chamber where the powder is dispersed and deagglomerated into primary particles and transported out of the device. Several other devices are available on the market. Worth mentioning is also the fluidized bed generator.

2.1.4 Generation characteristics

Before a dry powder can be used in inhalation studies it must be micronised in a jet mill down to a particle size of 1-2 µm (mass medium aerodynamic diameter, MMAD). Dry powder generators produce aerosol in the size range of 1 to 15 µm. Occasionally larger particles can be produced and they are often the result of incomplete deagglomeration of the initial powder. The aerosol output varies from µg/h to mg/min giving aerosol concentrations below 1 µg/l up to 2 mg/l. The amount of powder needed to run a powder generator varies from below 1 gram to more than 50 grams depending on size of the canister (WDF) or type of piston system (RGB 1000) used. Typical air flows used in inhalation is between 4-10 l/min. Powder generators can give momentary variations in aerosol concentration but are fairly stable in long term operation. This phenomena is not seen when using a nebuliser.

2.1.5 Pressurised Metered Dose Inhalers (pMDI)
In our studies pMDI devices are not as frequently used as the liquid/powder devices described previously. The decline in their further development is partly due to the banning of CFC's during the late 1980's and more efforts were put into dry powder inhalers such as Turbuhaler®. Replacement of propellant gases are now available on the market, and we can expect an increase in the use of pMDI devices in inhalation studies. Care has to taken when designing the inhalation systems as the exit velocity of the generated aerosol often is above 30 m/s as well as rapid evaporation of the propellants takes place at the same time. With pMDI devices, carefully designed collection and stabilising chambers have to be used in order to achieve well defined exposure conditions (7).

2.1.6 Generation characteristics
The initial droplet size (MMD) of an pMDI aerosol is in the region of 50 µm to 100 µm. They rapidly decrease in size as the propellant evaporates and form particles in the size range of 1 to 15 µm depending on the mass fraction of drug in the initial droplet. The droplet velocity decreases from above 30 m/s down to zero within a half meter distance (8). The final particle also contains lubricant used by the valve system, which in some persons can cause irritation.

3. *Characterisation of pharmaceutical aerosols*

When using aerosols of pharmaceutical formulations for inhalation purposes in animal research it is important to get the drug to the lung, the target organ. Characterisation of the aerosol properties is a criteria in order to be able to determine the lung deposition as accurate as possible. If the particle size distribution is too wide or too large there is a risk of overestimating the amount of drug entering the respiratory tract. There is also a possible danger of getting unwanted side effects due to large particle deposition in the nose and oropharyngeal region resulting in increased uptake from the gastrointestinal (GI) tract. It is necessary to assess the particle size characteristics, and comparing it with the lung deposition probabilities of the species used. If the particle size distribution is within the range of high lung deposition probability, there is a good reason to believe that a favourable deposition in the respiratory tract will occur. If not, the particle size characteristics must be changed. To assess the particle size characteristics a measure technique must be employed. Some of these will be discussed below.

3.1. Instruments measuring particle size distributions
Particle size characterisation of aerosols is done with several different techniques depending mainly on the size of the particle in question. Instruments used for characterisation of pharmaceutical aerosols have to operate in the size range of 1 to 20 µm as mentioned above. The instrument should also classify the aerosol from the aerodynamic point of view as aerodynamic behaviour of the aerosol within the airways is essential for deposition. This limits the instrumentation choice to a small number of types. Traditionally, within the pharmaceutical industry, instruments such as the liquid impinger, impactor and light scattering devices like the Mastersizer (Malvern Industries Ltd, England) and Aerodynamic Particle Sizer (APS 33B, TSI Inc., USA) are used to characterise the generated aerosol (9). The liquid impinger, impactor and Mastersizer can all be used for characterisation of droplet distributions generated by nebulisers. The most direct method for measuring a wet aerosol is the Mastersizer as the measurement is done close to

the nebuliser mouthpiece (10). The technique is fast and, if correctly used, gives a good measurement of the droplet size distribution. The liquid impinger and impactor both have to be used with precaution as the aerosol is measured further away from the generation site due to the need for adapters between the two devices and the nebuliser. The devices also need high air flows to operate. The air flow rate is often higher and less humid than the nebuliser flow rate. This leads to initial drying of the droplet size distribution and assessing an unstable size of the generated aerosol. Ideally particle size instrument always measures dry aerosols. The ultimate exposure situation should be to constantly try to expose animals to a stable dry aerosol. If so, then these instruments can be adequate to use. The impinger and impactor both rely on chemical analysis in order to give results. This often takes time, suggesting that these techniques are not suitable for real time measurements. If real time measurements of dry aerosols are needed, the APS is best suited for monitoring the particle size distribution within the exposure chamber. Again, caution should be exercised, so that an aerosol is always sampled in such a way, that the true distribution is measured.

4. *Administration of pharmaceutical aerosols*

The choice of delivery system used depends on the answer wanted from the pharmacological test system used. The questions asked, and their solutions, mainly depend on animal species, drug formulation and the type of generator used. For this reason different types of exposure chambers have been developed in the past years. The following paragraphs describes the exposure systems used in our laboratory.

4.1 Exposure systems

Exposure systems used today are designed in many different ways. The two main types of exposure systems used for awake animals are 1) the whole body exposure chamber and 2) the nose-only exposure chamber. In both systems the aerosol will be deposited in the upper airways and naso-pharyx, swallowed and later absorbed via the GI tract. An example of an inhalation chamber used for small animals like mice, rats and guinea pigs is the Battelle nose-only exposure chamber (11). Due to the design, 8 animals on each level and 5 levels, can be assembled to form one Battelle chamber. Thereby can 1 to 40 animals be exposed at the same time, see Figure 3a.

Selective aerosol deposition to the same lung can also be achieved in unconscious and anaesthetised animals. The aerosol is inhaled through an endotracheal tube. Even small animals like rats and guinea pig can successfully be intubated when anaesthetized. An example is an inhalation intubation chamber where up to eight small animals can be exposed to aerosol simultaneously, see Figure 3b.

Larger animals, like dog, pig and sheep, can also be exposed to aerosol either with a nose mask or via endotracheal intubation. A hand held dry powder delivery system can then be connected to the nose mask or endotracheal tube.

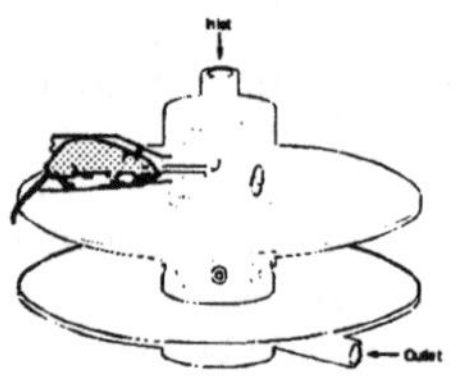

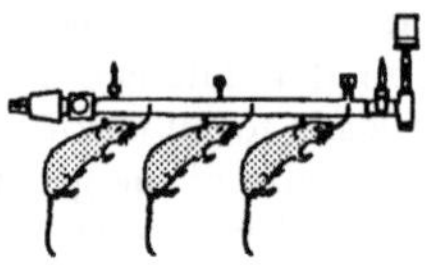

Figure 3a, Battelle chamber

Figure 3b, Intubation chamber

4.2 Filter correlation

During aerosol exposure of experimental animals it is essential to assess the inhaled dose in order to estimate the deposited dose in the animal. A filter sample taken in the breathing zone of an animal during the whole exposure period is used to estimate the inhaled dose. The exposure concentration of the aerosolised drug is calculated from the analysis of the mass collected on the filter at a known air flow rate. The filter should be connected in such a way that what is deposited corresponds to the aerosol concentration inhaled by the animal. Connecting the filter, for example, to a free port of the exposure chamber, will assure a correct sampling of the aerosol mass that the animal is exposed to. It is also vital that the chosen filter flow rate is low so it corresponds to the minute volume of the animal used. The disadvantage of this approach is that it will only give the aerosol mass concentration the animal has been exposed to, after the exposure has ended. Thus, there is no way to correct for high or low aerosol concentrations during the exposure.

4.3 Real time monitoring

A way of solving this dilemma is to monitor the aerosol concentration during the exposure period in real-time with an light scattering instrument. There are several real time aerosol monitoring systems available today. Two examples are the Real-time Air Monitor (RAM-1 or MiniRAM, MIE, England) and the Casella AMS 950 (Casella Ltd, England). The Casella AMS 950 is preferable as it is simple and easy to use. It has a wide concentration range with a high maximum aerosol concentration (0-2000 mg/m^3) and does not contaminate easily. The monitoring system can be connected to the exposure chamber on line with the filter or to an adjacent port. Running the monitoring system during an exposure will enable the operator to adjust for variations in the aerosol concentration. Together with the filter there is an option of later correlating the monitor mean reading during the exposure to the filter mass concentration assay. By setting up the exposure system prior to the actual animal study and performing aerosol exposures as in the study, the ratio of the aerosol monitor mean readout/filter mass can be estimated, and thereby the dose that the animal inhales calculated. This can be done for all dose levels involved in the study. The advantage of this procedure is obvious: the operator has the possibility of adjusting aerosol concentrations to the expected mass on filter and the monitoring systems sensitivity for different formulations and aerosol distributions can be compensated for.

5. *Deposition characteristics*

5.1. *Deposition studies*

We characterise our inhalation delivery systems by performing deposition studies using different tracer aerosols. Nebulised Evans blue dye (EB, solution) or nebulised monodisperse fluorescent polystyrene spheres, (PS) in suspension, are useful tools when trying to estimate the actual deposited dose in the lung or total body of an experimental animal. Deposition studies have been performed in guinea pigs and rats and we have used the two delivery systems described above, nose only and/or intubation (12, 13, 14).

5.1.1 *Methods*

In our deposition studies the animals are exposed for 5 or 10 minutes, sacrificed immediately thereafter and the lungs are taken out. The EB or PS aerosol is extracted from different parts of the lungs and from different parts of the body where it could have been deposited. The lungs are divided into central and peripheral part by scraping the parenchymal tissue from central airways > 0.5 mm.

5.2. *Dosimetry*

5.2.1 *Estimation of the Inhaled dose, Lung and Body burden*

To estimate the inhaled dose (ID = CC x RMV x ET / BW, μg /kg) from the above mentioned deposition studies, the following parameters are necessary to know. The aerosol concentration (CC, μg/l) in the delivery system is estimated from filter sampling during each exposure. The respiratory minute volume (RMV, ml/min, $4.19 \times BW^{0.66}$, body weight in g) is calculated according to McMahon *et al* (15). The exposure time (ET, min) and body weight (BW, kg) is also needed. Deposition studies with a wide size range of monodisperse particles have been published for mice and rats by Raabe et al (16, 17) and for dogs by Schlesinger (18). From these data deposition probability curves can be drawn and we have plotted smoothed curves of both lung and total body deposition for rats and dogs, see Figure 4 and 5. The particle size distribution from a dry powder aerosol is used in conjunction with the deposition probability curve when calculating the actual deposition probability factor for lung burden (DF_{LB}) and body burden (DF_{BB}). These factors are used when calculating lung burden (LB =ID x DF_{LB}) and body burden (BB = IDx DF_{BB}).

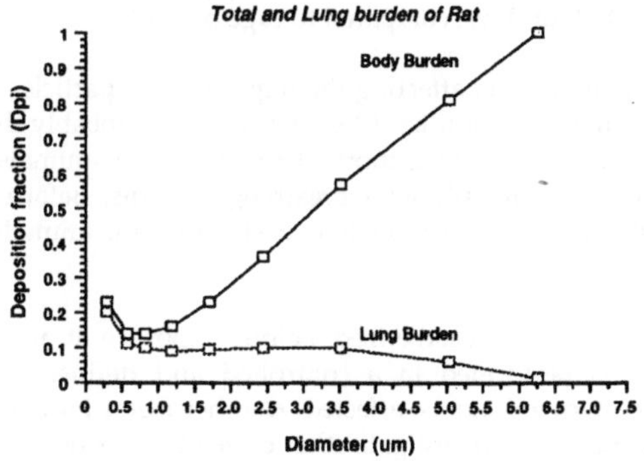

Figure 4

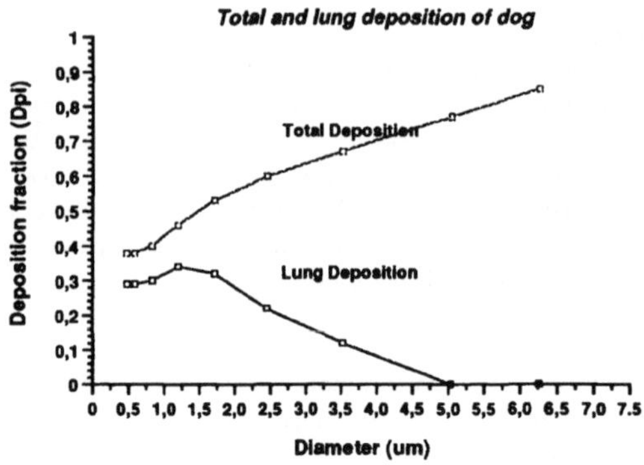

Figure 5

5.2.2 Results and discussion

After nose-only exposure of EB the fraction of the inhaled dose deposited in the lung (LB) is 8.5 % in rats and 18 % in guinea pigs. Total deposited dose (BB) after inhalation are 43 % and 32 % of the inhaled dose for rats and guinea pigs, respectively (12).

The inhaled aerosol is more efficiently deposited in the upper part of the respiratory tract in the rat than in the guinea pig. The high deposition is due to the nose anatomy of the rat which favours high deposition compared with guinea pigs (19). The fraction deposited in the nose and in the mouth will be swallowed. Of the lung burden, 15% is deposited centrally in the lung and 85% peripherally in the lung. In guinea pigs, 20 % is deposited in the central airways, and 80 % in the peripheral part of the lung. The deposition efficiency is equal when rats are exposed nose-only or via an endotracheal tube. When rats are intubated and exposed to EB aerosol, about 9 % of the inhaled dose is deposited in the lung. The ratio between central and peripheral deposition of the lung is the same as when rats are intubated.

The effect of repeated anaesthesia and intubation on rats have also been investigated. There are no significant effects on body weight or thymus weight in rats when handling them in these delivery systems (nose only or intubation) daily for up to four days, compared to cage control.

There are many factors affecting the deposition of particles in different part of the body in inhalation studies. Most important is probably the breathing pattern of the animal. It is always important to handle the animals carefully to avoid stress, that can eventually alter breathing patterns. Before starting the actual experiment it is often worth while to acclimatise the animals to the experiment situation.

When animals are anaesthetized before exposure it is important that the anaesthesia is performed in a controlled and defined manner so that the breathing pattern will not be affected. Another factor influencing the deposition is particle size. To investigate the correlation between particle size and deposition in the lung and the rest of the body, we use suspended PS of different particle sizes as a tracer aerosol. We found the following: The deposition of particles between 0.6 and 1.2 μm is 15 to 20 percent of the inhaled dose. About 10 percent of the inhaled dose is deposited in the lung when the particles are about 2 µm. One percent of the inhaled dose of particles that are 3.5 µm or larger (up to 10 µm) are deposited in the lung.

It is difficult to estimate the body burden of an inhaled dose. Due to high deviations in our studies we often found body burden values of more than 100 % of the inhaled dose . This could be caused by difficulties in estimating the inhaled dose. We believe, derived from our studies, that the body burden rises up to 100 % of the inhaled dose when the particle size range increases from 0.5 to 6 µm.
Lung burden in the studies above performed on rats correlates well to particle size. The relation between particle size and deposition could be estimated. The deposition probability for a polydisperse EB aerosol will be about 0.09 (that is; 9 % of the inhaled dose will be deposited within the lungs).

5.2.3. Comparison to other studies
The results from the deposition studies of inhaled EB in rats were compared to the deposition probability calculations in the study on inhaled monodisperse particles and with the two studies performed by Raabe et al. (16, 17). Using the droplet/particle size distribution from nebulised EB deposition study in conjunction with the deposition probability curve by Raabe et al. we found a calculated LB of 8.9% of the inhaled dose of EB aerosol. The probability of the aerosol being deposited in the total body (BB) was calculated to be 0.3 (30 % of the inhaled dose).The inhaled dose EB in the study described above was 460 µg/kg.

Deposition site	Actual deposition Dahlbäck, 1989[12]	According to Raabe, 1977, 1988[16, 17]	According to Dahlbäck, 1994[14]
Lung burden (µg/kg)	39 (8.5%)	41 (8.9%)	41 (8.9%)
Body burden (µg/kg)	200 (43%)	140 (30%)	460 (100%)

The calculated lung burden probability from three different studies (14, 16, 17) seems to correlate well to the actual deposition in one of our earlier studies (12).

6. *Effects of inhaled glucocorticosteroids*

Glucocorticosteroids (GCS) are often used in the treatment of inflammatory diseases in the human respiratory tract (e.g. asthma and rhinits). In the development of new GCS it is important to have animal models that can predict the effect in humans. Also we must know the kinetics of new GCS preparations. The route of administration should also mimic the clinical situation. We have compared the effect of budesonide when given by three different routes of administration in the Sephadex-induced lung oedema model in rats (20). The bioavailability of inhaled budesonide has also been studied in dogs.

6.1. *Effects of inhaled GCS on Sephadex-induced lung oedema in rats*

6.1.1. *Method.*

Rats were treated with budesonide on day one of the study. Two hours later Sephadex was given by intra tracheal (i.t.) instillation in order to induce lung oedema. The rats were then treated once daily for three consecutive days (day 2-4). On the fifth day the rats were sacrificed, lungs and thymus were taken out and weighed.

Budesonide was administered by i.t. instillation, nose-only exposure or aerosol via an endotracheal tube. The effect of budesonide was calculated from dose-response curves as an effective dose (ED) that decreases the lung oedema to 50 % (Lungs ED_{50}), and the side effect as a decrease in thymus weight of 25 % (Thymus ED_{25}), compared to control animals. The ratio Thymus ED_{25} /Lungs ED_{50} (T/L) is calculated to describe the "therapeutic profile" of budesonide.

6.1.2. *Results and discussion.*

Administration	Lung ED_{50} (μg/kg)	Thymus ED_{50} (μg/kg)	Thymus/Lung
I.T Instillation	43 (32, 37)	24 (20, 27)	0.55
Nose Only	4.1 (2.5, 6.1)	2.6 (2.0, 3.2)	0.63
Endotracheal tube	7.1 (4.2, 10)	6.3 (5.0, 7.6)	0.89

Budesonide is 6 times more potent when administered as an aerosol via endotracheal tube as compared with i.t. instillation of a budesonide suspension. This can be due to a more even distribution of budesonide in the lung reaching a larger surface area compared to the surface area covered by i. t. administration. After nose-only exposure budesonide is 10 times more potent as compared with i.t. instillation and almost 2 times more potent as compared with aerosol exposure via endotracheal tube. The higher potency can be due to a small contribution from budesonide deposited in the nose and GI tract after nose-only exposure. The T/L ratio is slightly higher after aerosol exposure via endotracheal tube as compared with the other two administration routes.

The study described above point out the importance of using a relevant route of administration and delivery system to be able to predict the effects of new inhaled GCS:s in humans.

6.2. *Bioavilability of inhaled GCS (budesonide).*

The kinetic profile of GCS after inhalation is an important factor that can affect the potency and the ratio between desired effects and side effects. Therefore it is important to know the time course of the uptake of GCS:s from the lung to the blood stream.

6.2.1. *Method.*

Beagle dogs were exposed to a dry powder aerosol of budesonide for 5 minutes via a nose mask. The inhaled dose was 40 μg/kg. To calculate bioavailability, inhaled budesonide was compared with an intravenous infusion of budesonide for 5 min, (20 μg/kg). Blood samples were taken before and up to 6 h after start of administration to determine budesonide in plasma.

6.2.2. Results and discussion.

The uptake of inhaled budesonide is rapid. The peak plasma concentration occurred immediately after end of inhalation (T_{max}= 6 min after start of exposure). The systemic bioavailability of inhaled budesonide is F= 60 ± 17 % of the inhaled dose. This means that when calculating systemic bioavailability a body burden probability 0.6 should be used. This is close to 100 % of body burden.

The deviations in plasma budesonide concentrations are small. Relative standard deviation of mean bioavailabilty was 28 % in this study. The small variation in plasma levels is probably due to the well defined way to administer the dose by inhalation and to the relatively similar breathing patterns of the dogs when inhaling.

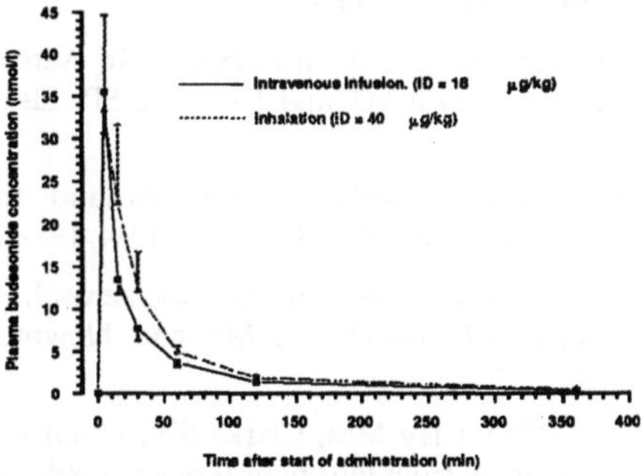

Figure 6. Plasma concentration of inhaled and intravenous administration of budesonide

7. *Summary*

When performing inhalation studies it is important to determine accurately the amount of inhaled drug deposited in the respiratory tract of the species used. Thus, when deciding on a specific pharmacological test system we select a generator and measure the particle size of the pharmaceutical formulations used as accurate as possible. The delivery systems, together with the different animal species used are then characterised, by doing deposition studies. This will indicate how the delivery system effects the deposition within the animal. Apparently, the deposited dose is important to know, but a major interest lies in where in the respiratory tract the drug is deposited as the site of deposition of a drug can play an important role in the way how a drug exerts its effect. It is therefore of considerable value to use new techniques to increase the knowledge of: 1) the actual dose deposited in the respiratory tract and 2) where in the lung the deposition has occurred. This can be done using either tracer aerosols or radiolabelling pharmaceutical dry powder formulations together with new non-invasive measuring techniques. When combining the site of specific action of a drug with selective deposition, it can be possible to increase the efficacy of a drug working locally in the lung and thereby reducing the systemic side effects.

8. References.

1. Dahlbäck M, Behaviour of nebulizing solutions and suspensions. J Aerosol Medicine, 1994, 7, S13-18.

2. Nerbrink O, Dahlbäck M, Hansson H-C, Why do medical nebulizers differ in their output and particle characteristics? J Aerosol Medicine, 1994, 7, 259-276.

3. Hinds WC, Aerosol technology, John Wiley & sons, 1982, 379-382.

4. Clay MM, Newman SP, Clarke SW, Factors Influencing the size distribution of aerosols from jet nebulisers, Thorax 1983, 38, 755-759.

5. Clay MM, Newman SP, Lennard-Jones T, Clarke SW, Assessment of jet nebulisers for lung aerosol therapy, Lancet, 1983, 592-593.

6. Smith EC, et al, Comparison of twenty-three nebulizer/compressor combinations for domicilary use. Eur Respir J, 1995, 8, 1214-1221.

7. Morén F, Aerosol dosage forms and formulations, In Aerosols in Medicine, Ed by Morén F, Dolovich MB, Newhouse MT and Newman SP, Elsevier Science Publishers, 1993, 321-350.

8. Newman SP, Therapeutic aerosols, In Aerosols and the Lung, Clinical and Experimental Aspects, Ed by Clarke SW and Pavia D, 1984, 197-224.

9. Hollworth GW, Particle size analysis of therapeutic aerosol, In Aerosols in Medicine, Ed by Morén F, Dolovich MB, Newhouse MT and Newman SP, Elsevier Science Publishers, 1993, 351-374.

10. Newman SP, Pellow PGD, Clay MM, Clarke SW, Droplet size and administration time of hypertonic saline aerosol from four brands of jet nebulizers. ICLASS, 1985, 1-10.

11. Cannon WC, Blanton EF, McDonald KE, The flow-past chamber: An improved nose-only exposure system for rodents. Am Ind Hyg Assoc, 1983, 44, 923-928.

12. Dahlbäck M, Eirefelt S, Karlberg I-K, Nerbrink O, Total deposition of Evan's blue in aerosol exposed rats and guinea pigs. J Aerosol Sci, 1989, 20, 1325-1327.

13. Dahlbäck M, Eirefelt S, Nerbrink O, Selective aerosol delivery to intubated rats; a comparison with nose only exposure, J Aerosol Sci, 1992, 23, 491-494.

14. Dahlbäck M, Eirefelt S, Total deposition of fluorescent monodisperse particles in rats. In inhaled particles IIV, Ann Occup Hyg, 1994, 38, Suppl 1, 127-134.

15. McMahon TA, Brain JD, LeMott S, Species differences in aerosol deposition. In inhaled particles IV, Edited by Walton WH, 1977, 23-33.

16. Raabe OG, Yeh H-C, Newton GJ, Phalen RF, Valasques DJ, Deposition of inhaled monodisperse aerosol in small rodents. In inhaled particles IV, Edited by Walton WH, 1977, 3-20.

17. Raabe OG, Al-Bayati MA, Teaque SV, Rasolt A, Regional deposition of inhaled monodisperse coarse and fine particles in small laboratory animals. Ann Occup Hyg, 1988, 32, Suppl 1, 53-63.

18. Schlesinger RB, Comparative deposition of inhaled aerosols in experimental animals and humans. A review. J Toxicol Environ Health, 1985, 15, 197-214.

19. Martonen TB, Yang Y, Simulation of aerosol deposition in extrathoracic and laryngeal passages of the laboratory rat. J Aerosol Sci, 1993, 24, 103-113.

20. Eirefelt S, Nerbrink O, Källström L, Axelsson B, Dahlbäck M, Anti-inflammatory effect of budesonide; differences between intratracheal instillation and aerosol administration in Sephadex-induced lung oedema in rats, J Aerosol Med, 1993, 6, 80.

Experimental Investigation of Deposition and Fate of Particles: Animal Models and Interspecies Differences

R. K. Wolff

Toxicology Research Laboratories
Lilly Research Laboratores, A Division of Eli Lilly and Company,
Greenfield, IN, USA 46140

Keywords

Aerosol; Deposition; Nasal Deposition, Tracheobronchial Deposition; Alveolar Deposition; Mucociliary Clearance, Alveolar Clearance; Susceptible Populations; Age; Cigarette Smoking; Chronic Obstructive Pulmonary Disease

Introduction

The purpose of this paper is to review the available information related to deposition and fate of particles in order to help interpret health effects data. As a part of this endeavor it is important to consider species differences in order to extrapolate from effects observed in animal toxicology studies to possible effects in humans exposed in the environment, usually at lower concentrations. A large body of information has been developed on deposition and fate of particles in humans. The vast majority of the data, however, have been obtained in normal healthy adults. There are limited data on susceptible populations or on factors that would modulate responses. The extent to which factors such as age, disease, exercise, and cigarette smoking might alter deposition and fate, and consequently health effects needs to be considered. Another issue dealt with in this review is the extent to which there are clues in the experimental data on deposition and fate of particles that might shed light on the apparently linear correlation of mortality with daily pollution levels even at relatively low ambient concentrations. It is also important to understand species difference so that toxicology studies in animals of pollutant effects can be used effectively to extrapolate to human health effects.

Available Deposition Data

There is a large body of deposition data that have been extensively reviewed previously (Schlesinger, 1985; 1989). Total deposition is defined as the amount of material deposited throughout the entire respiratory tract - from nose or mouth down to the alveolar region of the lung. Regional deposition is defined as the amount of material deposited in specific anatomical sub regions of the respiratory tract. Regional deposition of aerosols has been reviewed by a number of authors (Morrow, 1966; Lippmann, 1977, 1984; Schlesinger,

J. C. M. Marijnissen and L. Gradoń (eds.), Aerosol Inhalation: Recent Research Frontiers, 247–263.

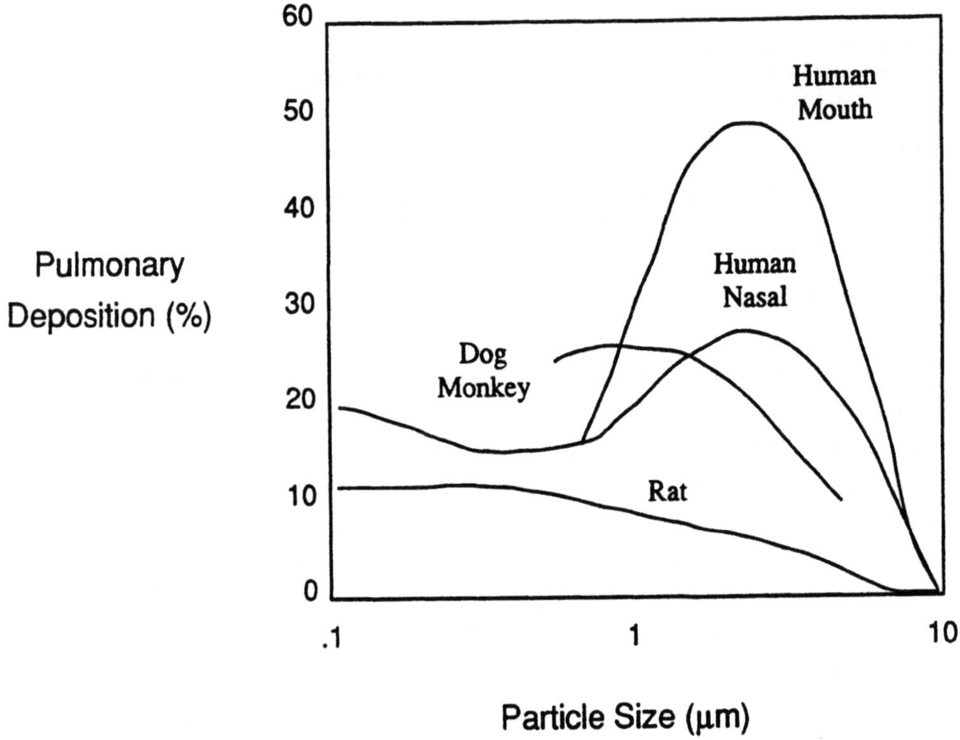

Figure 1. Pulmonary deposition in nose-breathing rats (Raabe et al., 1988; Mauderly et al., 1987), dogs (Schlesinger, 1985), and in nose-breathing and mouth breathing humans(Schlesinger, 1985). The deposition curves are hand-fit curves through the data noted above reviewed by Schlesinger (1985).

1985, 1989; Phalen, 1988) mainly from the perspective of assessing effects from inhaled environmental or occupational agents. Three major regions have been designated - the head airways (nose, mouth, naso- and oropharynx, larynx), the tracheobronchial region (conducting airways from the trachea to terminal bronchioles), and the pulmonary region (alveoli) (Morrow et al., 1966; Lippmann, 1977, 1984; Schlesinger, 1985, 1989; Phalen et al., 1988).

Aerosol deposition is an important parameter in determining delivered dose. It is affected by particle characteristics, respiratory tract anatomy, and breathing patterns. It is also important to distinguish between total respiratory tract deposition and regional deposition. Variations in toxicity may reflect regional deposition pattern as well as total deposition. Frequently, a toxic effect is determined by the local concentration of a toxic agent at the target tissue site. Therefore, comparative respiratory tract deposition between experimental animals and man is of central importance in interpreting toxicology studies.

In spite of differences in respiratory tract anatomy, ventilation and body size, there are some general qualitative similarities between deposition in experimental animals and in man (Figure 1). Pulmonary deposition is at a minimum at approximately 0.5 μm, and increases from this minimum to a relative maximum at about 1 μm for rats and 2-4 μm for man (Schum, 1980; Schlesinger, 1985).

There are, however, important qualitative differences in deposition among species (Figure 1). Pulmonary deposition in rats is considerably less than human deposition(Yeh, 1976; Schum et al., 1980; Schlesinger, 1985; Raabe, 1988; Jarabek, 1989). The deposition curve in rodents decreases dramatically for particles over 4 μm.(Raabe et al., 1988) . Deposition in larger species such as dogs and monkeys is similar to that in people (Figure 1). Therefore, toxicology studies in the larger species are sometimes of more direct applicability in making safety assessments for people, however they are more expensive to conduct.

Delivered Dose Determinations

In conducting adequate toxicology studies, it becomes important to quantitate the delivered dose so that it can be compared to projected or measured human doses. Measuring lung burden has been a very useful approach in the studies of low solubility particles of environmental and occupational concern (McClellan, 1986; Mauderly, 1990) such as diesel

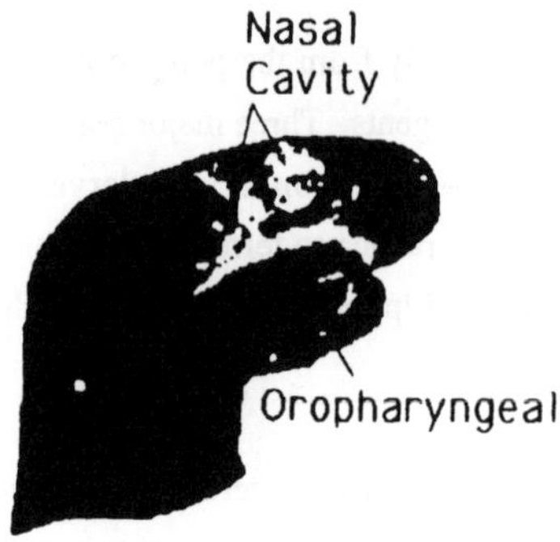

a) Head

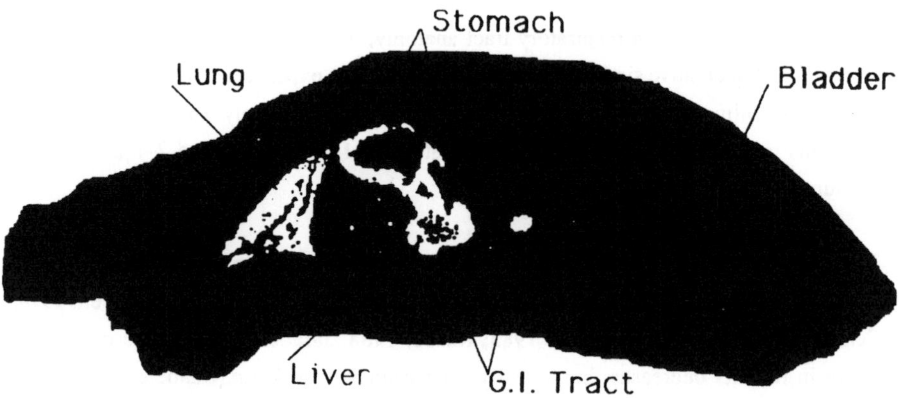

b) Torso

Figure 2. Whole body autoradiographs of a rat immediately following inhalation of a 3 μm MMAD ^{14}C radiolabeled aerosol. Panel a) shows the substantial deposition of aerosol in the nasal region. Panel b) shows the large amount of material in the gastrointestinal region cleared from the nose by mucociliary clearance. Deposition shown in the lung was only approximately 1/10 that seen in the nasal region as estimated by the amount in the head + gastrointestinal regions.

exhaust (Wolff, 1987; Strom, 1988; Muhle, 1990), carbon black (Muhle et al., 1990), toner (Muhle et al., 1990), titanium dioxide (Lee, 1985), and shale dusts (Mauderly, 1987).

Calculations of lung burdens can be made, but these values are subject to considerable uncertainty. Inferences of deposited dose from blood or urine levels can be made for agents that are soluble and absorbed well into the systemic circulation from the lung. However, this is an indirect determination and great care must be used with this approach. Measurements of material deposited in lung provide the most direct determination of deposited dose for soluble as well as insoluble agents. Lung tissue assays using analytical chemical methods can be developed in some cases. However, it is frequently difficult to achieve sufficient sensitivity to measure lung levels with confidence following inhalation exposure, which often results in deposition of relatively small amounts of material.

If a ^{14}C radiolabel is available for the compound then tissue counting and also whole-body autoradiography techniques (Ullberg, 1982; Sainio, 1991; Chay and Pohland, 1994) can provide considerable detail on deposition. This method has been used to delineate lung deposition and to ascertain if there were any other organs of accumulation following inhalation exposure to an anti-asthma candidate (Pohland, 1991). Rats were exposed nose-only for 1 hr to an aerosol of ^{14}C-LY170680 generated from a Lovelace nebulizer. The particle size distribution had a mass median aerodynamic diameter of 3 μm with a geometric standard deviation of 1.8. The autoradiographs (Fig. 2) clearly showed that the largest percent of the dose was delivered to the nose and cleared to the gastrointestinal tract. The second highest deposition site was in the lung, but the level was considerably lower than that seen in the nose; no significant accumulation was noted in any other organ for this compound. The disadvantage of the rat as a toxicology model is clearly seen, in that nasal deposition predominates in rats. The disproportionate gastrointestinal dose in rats compared to people may also complicate interpretations. These differences must be taken into account when extrapolating from data in rats to that in humans. Similar data have also been generated in the study of inhaled polycyclic hydrocarbons (Bond, 1986). If nasal toxicity is the predominant finding, there needs to be careful consideration of the possible relevance to man because of the higher nasal deposition in rats than humans.

The above description is a relatively simplistic approach to interspecies dosimetry. A more detailed approach has been used in the EPA reference concentration methodology (Jarabek et al., 1989). Oberdorster (1991) has also shown how such an interspecies dosimetry

a) b)

Figure 3. Deposition a) in an asthmatic subject breathing normally and b) in the same subject during an experimentally induced asthma attack produced by inhaling methacholine aerosols. The narrowed airways induced by bronchoconstriction in case b) results in much higher relative deposition of particles (indicated by the white areas) in the tracheobronchial region or central region than when breathing normally (Smaldone, 1994).

approach can provide useful quantitative insights when making extrapolations from rodent studies to possible human effects for agents such as diesel exhaust. These approaches become more useful with the aid of increased knowledge and data about the chemistry of the agents involved, interaction with lung cells and fluids, and improved comparative morphometry between laboratory animals and man. Miller (1978) has demonstrated that this can be a powerful approach when applied to reactive gases such as ozone. It is not yet possible to make such detailed predictions for inhaled particles but the information on morphometry, lung cell types, and airway fluid linings is available to aid future developments (Miller, 1993).

Deposition predictions in both man and animals require additional refinements (Swift, 1993). Most current deposition models assume tidal breathing of stable aerosols by normal healthy adults. There are limited data available related to effects of factors such as age, disease, exercise, or cigarette smoking. These are factors that might be of importance in considering sub populations that might be particularly at risk when exposed to high levels of pollutants.

Modifying factors

There are very few data on the effects of age on deposition. Phalen (1985) has provided valuable data on lung morphometry from lung casts of infants, children, and young adults to compare with those obtained previously from mature adults . Phalen (1988) has also used the lung cast data in a mathematical lung modeling approach to predict that tracheobronchial deposition of fine aerosols (0.1 to 2.5 μm) will be increased compared to that for normal healthy adults while pulmonary deposition will be decreased.

There are some data on the effects of respiratory disease on deposition. Newhouse (1978) and others (Thomson, 1974; Ramanna, 1975; Smaldone, 1994) have shown that chronic obstructive pulmonary disease, increases total deposition by increasing tracheobronchial deposition. Figure 3 shows a gamma camera scintigraphic demonstration of this phenomenon that occurs primarily because of airway narrowing in these disease states (Smaldone, 1994).

Very few studies have been conducted on the effects of exercise. Morgan et al (1984) showed that exercise increased tracheobronchial deposition. This effect is consistent with what would be predicted from the effects of increased flow rate (Heyder, 1975).

Cigarette smoking produces increased tracheobronchial deposition as a result of inducing respiratory disease and so the effect is similar to that shown above because the net effect is the same, airway narrowing. Deposition of cigarette smoke itself is quite interesting. Although the particle size of cigarette smoke is approximately 0.4 - 0.6 μm which according to standard predictions should be at a near minimum for deposition, the measured values are approximately 70-80%. Martonen (1992) and Phalen (1992) have both provided recent data that indicates that the enhanced tracheobronchial deposition is primarily because of the "cloud" effect of the extremely high particle number concentrations of cigarette smoke that results in the particles acting as an ensemble with a considerably higher effective particle size than the individual particles. Martonen (1992) has shown in a cast model, that 6 μm particles have an equivalent deposition pattern to that of cigarette smoke particles.

Taken in aggregate, most of the modifying factors would result in an increase in tracheobronchial deposition compared to that found in normal healthy adults. The magnitude of the increase is generally restricted to factors of the order of 2 or 3 and not orders of magnitude. There is some increase in dose in compromised subjects but not an overwhelming difference compared to normal subjects.

Fate of Particles

The fate of particles within the lung and body is also a very important determinant of health effects. If particles are relatively insoluble they may be retained in the lung for long periods (months to years) and so exhibit their effects over a long time period. If particles are more soluble, their residence time in the lung is likely to be short (minutes or hours) and their effects are likely to be either at or near the original deposition site or in other organs if they produce effects systemically after being transported via the blood to other organs. If the latter is the case, then deposition site and local concentration are likely to be the main determinants of respiratory tract toxicity. Clearance effects are usually described in terms of effects on upper airway or mucociliary clearance that takes place over hours and long-term or alveolar clearance, that takes place over weeks to years (Snipes, 1989).

Mucociliary Clearance

Mucociliary clearance is a lung defense mechanism by which inhaled and deposited particles, including toxic and infectious agents, are removed from the conducting airways of the lung. Beating cilia propel the overlying mucus, containing deposited particles, up the airways to the trachea and larynx and the mucus is swallowed (Newhouse, 1976; Wanner, 1977). Impairment of this function could act to produce accumulations of secretions in airways or to give toxic materials a longer residence in the airways. Abnormal elimination of mucus is associated with, and is perhaps a factor in, the pathogenesis of chronic obstructive pulmonary disease, especially chronic bronchitis (Hilding, 1965; Newhouse et al., 1976). It is important to determine the response of the mucociliary system to general community and occupational exposures.

A large range of studies have been conducted on the effects of pollutants on lung defense mechanisms. These have been reviewed by Wolff (1986) and Schlesinger (1989). Many pollutants such as sulfuric acid, sulfur dioxide, ozone, and nitrogen dioxide depress mucociliary clearance at high concentrations compared to those encountered in urban atmospheres. However, at lower concentrations there may be only subtle changes detected. Sulfuric acid is undoubtedly the most extensively studied pollutant. At high concentrations approaching 1000 $\mu g/m^3$ mucociliary clearance is depressed. However, at concentrations in the range of 100-300 $\mu g/m^3$, mucociliary clearance is actually increased somewhat, probably as an adaptive response to increased secretions. At concentrations below 100 μg/m3 changes in mucociliary clearance are difficult to detect. It has been observed, however, that bronchial walls are thickened and the number of secretory cells is increased. Thus, measurable effects on airways can be seen at exposure concentrations approaching those seen in episodes of high air pollution in U. S. urban centers.

These studies suggest that short-term pollutant exposures can produce subtle damage in normal healthy populations. The prime mechanism for observed changes appears to be an increase in secretions produced by reflex effects on upper airway receptors (Nadel, 1979), although direct effects on epithelia may also play a role (Lipton, 1980). Only at exposure concentrations much higher than ambient are sufficient increases in clearance or direct ciliary damage produced, resulting in depressed clearance. The results are consistent with a picture of modest increases in secretions produced at low levels of exposure that in the normal state are dealt with effectively by mucociliary clearance, not resulting in impaired clearance. However, in people already compromised by pre-existing lung disease it may be more difficult to deal with an additional insult. This is a difficult area to study and there are few experimental data. One of the major challenges to experimental study is that the deaths

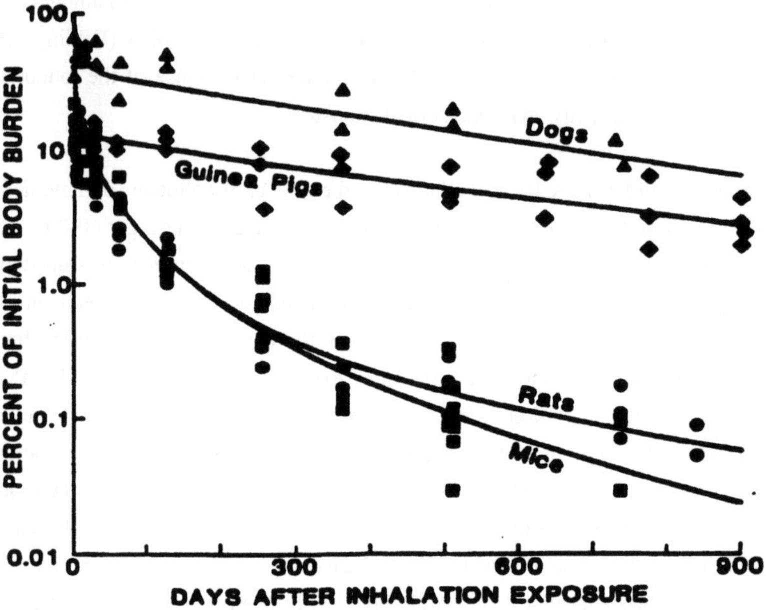

Figure 4. species comparison of long-term clearance. Rats and mice have much faster long-term clearance than large animals. The long-term clearance in humans is very similar to that seen in dogs.(Snipes, 1989)

and morbidity attributed to pollution are low probability events in the population. Subtle changes in respiratory impaired individuals may trigger a cascade of events, or act synergistically with respiratory infections or exposure to other agents, producing respiratory failure.

Long-term clearance

Studies of pollutant effects have been made on long-term clearance as well as mucociliary clearance. In general, the pollutant gases do not show marked effects on long term clearance. Small depressions in clearance have been observed for ozone and nitrogen dioxide and variable effects for sulfuric acid. The effects are generally less sensitive than those observed on mucociliary clearance. This observation is not surprising in the sense that the effects of pollutant gases, particularly the soluble gases, tend to occur more frequently in conducting airways and effects on long-term clearance are more generally produced by effects of the pollutants in the deep lung. For instance, exposure to insoluble particles produces impairments in long-term clearance at high concentrations in the range of several mg/m^3. This has been a common feature observed in rodent exposures to most insoluble particles, even those of low inherent toxicity such as titanium dioxide, and carbon black. This has been referred to as a "particle overload" response. Particle accumulation occurs in macrophages and is accompanied by clearance impairment and a chronic inflammatory response that often leads to lung tumor formation in rats (Hahn, 1979; Morrow, 1982, 1988). The tumor formation appears to be a rat specific phenomenon (Hext, 1994) while particle clearance impairment is seen in all rodents tested. Other, more cytotoxic, particles such as silica also produce clearance impairments but after considerably shorter exposures with much lower lung burdens of particles than for the low toxicity particles.

Long-term retention of insoluble particles is highly species dependent. As shown in Fig. 4, large animals and man have relatively slow clearance of insoluble particles while rodents have relatively fast clearance of insoluble particles (Snipes, 1989). This difference must be considered and factored into dosimetric comparisons. Oberdorster and Yu (1990) have made careful quantitative dosimetric adjustment to use rodent data from diesel exhaust exposures to predict possible responses in humans. Their findings suggest that particle clearance impairments might be produced at lower airborne concentrations in humans than in rats, if it is assumed that a similar accumulated lung burden results in similar lung injury in the two species. The reasons for the difference are the slower long-term clearance in humans and also the fact that exposures might take place over several years in humans whereas lifetime exposures in rats are restricted to approximately 2 years.

Susceptibility Factors

There have been few studies of the effects of susceptibility factors on long-term clearance of particles. There is some suggestion in rodent studies that age and the development of emphysema prolong long-term clearance (Hahn et al., 1979). Cigarette smoking has dramatic effects on particle clearance. Chronic cigarette smokers have both substantially slower mucociliary clearance (Camner, 1972; Goodman, 1978) as well as much slower long-term clearance of particles than normals (Cohen, 1979).

Air pollutant effects on long-term clearance are modest and would appear to be related mainly to chronic changes. As such, it seems unlikely that these types of changes would lead directly to increased mortality or morbidity in acute situations. However, changes in either mucociliary clearance or long-term clearance might signal effects that could produce increased susceptibility to respiratory tract infections. Such effects could occur simply by allowing greater residence time in the lung for pathogens and so a greater probability for toxic effects. Other possibilities are impaired alveolar macrophage function that might lead to reduced bactericidal effects. There are data that show altered macrophage function with exposure to some pollutants (Gardner, 1984). Pollutant exposure can also produce epithelial damage and increased permeability. Such effects could increase the possibility of viruses or bacteria multiplying in the airways or lung and colonizing the respiratory tract. There is extensive evidence from previous reviews that air pollutants can increase susceptibility to bacterial infections (Ehrlich, 1979; Gardner, 1979). Effects on immune function could also contribute to adverse effects. Increased susceptibility to infections is one mechanism by which pollutant effects could result in increased mortality and morbidity as a result of acute exposures at relatively low levels. The challenge is to determine what mechanisms are particularly critical to induction or exacerbation of respiratory tract infections and then to determine in appropriate models what the sensitivity is to air pollutants.

As mentioned previously, the question of sensitive sub populations is also an important issue with respect to susceptibility to infections. Possibly pollutant exposure and concomitant exposure to infectious agents is more likely to produce adverse effects in those with underlying respiratory disease or in people who are immunocompromised. Additional epidemiological data on such sub populations would be useful. Experimental approaches that could be envisaged might include the use of experimental animal models that have some of the features of the sensitive sub populations. For instance, the hairless mouse might be used to explore if reduced immune function predisposes to increased susceptibility. Also, it may prove fruitful to use transgenic or knockout gene models as they are developed to mimic human respiratory disease conditions.

Emerging Concepts of the Fate of Materials in the Lungs

More basic knowledge on the fate of particles in the lung will provide information on mechanisms of disease initiation and potential subtle effects of pollutants. There are a few examples that deserve mention. Geiser and colleagues (1995) have done extensive work in examining the fate of particles in conducting airways. They have shown that particles frequently are not carried along on the top of mucus like rafts, but rather can become submerged in the mucus. They have also shown that particle containing macrophages can be observed directly on the epithelium under overlying mucus. These observations indicate that airway responses to particles are more complex than previously thought and suggest mechanisms by which particles may interact with epithelial cells. Another area of study that has produced interesting results are current investigations into the fate of proteins introduced into the lung. Patton (1992) and others (Laube, 1993) have done extensive studies on lung delivery of proteins related to potential therapeutic use of proteins as inhaled pharmaceuticals. It has been found that when proteins are delivered to the alveolar region of the lung, absorption into the blood is quite high. The bioavailability of proteins deposited in lung relative to subcutaneous delivery has generally been in the 20 -50 % range, even for proteins as large as 20,000 daltons in molecular weight. The mechanisms that allow this efficient transfer of large molecules across the alveolar epithelium have been speculated to be dissolution of the proteins from the deposited particles into the alveolar surfactant layer followed by transcytosis occurring by vesicular uptake at the alveolar side of the Type I cells and transfer to the blood capillary side. It is also possible that there is some passive diffusion of the proteins as well as intracellular transfer. These studies do not directly shed light on pollutant effects but they do show the complex interactions that take place in the lung in processing inhaled particles and the current imperfection of our understanding of these mechanisms. As additional information becomes available on mechanisms involved in the fate of inhaled materials, particularly biological agents such as viruses and bacteria, it will become easier to understand pollutant influences on events such as susceptibility to infection.

Summary

Although there is a considerable body of data on deposition and fate of particles, there are substantial additional research needs to help define the effects of pollutants at ambient concentrations. Most of the available data are derived from studies on normal healthy adult humans or animals. There is a need to provide data relevant to susceptible sub populations, such as the aged, and those with underlying respiratory disease. There is also a need to understand much more fully the intricate mechanisms responsible for processing particles in the lung so that it can be determined if pollutant effects are perturbing these mechanisms and perhaps increasing susceptibility to infections. As with most other considerations of lung disease, cigarette smoking has profound effects on deposition and fate of particles and these must be accounted for before addressing effects of other pollutants.

References

Bond, J. A., Mitchell, C. E., Dutcher, J. S., Mauderly, J. L., Wolff, R. K. and McClellan, R. O. , Biological fate of inhaled organic compounds associated with particulate matter. Aerosols. S. D. Lee, T. Schreider, L. D. Grant and P. J. Verkerk. Chelsea, MI, Lewis: 579-592, (1986)

Camner, P. and Philipson, K. . "Tracheobronchial clearance in smoking discordant twins." Arch. Env. Health **25**: 60-63, (1972)

Chay, S. H., and Pohland, R. C. " Comparison of quantitative whole-body autoradiographic and tissue dissection techniques in the evaluation of the tissue distribution of [^{14}C]daptomycin in rats" J. Pharm. Sci. **83**: 1294-1299, (1994)

Cohen, D., Arai, S. and Brain, J. D. . "Smoking impairs long-term dust clearance from the lung." Science **204**: 514-517, (1979)

Ehrlich, R. . "Interaction between environmental pollutants and respiratory infections." Env. Health Perspec. **35**: 89-100, (1979)

Gardner, D. E. , Alteration in host-bacteria interaction by environmental chemicals. Assessing Toxic Effects of Environmental Pollutants. S. D. Lee and J. Bryan. Ann Arbor, Ann Arbor Publishers: 87-103, (1979)

Gardner, D. E. . "Alterations in in macrophage functions by environmental chemicals." Env. Health Perspect. **55**: 233-239, (1984)

Geiser, M., Cruz-Orive, L. M., Waber, U., Hof, V. I. and Gehr, P. . "Particle retention in airways estimated by unbiased syereology and photometry." J. Aerosol Med. **8**: 149-165, (1995)

Goodman, R. M., Yergin, B. M., Landa, J. F., Golinvaux, M. H. and Sackner, M. A. . "Relationship of smoking history and pulmonary function tests to tracheal mucous velocity in non-smokers, young smokers, ex-smokers, and patients with chronic bronchitis." Am. Rev. Resp. Dis. **117**: 205-214, (1978)

Hahn, F. F. and Hobbs, C. H. . "The effect of enzyme-induced pulmonary emphysemain Syrian hamsters on the deposition and long-term retention of inhaled particles." Arch. Env. Health **34**: 203-211, (1979)

Hext, P. M. . "Current perspectives on particulate induced pulmonary tumors." Human and Exper. Toxicol. **13**: 700-715, (1994)

Heyder, J., Armbruster, L., Gebhart, J., Grein, E. and Stahlhofen, W. . "Total Deposition of of aerosol particles in the human respiratory tract for nose and mouth breathing." J. Aer. Sci. **6**: 311-328, (1975)

Hilding, A. C. . "Mucociliary insufficiency and its possible relation to chronic bronchitis and emphysema." Med. Thorac. **22**: 329-345, (1965)

Jarabek, A. M., Menache, M. G., Overton, J., J. H., Dourson, M. L. and Miller, F. J. . "Inhalation reference dose: An application of interspecies dosimetry modelling for risk assessment of insoluble particles." Health Physics **57**(Suppl. 1): 177-183, (1989)

Laube, B. . "Preliminary study of the efficacy of insulin aerosol delivered by oral inhalation in diabetic patients." J. A. M. A. **269**: 2106-2109, (1993)

Lee, K. P., Trochimowicz, H. J. and Reinhardt, C. F. . "Pulmonary response of rats exposed to titanium dioxide (TiO_2) by inhalation for two years." Toxicol. Appl. Pharmacol. **79**: 179-192, (1985)

Lippmann, M. , Section A: Reactions to Environmental Agents, Chapter 14. Handbook of Physiology. D. H. K. Lee, The American Physiological Society: 213-232, (1977)

Lippmann, M. and Schlesinger, R. B. . "Interspecies comparison of particle deposition and mucociliary clearance in tracheobronchial airways." J. Toxicol. Environ. Health **13**: 441-469, (1984)

Lippmann, M., Schlesinger, R. B. and Leifauf, G. . "Effects of sulfuric acid aerosol inhalations." Am. J. Ind. Med. **1**: 375-381, (1980)

Martonen, T. B. . "Deposition patterns of cigarette smoke in human airways." Am. Ind. Hyg. Assoc. J. **53**: 6-18, (1992)

Mauderly, J. L., Bice, D. E., Carpenter, R. L., Hanson, R. L., Henderson, R. F., Jones, R. K., McClellan, R. O. and Wolff, R. K. , Effects of 4-week inhalation exposure of rats to oil shale dust and diesel exhaust, available from NTIS, (1987)

Mauderly, J. L., Cheng, Y.-S. and Snipes, M. B. . "Particle overload in toxicological studies: Friend or foe?" J. Aerosol Med. **3**: S169-187, (1990)

McClellan, R. O. . "Health effects of diesel exhaust: A case study in risk assessment." Am. Ind. Hyg. Assoc. J. **47**: 1-13, (1986)

Miller, F. J., Menzel, D. B. and Coffin, D. L. . "Similarity between man and laboratory animals in regional pulmonary deposition of ozone." Environ. Res. **17**: 84-101, (1978)

Miller, F. J., Mercer, R. R. and Crapo, J. D. . "Lower respiratory tract structure of laboratory animals and humans: Dosimetry implications." Aerosol Sci. Technol. **18**: 257-271, (1993)

Morgan, W. K. C., Ahmad, D., Chamberlain, M. J., Clague, R. W., Pearson, M. G. and Vinitski, S. "The effect of exercise on the deposition of an inhaled aerosol." Respir. Physiol.. **56**: 327-338, (1984)

Morrow, P. E. . "Possible mechanisms to explain dust overloading of the lungs." Fundam. Appl. Toxicol. **10**: 369-384, (1988)

Morrow, P. E. . "Dust overloading of the lungs: Update and appraisal." Toxicol. Appl. Pharmacol. **113**: 1-12, (1992)

Morrow, P. E. and et al. . "Task group on lung dynamics: Deposition and retention models for internal dosimetry of the human respiratory tract." Health Physics **12**: 173-208, (1966)

Muhle, H., Creutzenberg, O., Bellmann, B., Heinrich, U. and Mermelstein, R. . "Dust overloading of the lungs: Investigations of various materials, species differences, and irreversibility of effects." J. Aerosol Med. **3**(Suppl): S111-S128, (1990)

Nadel, J. . "Control of mucus secretion and ion transport in airways." Ann. Rev. Physiol. **41**: 361-381, (1979)

Newhouse, M. T. and Ruffin, R. E. . "Deposition and fate of aerosolized drugs." Chest **73**(Suppl): 936-943, (1978)
Newhouse, M. T., Sanchis, J. and Bienenstock, J. . "Lung defense mechanisms." N. Engl. J. Med. **295**: 9990, (1976)
Oberdorster, G. . "Lung dosimetry and extrapolation of results from animal inhalation studies to man." J. Aerosol Med. **4**: 335-348, (1991)
Oberdorster, G. and Yu, C. P. . "The carcinogenic potential of inhaled diesel exhaust: a particle effect?" J. Aerosol Sci. **21**((suppl 1)): s397-3407, (1990)
Patton, J. S. and Platz, R. M. . "Pulmonary delivery of peptides and proteins for systemic action." Advanced Drug Delivery Reviews **8**: 179-196, (1992)
Phalen, R. F. . "Particle deposition predictions for infants, children, and adolescents." Ann. Occup. Hyg. **32**(Suppl 1): 11-21, (1988)
Phalen, R. F., Oldham, M. J., Beaucage, C. B., Crocker, T. T. and Mortensen, J. D. . "Postnatal enlargement of human tacheobronchial airways and implications for particle deposition." Anat. Rec. **212**: 368-380, (1985)
Phalen, R. F., Oldham, M. j. and Mannix, R. C. . "Cigarette smoke deposition in hollow upper airway models representing adults and children." Am. Assoc. For Aerosol Research Annual Meeting: 347, (1992)
Phalen, R. F., Stuart, B. O. and Lioy, P. J. (1988). Rationale and implications of particle size-selective sampling. American Conference of Governmental Industrial Hygienists. Chelsea, MI, Lewis Publishers: 3-15.
Pohland, R. C., Beck, J. M., Carlson, K. H., Herman, D. R., Hoppes, J. L., Vavrek, M. T. and Wolff, R. K. . "Tissue distribution of a leukotriene antagonist, 14C-LY170680, following inhalation exposure in the rat." Toxicologist **11**: 95, (1991)
Raabe, O. G., Al-Bayati, M. A., Teague, S. V. and Rasolt, A. . "Regional deposition of inhaled monodisperse coarse and fine aerosol particles in small laboratory animals." Ann. Occup. Hyg. **32**: 53-63, (1988)
Ramanna, L., Tashkin, D. P., Taplin, G. V., Elam, D., Detels, R., Coulson, A. and Rokaw, S. N. . "Radioaerosol lung imaging in chronic obstructive lung disease." Chest **68**: 634-640, (1975)
Sainio, E. L. and Sainio, P. . "A quantitative method of measuring labeled compounds with whole-body autoradiography in tissue sections." J. Pharmacol Methods **26**: 53-59, (1991)
Schlesinger, R. B. . "Comparative deposition of inhaled aerosols in experimental animals and humans: A review." J. Toxicol. Environ. Health **15**: 197-214, (1985)
Schlesinger, R. B. , Deposition and clearance of inhaled particles. Concepts in Inhalation Toxicology. R. O. McClellan and R. F. Henderson. New York, Hemisphere: 163-192, (1989)
Schum, M. and Yeh, H.-C. . "Theoretical evaluation of aerosol deposition in anatomical models of mammalian lung airways." Bull. Math. Biol. **42**: 1-15, (1980)
Smaldone, G. C. . "Deposition patterns of nebulized drugs: Is the pattern important?" J. Aerosol Med. **7**(Suppl 1): S25-S32, (1994)
Snipes, M. B. . "Long-term retention and clearance of particles inhaled by mammalian species." Crit. Rev. Toxicol. **20**: 175-211, (1989)
Snipes, M. B. , Species comparisons for pulmonary retention of inhaled particles. Concepts in Inhalation Toxicology. R. O. McClellan, and Henderson, R. F. New York, Hemisphere: 193-227, (1989)
Strom, K. A., Chan, T. L. and Johnson, J. T. . "Pulmonary retention of inhaled submicron particles in rats: Diesel exhaust exposures and lung retention model." Ann. Occup. Hyg. **32**: 645-658, (1988)
Swift, D. L. . "Dose distribution of pharmaceutical aerosols." Aerosol Sci. Technol. **18**: 272-278, (1993)
Thomson, M. L. and Pavia, D. . "Particle penetration and clearance in the human lung." Arch. Env. health **29**: 214-219, (1974)

Ullberg, S., Larsson, B. and Tjalve, H. , Autoradiography. Biological Applications of Radiotracers. H. J. Glenn. Boca Raton, FL, CRC Press: 55-108, (1982)
Wanner, A. . "Clinical aspects of mucociliary transport." Am. Rev. Respir. Dis. **116**: 73, (1977)
Wolff, R. K. . "Effects of airborne pollutants on mucociliary clearance." Environ. Health Perspec. **66**: 222-237, (1986)
Wolff, R. K., Henderson, R. F., Snipes, M. B., Griffith, W. C., Mauderly, J. L., Cuddihy, R. G. and McClellan, R. O. . "Alterations in particle accumulation and clearance in lungs of rats chronically exposed to diesel exhaust." Fundam. Appl. Toxicol. **9**: 154-166, (1987)
Yeh, H.-C., Phalen, R. F. and Raabe, O. G. . "Factors influencing the deposition of inhaled particles." Environ. Health Perspect. **15**: 147-156, (1976)

Pollen Grains, their Role in Allergy

Gabriel Peltre*

Keywords

Allergy, Atopy, Allergens, Immune Response, Pollen grains, Environment, Pollution.

Introduction

Pollen allergens are mainly water soluble proteins representing up to 1% of the pollen dry mass. By themselves they are able to induce allergenic sensitization (allergenicity) in experimental animals as well as symptoms in sensitized individuals.

Pollen grains are the most common carriers of their allergens on the patient's exposed mucosa. They are protecting very efficiently the fragile allergens during there long journey from their emiting flower to their animal targets. Clean pollen grains passively exposed to different urban sites have been shown to be excellent pollution markers. They carry on their surface particles representative of their exposure together with some altered allergens that are still able to induce an experimental allergic sensitization. Pollen grains and smaller particles carrying allergens present in the atmosphere have a facilitated penetration in allergic individual's upper airways by an actively suppressed tracheal ciliary motion.

Furthermore experimental animal exposure by allergen aerosols, or intranasely, have been shown to be very efficiently allergy inducing immunization routes as compared to conventional injections which usualy require the use of adjuvants.

Allergy - Atopy

Allergy is a hypersensitivity state which consists in an exagerated immune reaction to some given antigens named allergens. This immune response is mainly characterized by the production of specific antibodies to allergens belonging to the IgE class. These immunoglobulins are able to bind receptors on basophils from the blood or on tissular mast cells. When allergens are in contact with the IgE bound to these cells they induce very rapidly a release of mediators accumulated in granules present in these cells such as the well known histamine. This small molecule acts on the surrounding blood vessels by inducing an abnormal permeability for the plasma they contain. This provokes local swelling associated with itching, pain, nose obstruction, eczema ... Among the severe symptoms are also vomiting, diarrhea for food allergens, asthma for inhaled allergens, anaphylatic chocs that may cause death when a great amount of allergen is introduced suddenly in the organism like a drug intake or an insect sting.

Atopy is a state of a human or an animal who is genetically "programmed" to be allergic to at least one allergen source. Although great progress has been made

*Immuno Allergy, Pasteur Institute, 28, rue du Docteur Roux F. 75015 Paris, France.

J. C. M. Marijnissen and L. Gradoń (eds.), Aerosol Inhalation: Recent Research Frontiers, 265–272.

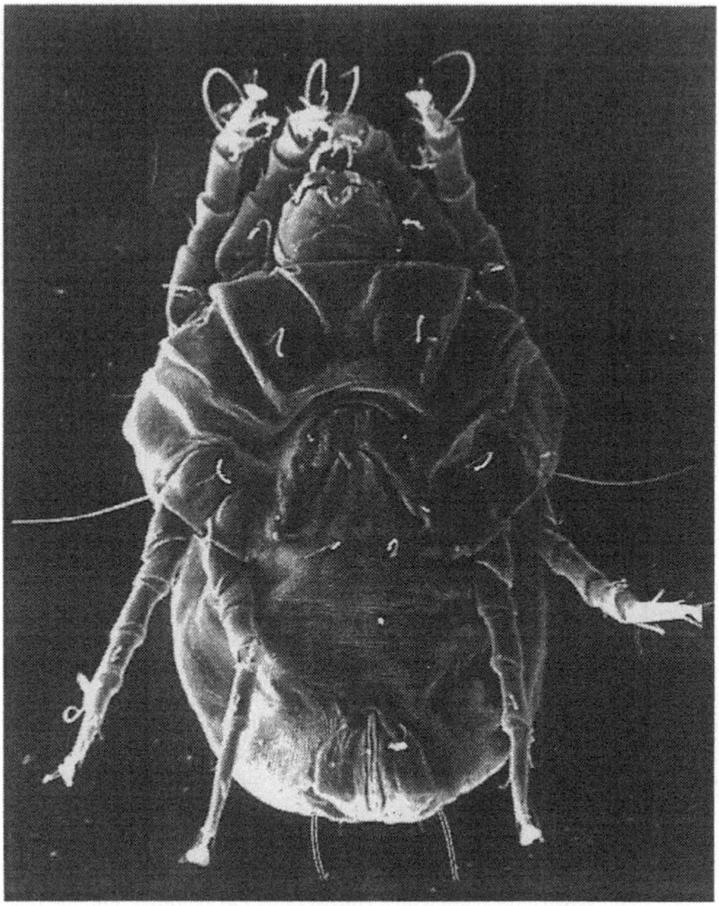

Dermatophagoïdes farinæ
(length 350µm)
from Guérin, B., Allerbio

in this field of research the genes coding for allergy are not yet found ! Some regions of the genome are obviously playing a role in this disease such as the HLA region on chromosome 6 responsible for the presentation of allergen fragments in the immune system. Besides the immune system the inflammatory and the central nervous systems are interacting to regulate the induction and the evolution of allergy.

Allergens

They are coming from very diverse sources : plants contain allergens in seeds (wheat flour, ...) in roots (carrots, ...) in fruits (apples, strawberries, ...) in leaves and pollen (trees, weeds or grasses ...) Animals are also producing allergens : housed dust mites, domestic animals such as cats or dogs, insects ... Allergens are mainly water soluble proteins . They are usualy quickly released when a pollen grain is in contact with a mucosa.

Approximately 10 groups of allergens, well characterized, are known from grass pollen [8]. Three major allergen groups are described from tree pollen, mainly from birch. In some weeds like Ambrosia 6 or 7 allergens are known. With the help of molecular genetic techniques more allergens are better characterized [6]. Besides their mass, their charge or isoelectric point, their sequence in amino acids can be determined. The major allergens are those which are more frequently recognized as allergens by patients. They usually form groups of isoallergens differing from each other by minor modifications in sequence and thus in physico-chemical properties. They can be found in many different but related plants. A molecular taxonomy based on the immune cross-reaction between allergens has been proposed with success [10].

Roles of pollen grains in allergy

Pollen grains are carrying allergens but also air pollutants that are bound to the pollen.

Allergen vectors

Approximately 10% of the dry mass of a grass pollen are water soluble proteins. One can estimate that an average of 10% of these proteins can be allergens. A maximum of 1000 pollen grains per m^3 of air can be observed during the pollen season, which means 30 μg of pollen per m^3 of air (Cour P. in [9]). D. Marsh estimates that a normal adult may inhale 30 to 300 μg of pollen during a 6 to 8 week pollen season that is less than 3 μg of allergens. This is far less than a usual immunization, such as a vaccination. This is also far less than what we use to induce, by subcutaneous injection with adjuvants, an allergic response to soluble pollen allergens in a mouse. These observations suggest that the airborne pollen is a very efficient allergy inducing agent. May be the water insoluble pollinic structure acts as a potent adjuvant in the animal mucosa and orients the immune response of the host towards the production of specific IgE and other immunoglobulins [11].

We know that due to their size (40 μm) the grass pollen grains are unable to enter deeply into the upper airways. They are stopped in the nose [3,5]. They may then release their allergens as these are trapped in the mucus layer that covers the microvillous surface of the inner nose epithelium. The nose barrier can be bypassed if, as the congested nose of an allergic patient offers too much resistance to the inhaled air, the respiration is performed by the open mouth. Pollen grains can then reach more deeply into the pulmonary airways. Experiments on allergic

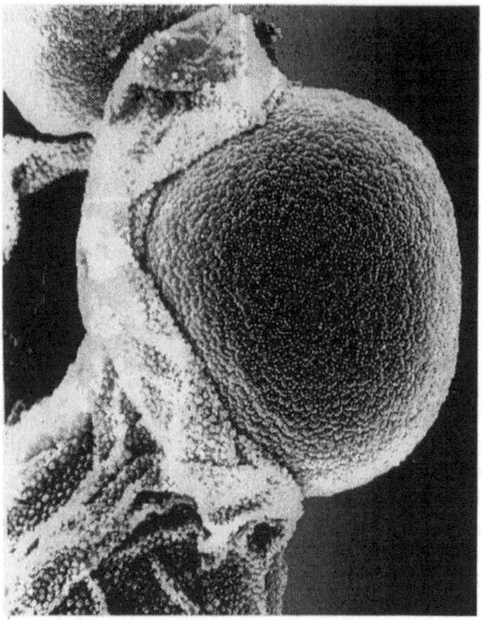

Scanning electron microscopy of grass pollen grain from *Dactylis glomerata L.*, on the anther tapetum.
Note the structure similarity between the pollen and the tapetum surface (pollen diameter : 40 μm)
Photo : M.T.Cerceau-Larrival, MNHN, Paris

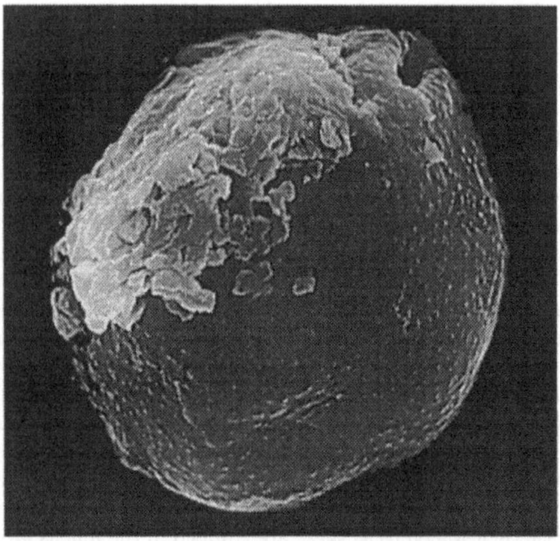

Scanning electron microscopy of a *Dactylis glomerata L.* pollen grain exposed for two days to urban pollution.
(from Bocquel, C., and Cerceau-Larrival, M.T., MNHN, Paris)

animals have shown that when a pollen grain reaches the tracheal epithelium it inhibits the beating microvilli which are the main physical mechanism to clean the pulmonary air [2].

It has been shown that besides intact pollen grains carrying allergens smaller particles (0,46 to 1,4 μm) from the plant flowers are also the support of airborne allergens [4,7]. Pollen grains emitted from their flowers may fall down frequently and be reemitted in the air, submitted to rain falls and eventually may burst before reaching the animal mucosa. Knox (in [6], p.31-38) has shown that starch granules from disrupted grass pollen grains are carrying allergens. These particles are able to form aerosols and persist in the air far longer than pollen grains.

Besides water soluble pollen allergens we have observed allergens that can be solubilized only by using detergents and even organic solvants. These allergens could be released in the mucosa by the local surfactants and/or an appropriate enzymatic activity coming from the animal or from the pollen grain itself. All these allergens entering into the respiratory tract may cause asthma, which is a very severe disease and is responsible for more than 2000 deaths per year in France.

Pollen grains are biomarkers of the environment

Before reaching an animal mucosa a pollen grain, or its fragments, can collect on their surface chemicals or particles representative of their origin or of the places where they have passed. We have exposed in the city of Mulhouse, in the east of France, grass pollen grains that were harvested in non polluted country side. We have chosen different exposure sites, in a residential area, in a pedestrian zone in the center of the city, a very busy car trafic zone and an industrial neighborhood. After 2 days of exposure the pollen grains were collected. By scanning electron microscopy, the pollen surfaces were observed. Their alteration was quite specific of the exposure site so that we really can say that pollen grains are good biomarkers of their environment.

By extracting the pollen allergen contents, we observed also that each exposure site was affecting rather specifically the allergen repertoire. This suggests that the air quality can act on the allergen content of the pollen grain, which also becomes the carrier of chemicals and particles. These latter components may act as adjuvants for an allergenic sensitization. We have exposed clean pollen grains to different model exposures in the laboratory such as to cigarette smoke, car exhaust pipe gas, heat, UV irradiation or ozone. We were very surprised to observe how the pollen grain was a good protection for the allergens they contain [12].

Only a temperature above 60 °C for more than one hour, a prolonged UV irradiation and ozone exposure were able to modify noticeably the amount and the nature of the allergens. We conclude that an outdoor exposure which is a complex exposure can be very dramatic for the pollen structure and for its role on animal mucosa.

Finally, the environment can act also on the plant which is producing the allergenic pollen. The best example is given by the birch tree. Jilek et al. (in [6] p.39-45) have found in the pollen of a birch tree exposed to the polluted air of Vienna in Austria 10 to 20 times more Bet v 1, the major birch pollen allergen, than in the same weight of a pollen coming from a birch tree enjoying the pure air of a distant country side. The Bet v 1 molecule is most likely a stress protein of the plant, produced to protect the plant from its agressive environment !

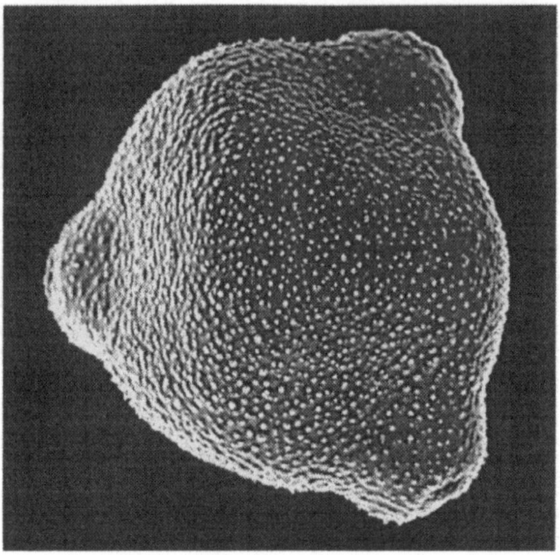

Scanning electron microscopy of
pollen grain from *Betula verrucosa E.*
(pollen diameter : 24 μm)
Photo : M.T.Cerceau-Larrival, MNHN, Paris

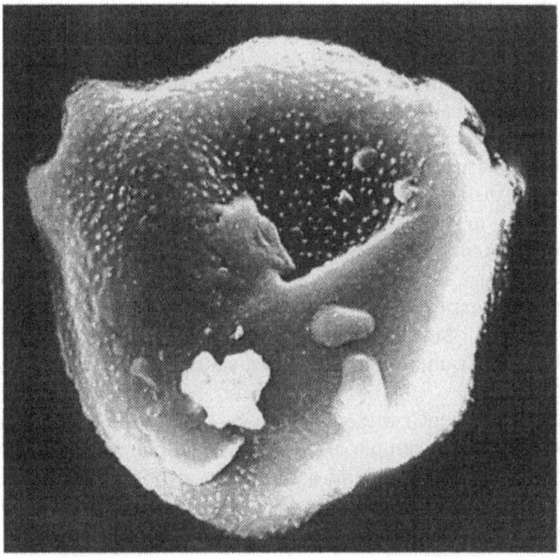

Scanning electron microscopy of
pollen grain from *Betula verrucosa E.* exposed for two
days to urban pollution.
(from Bocquel, C., and Cerceau-Larrival, M.T., MNHN, Paris)

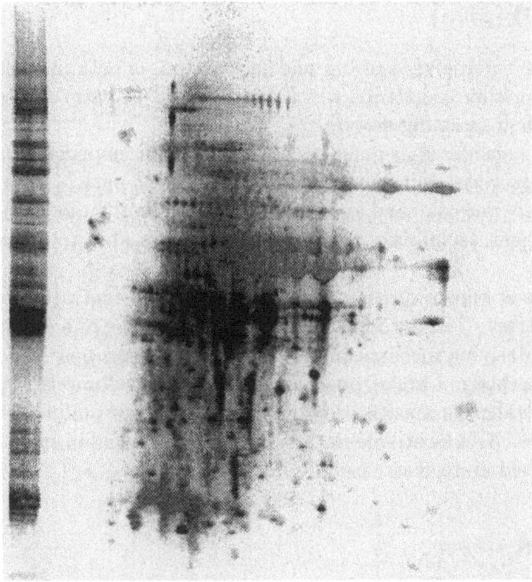

A grass pollen extract of *Dactylis glomerata L.* analysed by two-dimensional electrophoresis.
(from V.Brodard & al., 1993)

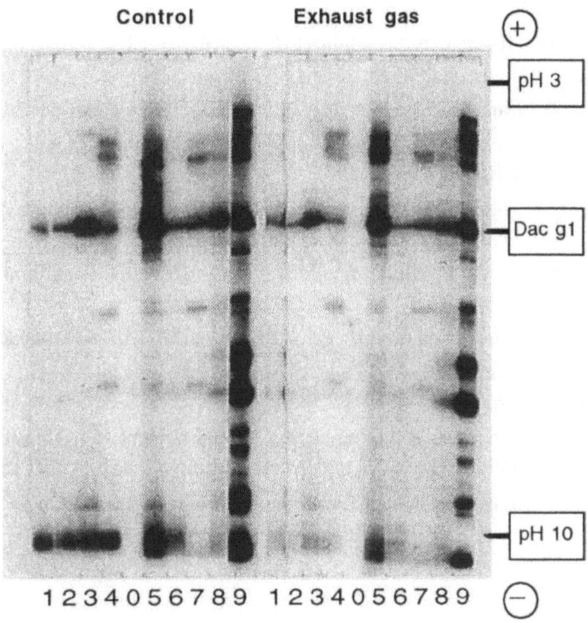

Dactylis glomerata allergens recognized by the IgE antibodies from 9 patient sera(1 to 9) analysed by isoelectric focusing in a 3 to 10 pH range followed by immunoprints.

On the left lanes are the allergens extracted from clean pollen. On the right lanes are the allergens extracted from pollen exposed to car exhaust pipe gas for 1h at 10°C.

Conclusion

Pollen grains by themselves are big and heavy components in biological aerosols. However, their smaller fragments, which are also able to carry allergens, can travel far away from their emitting flowers.

They can be considered as good biomarkers of their environment as they collect chemicals and particles representative of their travel path. In contact with an animal or human mucosa, and especially in the upper airway, they act as very potent immunogens as minutes amounts of them are able to induce specific IgE response.

Far too little is known on the mechanism of their penetration in the nose and the respiratory tract. The aerosol science will certainly be of a great help in understanding the underlying mechanisms, a prerequisite to propose efficient protection or prevention methods. I am strongly convinced that a domestic exposure plays a major role in the allergic sensitisation by aerosols ,even of pollinic origin ,in our developed countries. All physico-chemical parameters involved in such an environment should be analyzed and taken carefully into account.

References

[1] Marsh, D.G., Blumenthal, M., "Genetic and Environmental Factors in Clinical Allergy", University of Minnesota Press, Minneapolis, USA (1990).

[2] Chevance, L.G., Acta Otolaryng. 72 (1971) 121.

[3] Wilson, A.F., Novey, H.S., Berke, M.D., Surprenant, E.L., New Engl. J. Med. 288 (1973)1056.

[4] Busse, W.W., Reed, C.E., Hoehne, J.H., J. Allergy Clin. Immunol. 50 (1972) 289.

[5] Michel, F.B., Marty, J.P., Quet, L., Cour, P., An. Rev. Resp. Dis. 115 (1977).

[6] Kraft, D., Sehon, A., "Molecular Biology and Immunology of Allergens", CRC Press, Boca Raton, USA (1993).

[7] Takahashi, Y., Sasaki, K., Nakamura, S., Miki-Hirosige, H., Nitta, H., Grana. 34 (1995) 45.

[8] Brodard, V., David, B., Görg, A., Peltre, G., Int. Arch. Allergy Immunol. 102 (1993) 72.

[9] Guérin, B., "Pollen et Allergies", Ed. Allerbio, 55270 Varennes, France (1993).

[10] Peltre, G., Cerceau-Larrival, M.T., Hideux, M., Abadie, M., David, B., Grana. 26 (1987) 158.

[11] Batard, T., Basuyaux, B., Lambin, P., Brémard-Oury, C., Hamilton, R.G., David, B., Peltre, G., Int. Arch. Allergy Immunol. 100 (1993) 68.

[12] Peltre, G.,Derouet, L.,Cerceau-Larrival, M.T., Grana. 30 (1991) 59.

THE ATMOSPHERE AT HIGH ALTITUDES: ANOTHER APPROACH

REIN A. ROOS

La Pichotière,
61340 Nocé, France

KEYWORDS

Pulmonary, altitude, atmosphere, particles, terpenes, electricity, photo-ionisation, radioactivity, ozone, health

INTRODUCTION

When we consider the effects on health of high altitudes as observed by various authorities we are generally confronted with statements like:

The climate at high altitudes has a remarkably stimulating effect on the sick. It is very good for certain illnesses such as asthma, etc.

Or more precise comments such as:

1. the climate at low altitudes (less then 400m) is generally stable as regards temperature and barometric pressure, and the medium or high level of humidity makes it sedative. 2. The climat at medium altitudes (500-800 m) is good for numerous illnesses, is tonic and well tolerated. 3. The climate at high altitudes (over 1000 m) is tonic and even exciting; it us beneficial to convalescents and people with poor health, provided they have enough physical reserves to support them in this situation. The absence of dust at an altitude of over 1500 m makes the climate there suitable for certain types of asthma, bronchitis and other allergic illnesses .

as well as remarks like:

If we have the conviction that we respirate better in the country side, in the mountains, at the sea side or near a waterfall, then this is due to the richness of the air and the negative ions. (Robert, 1990)

Although information of this kind can frequently be found in numerous sources, it does not necessarily mean that it can be taken at face value. Perhaps it was correct at a certain time, or perhaps it is a compilation of observations not well investigated. Another general problem is that science and its results are used and manipulated in our enlightened epoque by numerous people - lobbyists, pressure

J. C. M. Marijnissen and L. Gradoń (eds.), Aerosol Inhalation: Recent Research Frontiers, 273–286.

groups and politicians etc. Open your paper or look at the TV and you will see that the metaphysics of the Middle Ages is still the order of the day.

It might be useful, therefore, to go back to the position taken by René Descartes (1596-1650):

...de ne recevoir aucune chose pour vraye que je ne la connusse evidemment estre tell: c'est à dire, d'eviter soigneusement la Precipation, et la Prevention; et de ne comprendre rien de plus en mes jugemens, que ce qui se presenteroit si clairement et si distinctement a mon esprit, que je n'eusse aucune occasion de le mettre en doute..... (Discours sur la Méthode).

A point of view concerning altitude and much more in keeping with the method of Descartes is the following:

There is a general belief that people living in the mountains become very old. Three regions, the Hunza or Karakoram, the Abkhazia in the Caucasus and Vilcomba in the Andes, are considered as places where people live longest. This is completely wrong. This legend is not invented by the people who live there but by us Westerners, who like to link, incorrectly, longevity and altitude. (Allan, 1994)

WHAT DO WE REALLY KNOW ABOUT THE ATMOSPHERE AT HIGH ALTITUDES

When we go up in the sky, for example in a balloon, the air becomes less dense, the temperature decreases, and the light of the sun increases in brightness. What goes almost unnoticed is that after some time we leave a region with a real but invisible border, namely the

EXCHANGE LAYER

and we enter the

FREE ATMOSPHERE

These are the two constituents of the atmosphere near the earth's surface. Let's take a closer look at both of them.

EXCHANGE LAYER

Height

The exchange layer surrounds practically the whole earth and its height varies from almost zero to about 3 kilometers. It acts as a buffer between the earth and the free atmosphere and in it we find most of the emissions produced by nature and mankind. The upper boundary of the exchange layer is usually easily visible as the level of fine-weather cumulus clouds or the top of the haze. (Chalmers, 1967)

Behaviour

Why does the exchange layer not spread out gradually in the free atmosphere, and present itself as a real border? At night no clear-cut exchange layer exists. What is present is more often a low level invisible cloud formed by the emissions from the earth: vapor from water and Volatile Organic Compounds (VOC's), waxy particles from plant surfaces, organic material in sea spray, and emissions caused by human activity. Do not underestimate the magnitude of the emission of plant hydrocarbons like VOC vapors. According to Hinds (1982), 75 000 000 to 200 000 000 metric tons of it are emitted into the atmosphere per year, compared to 10 000 000 to 90 000 000 metric tons caused by direct man-made emissions.

Because the earth is charged negatively and so has an excess of electrons, almost everything emitted from the earth carries with it a negative charge, helping it immediately to gain altitude, once airborne, due to mutual electrostatic repulsion. This means that these small particles will remain aloft for quite a time under night-time conditions. However the cloud stays at low levels due to the opposing temperature gradient generally present.

At dawn everything changes: the capping temperature gradient leaves, and makes it possible for emitted particles to drift upwards into the atmosphere. These particles are helped by the positively charged electrosphere that surrounds the earth at an altitude of about 50 km and the thermal buoyancy effects caused by the increased heating of the earth's surface from the sun's radition. The night emission will therefore leave the earth's surface some time after sunrise.

Studies carried out at the Atmospheric Aerosol Observatory (AAO) at Nocé, (61) in cooperation with the Institut Pasteur on this morning emission showed that the behaviour of this emission was strongly affected by its composition. The presence of pollen often inhibited the release of fine biological particles like VOC's. The slowly falling particles of pollen act as if they were fibers in a kind of invisible filter for the fast upwards-moving morning-emission particles. This is a very favourable condition, because the more negatively charged particles they intercept, the more buoyancy they obtain from both the earth and the electrosphere. A part of the effect is schematized in Figure 1

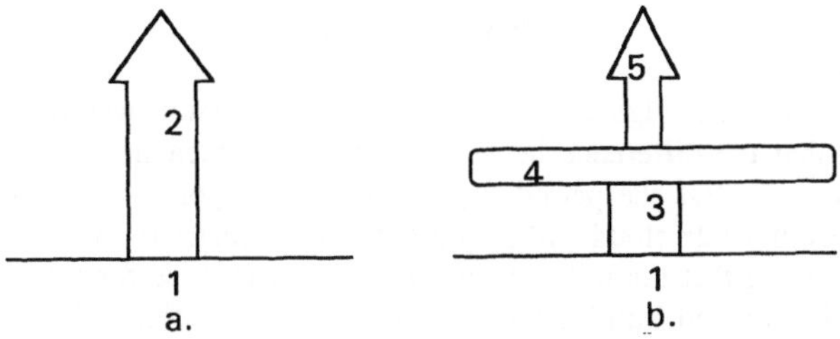

Figure 1. Mixing patterns in the exchange layer.
a. normal emission from active vegetation. 1. The earth's surface. 2. Emission composed of water vapor and VOC's. **b.** Emission when particles of pollen are present. 1. The earth's surface. 3. Emission composed of water vapor, VOC's, and pollen. 4. Filter layer composed of pollen and intercepting VOC's. 5. Water vapor.

These observations of non-uniform mixing clearly contrast with Chalmers' remarks: "Within the exchange region, there are small-scale convection currents which carry the surface air, together with the ions, nuclei, etc., which it contains, up to the top of the layer, and ensure a *thorough* mixing".

Daily variations in the exchange layer

One of the ways to measure the activity of the exchange layer is to observe the vapor pressure. When enough water vapor is present near the observation point, as is the case over the sea, an oscillation with one period per day approximately following the daily temperature curve, will be observed.

Over land this single oscillation profile will become an oscillation with two periods per day, one due to the effect already discussed of emission at night and one due to the fact that the vapor pressure early in the morning largely follows the temperature. With rapid heating after sunrise, evaporation also rises, but, simultaneously, convective air flow sets in with vertical diffusion and a consequent exchange between the vapor-poor layers at high altitudes with those rich in vapor near the ground.

This soon reduces the water vapor content near the ground, since that supplied by evaporation cannot compensate for the rapidly rising convection. Evaporation begins to predominate only after convection has reached its peak, i.e., the vapor pressure rises toward the evening until saturation recurs at the lowest layers when the temperature drops and condensation ensures further reduction (dew, rime, mist).

Figure 2 gives the daily (diurnal) variations of the vapor pressure for the month of July at Potsdam, Germany. The dotted line indicates the vapor pressure curve as it would appear in the absence of vertical exchange by convection, and indicates the pattern found over the oceans. (Israël 1973.)

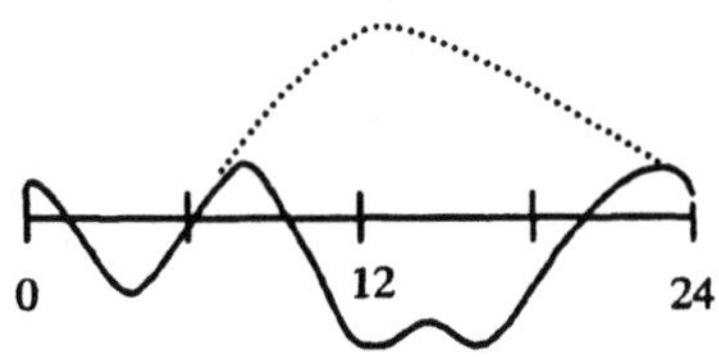

Figure 2. See text.

The solid line in Figure 2 represents the vapor pressure near the earth's surface. This means that the difference between it and the broken line in the same figure represents more or less the periodicity of the water vapor emission towards higher altitudes. This invisible cloud will only reach limited height for a number of reasons, one of them being that the fall in temperature will cause ice formation, intercepting the heat of the sun and limiting the buoyancy of the emission, adding also to cloud formation. This process reduces significantly the amount of water vapor at high altitudes. Another reason why this cloud will only reach in limited height is because its top will be submitted to the very energetic part of the sun's light that produces photo-ionization and will be forced back to the earth, as we will see later on.

These observations make clear that the process of thorough mixing in the exchange layer is considerable. The reparticipation of the different constituents will in principle be inhomogeneous, so that complete mixing is something hypothetical.

Because of the fundamental differences in the constituents of the exchange layer, it would seem logical to indicate the part of the layer close to the earth as the inferior EXCHANGE LAYER (INF), and the one in contact with the free atmosphere as the superior EXCHANGE LAYER (SUP). From the literature it is known that the free atmosphere is also not homogeneous, so for this reason we propose that the part that touches the *exchange layer (SUP)* be called the inferior FREE ATMOSPHERE (INF)

FREE ATMOSPHERE (INF)

Particle behavior

Here the sun possesses an impressively large spectrum containing high energetic rays such as the ultraviolet "C". The energy of these rays is such that it can remove one or more electrons from most of the elements and their compounds present in the atmosphere.

This means that the initially negatively charged particles, due to the attachment of electrons supplied by the earth, will lose them and end up in general as neutral or often positively charged. They will then take a return ticket and drift back to the negatively charged earth, helped by gravitation. It should be clear, then, that the height of this layer caused by the emission of particles emitted from the earth's surface is *limited* by this ionisation effect from the sun.

Transparency

Clearly the transparency of the layer plays an important role in the effect solar rays have on it. In the case of a good transparency, a large part of the top of the layer will be affected by the sunlight and start to drift downwards; this can be seen on sunny days in the afternoon as a bluish layer because of the polymerisation by the sun of organic matter, especially VOC vapors.

Things are quite different when the top of the layer consists of ice crystals that reflect the light of the sun or when the layer is formed by highly light-absorbing particles such as volcano dust or dust generated by deserts. In both cases, sunlight can penetrate only to a very limited depth in the layer, discharging thus a small quantity of particles and leaving the cloud to continue its journey for a long time due to the repulsive force that exists between it and the earth's surface.

Particle treatment

Certain particles entering this part of the free atmosphere will not only modify their electrical charge content and have their shape changed by coagulation and polymerisation, but will also have completely modified their behaviour through bond rearrangements and oxidation. The *VOC's* especially are very sensitive to this process.

FREE ATMOSPHERE (INF.)

"The upper boundary of the exchange layer ... is shown very markedly by electrical measurements" (Chalmers, 1967). This effect is clearly illustrated in Figure 3, where the results are shown of a number of flights carried out with an aircraft equipped with a measuring system for coarse charged particles.

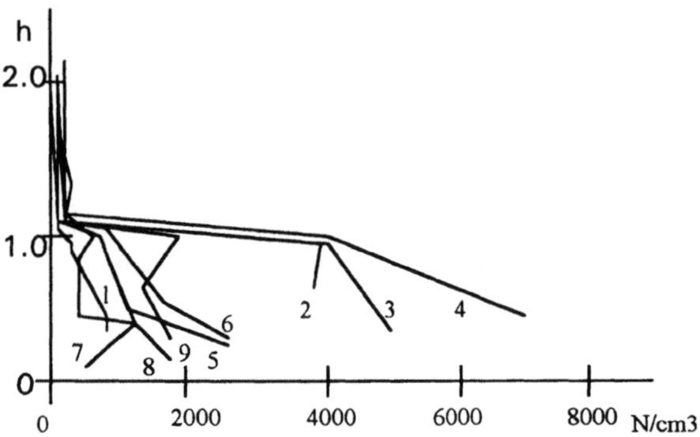

Figure 3
Result from large charged particle counting (N) in various aircraft flights effected between Dec. 1952 and Aug. 1953. H = 1.0 is the level of the top of the exchange layer. (After Sagalyn et al., 1954)

Conductivity is high in the free atmosphere due to the absence of particles through the photo-ionisation process described as well as through the formation of ions though the same process in the various gases and compounds constituting the earth's atmosphere such as:

$$H^+(H_2O)_n,\ H^+(CH_3CN)\ (H_2O)_m,\ NO^-_3(HNO_3)_n,\ HSO^-_4(H_2SO_4)_n\ (HNO_3)_m.$$

A high conductivity means that such an atmosphere acts more like a metal than an insulator; thus, for example, charged surfaces will be discharged faster. The consequence of a highly conductive atmosphere is that electrical field strengths are low. The mean value observed in the free atmosphere (inf.) is in the order of 5V/m.

THE EXCHANGE LAYER (INF.)

Ionisation

The energy-rich part of the sunlight does not penetrate deeply in the exchange layer (inf.): e.g. the Ultraviolet "C" component is non-existent near the earth's surface. By removing this powerful source of ionisation from the earth's atmosphere, then, only a few other sources of low-level ionizaton remain.

One of these sources is the ionisation caused by the decay of natural and artificial radioactive sources in the lower atmosphere and in the earth's crust. Another source is the effect of powerful particles related to so-called cosmic rays. The number of gaseous ions produced in this way is quite low, namely some 10-20 ion pairs per cm^3 per second.

The presence of numerous particles in the earth's atmosphere caused by natural and man-made pollution means that these ions have a very limited lifetime. This is due to the fact that these particles act like fibers in a filter as they are almost stationary compared to the highly mobile ions, thus cleaning the air from ions of both kinds. So in industrialized areas the number of free gaseous ions varies from some tenths to a few hundreds per cm^3. It is only over low polluted areas like the oceans where the often-cited number of 2000-3000 per cm^3 can be observed, a number limited by the presence of water vapor.

Conductivity, Electric Field

The absence of mobile gas ions near the earth's surface means that conductivity is low there. This is very important, because it means that the value of the electric field close to the earth is far higher than that of the free atmosphere and easily reaches values between 150-450V/m under fair weather conditions. In the case of extra loss of electrical conductivity, electrical activity values, as with pollution, fog, or thunderstorms, easily become higher by a factor of *10* at the exchange layer (inf.), and so a thousand times higher than the values found in the free atmosphere (inf.).

It should be clear that the electrical field value is dependent on the water vapor present. The higher the level of water vapor, the higher the field becomes. This seems strange because the presence of water is normally linked to a higher conductivity and thus a reduced electric field. However, in the case of the atmosphere things are different. Gaseous ions, especially negative ones, serve as favorite condensation sites for water molecules. By increasing in size they reduce their mobility and increase their filtering properties for other gas ions. Therefore water vapor tends to build an insulating layer and so creates high electric fields. This is illustrated in Figure 4.

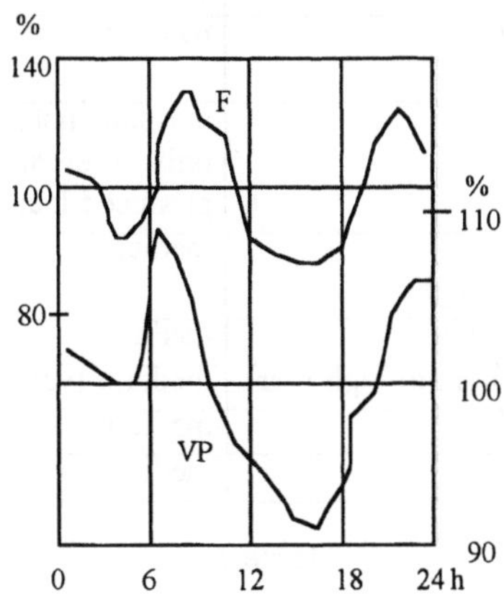

Figure 4
Variations observed in the electric field (F) and the level of water vapor (VP) at Potsdam, Germany. (After Israel, 1973)

The consequence of the low conductivity of the exchange layer (inf.) is that a surface that has obtained an electrical charge, (e.g. a carpet in your office, the leaves of the tree in your garden, or yourself with your suit à la mode), will retain this charge for a long time due to this low air conductivity This means that these surfaces will influence the propagation, interception, and deposition of the fine particles, sometimes allergenic, found throughout the exchange layer. Additionally, the generation of charged particles is highly affected by the often low conductivity and the high electric field of the atmosphere near the earth.

Effects of conductivity of the exchange layer (inf.).

In this context it is little known but important to mention that, after the passage of the radioactive cloud emitted by the explosion of the Chernobyl nuclear power plant over Poland fruit trees carried some months later more fruit than normal. Moreover, the fruit were much larger, and sometimes even gigantic. Assuming mutation effects caused by the radioactive constituents carried by the cloud and a continuing large scale ion presence implies that the consumption of these fruit by the population was very low. However the assumed mutation and eventual health effects on man and animal have not been not reported. A very plausible explication of the effect may be as follows. The presence of radioactive material made the atmosphere more conductive than normal. The consequence of this was that the often existing high electric fields with their negative influence on the vegetation diminished, allowing for a better use of its sap, as a consequence, allowed the vegetation to yield a higher production.

Summary of characteristics

In the following table a summary is given of a number of characteristics of the different layers near the earth's surface.

characteristic	exchange layer (inf.)	exchange layer (sup.)	free atmosphere (inf)
water vapor level	high-medium	low + clouds	no
presence of VOC's	yes	yes	no
chem modif. activity	low	medium	high
pollen	present	normally not present, only present when produced locally	not present
number of particles	very high	medium	low
number of ions	very low	medium	high
E field	medium to high (var)	medium (variable)	low (constant)
Temp	high	medium	low
Solar activity	low	medium to high	high

Table 1

THESIS: SOME POSSIBLE CHARACTERISTICS OF THE ATMOSPHERE AT HIGH ALTITUDES

On the basis of these parameters, let us see if we are able to reconstruct some characteristics of the atmosphere at high altitudes for the period of the year that the three basic components are reguarly present, which will be the case in general in spring and summer. In wintertime, the thermal effects will mean that the exchange layer will remain at low altitude Only the characteristics in Table 1 will be taken into account, so no reference will be made to the often more powerful thunderstorm activity or the regular presence of mist and fog at high altitudes.

1) The basic ingredient will be formed by the dry, good conductive air of the free atmosphere (inf), allowing for low level electrostatic fields. As a consequence generation, transport, and deposition of particles become more predictable, which means that counter measures become effective.

2) The number of natural and anthropogenic particles that are present will be limited due to the reduced influence of the exchange layer (inf.). Polluted pollen material of lower altitudes will be scarce.

3) On the other hand the presence of modified VOC's becomes very likely. The presence of micro-organisms will be limited by the powerful bactericidal action of the sun's ultra violet light.

4) The dominating character of the free atmosphere with its high conductivity state will reduce the stress on the vegetation as is the case at lower altitudes. Some similar aspects as observed in Poland concerning the activity of the vegetation after the Chernobyl accident could take place.

5) Another consequence is that the air at high altitudes will contain a greater number of positive and negative ions than regions at lower altitude.

VERIFICATION

1) The basic ingredient will be formed by the good, although dry, conductive air of the free atmosphere (inf.), allowing for low level electrostatic fields. As a consequence the generation, transport, and deposition of particles becomes predictable, which means that counter measures become effective.

1) Unfortunately no documents of scientific value could be found to support this part of the hypothesis. Only observations of intense nervousness in people and animals in high electric fields have been reported.

2) The number of natural and anthropogenic particles present will be limited due to the reduced influence of the exchange layer (inf.). Polluted pollen material of lower altitudes will be scarce.

2) In this context let us observe the results obtained by Pani et al. (1990), who measured the weekly average of mass and composition during one year at altitudes of 450, 1020, and 1780 m. on mount Rigi in a prealpine region of central Switzerland. Figure 5a,b shows the variations of the total suspended particulates (TSP), at the highest and the lowest sampling site for about a year-long period. The observations at the middle site have been left out for the sake of clarity. During the winter, a pronounced altitude dependence could be detected (see Figure 5a), especially when stable temperature inversion situations existed, such as at the beginning of February 1989. It is interesting to note the high TSP concentration at 1780m just after the

middle of February, due to a Sahara dust period. In spring and summer, the exchange layer is higher than the top of mount Rigi, leading to similar particle concentration patterns at different altitudes. However, as could be expected, particle load is less at high altitudes.

The progression in the free atmosphere (inf.) during the Sahara dust event, normally difficult to explain, is something which is predictable when electrical effects of the atmosphere are taken into account.

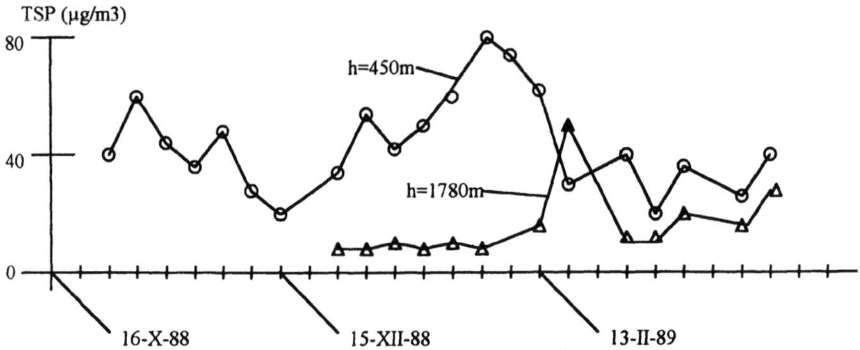

Figure 5a Weekly total suspended particulate matter concentrations (TSP) at two different altitudes for winter period. (after Pani et al. 1990)

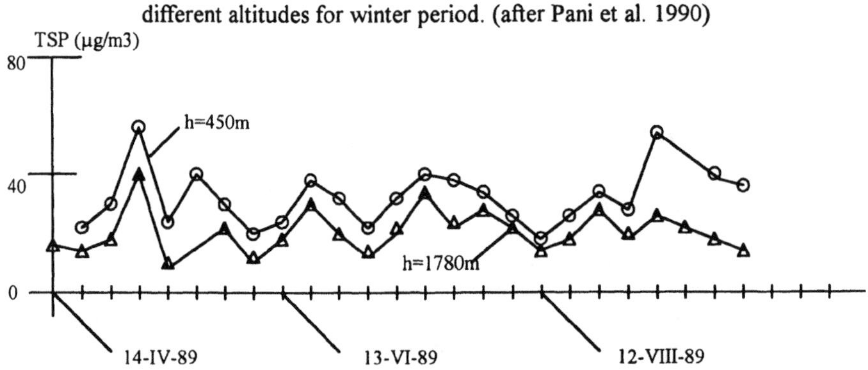

Figure 5b. Weekly total suspended particulate matter concentrations (TSP) at two different altitudes for the summer period. (after Pani et al. 1990)

No information about the presence of pollen has been obtained from this study. However, Rufin (1995) indicates that "In the mountain the pollen season is delayed. Its maximum is often between July and August". This shift indicates that the pollen from low altitudes is effectively screened off and that the pollen input at high altitudes is from local origin.

3) On the other hand the presence of modified VOC's becomes very likely. The presence of micro-organisms will be limited by the powerful bactericidal action of the sun's ultra violet light.

The ultraviolet part of the sun's spectrum at high altitudes is powerful enough to produce ozone. Although the maximum production is related to the so-called ozone hole situated between 12 and 17 km altitude, its presence is already noted at the level of the free atmosphere (inf.) It is well known that ozone is a strong bactericidal agent, which is thus coherent with our statement.

Another less well known source of ozone, is described in the Würtz of 1864. When we put a quart-sized bottle filled with turpentine oil (a typical VOC) in the sun and shake it violently to get it into contact with the air, the turpentine becomes charged with ozone, because it has highly energetic oxydizing characteristics. Left alone the turpentine oil becomes oxidized because ozone polymerizes it. The slow oxidation of this VOC is thus preceded during a certain time by the formation of ozone.

Summertime haze is a result of such a polymerization, which creates condensation nuclei, which together with high summer humidities produce particles that can reach droplet sizes of about 0.3μm in diameter, a size which is very effective in scattering light and causing visibility effects. However, under such circumstances, the formation of turpine is likely. Turpine is the result of the hydratation of turpentine oil and is used as a treatment for pulmonary related illnesses; therefore, it is possible that certain health effects at altitude are caused by this natural formation of a drug together with a bactericidal agent.

4) The dominating character of the free atmosphere with its high conductivity state will reduce the stress on the vegetation which applies at lower altitudes. Some similar aspects as observed in Poland in regard to the activity of the vegetation after the Chernobyl accident could take place.

Although this hypothesis doe not seem very obvious, it is a pleasure to discover a text in which the described phenomena coincide remarkably with our statement, although the explication is of course not the same.
"In the equatorial region where similar different levels of vegetation at high altitudes also exist, the biggest surprise is found at the highest level. With a temperature quite stable and favorable for growth there is often as well a very nebulous cause of the gigantic development of non-shrub-like species like the giant sénonçons of the Kilimandjaro, the *Puya Raimondii* of the Peruvian Cordillera Blanca, the lobélies of Zaïre and the rhodondendron, normally a bush, but which is a real tree in the Himalayas". (Encyclopedia Universalis, 1980)

The explication given in this citation is, as is often the case when it concerns questions of climate, very imprecise and can without problems be applied to a lower altitude where even more favorable conditions exist, without leading to something really astonishing, namely the gigantic size of certain plant species.The reduced influence of high electric fields together with favorable circumstances seems to be a better explication.

5) Another consequence is that the air at high altitudes will contain a greater number of positive and negative ions than regions at lower altitudes.
Here it is in our opinion necessary to point out a possible consequence of the presence of negative ions at high altitudes. This point of view is far from the ill-founded idea that these ions should have a very good influence upon the health of those who inhale

them. We have to point out that the quantity of positive and negative ions in the air even at high altitudes has a homeopathic level. Assuming an amount of 2 000 per cm^3, this means that we have

1 ion for 13.3 000 000 000 000 000 molecules of air.

It will be clear that the effect of ions seems to be marginal under these conditions.

However at the beginning of this century, the Nobel prize winner C.T.R Wilson, known for Wilson cloud chamber, showed that there is a clear difference between positive and negative ions. Condensation of water molecules around a positive ion is not possible under normal humidity conditions, and this is also the same for neutral molecules. However condensation around a negative ion is possible below a humidity level of 100%, which is also the case around fine atmospheric particles, so called Aitkin nuclei.

This means that a cloud of condensation around these two types of particles will result when you exhale at a high altitude. The fact that in a cold atmosphere, you are able to see this cloud as a form of mist, means that its size is larger by far than that of the wavelength of visible light, which is in in the order of 0.5μm. This means that the exhaled particles must have diameters in the micrometer region. At higher altitudes, where the air is so much colder, this process partly takes place in the airways.

The pulmonary system, when it comes to particle retention, has similar characteristics to that of a filter. Smaller particles are intercepted mainly by their Brownian movement, while large particles are filtered out because of interception and impaction. The intermediate sized particles with diameters of 0.2-0.3μm are hardly retained at all, the so-called Greenfield gap.

In Figure 6 the consequences of altitude for lung retention are schematically shown. At low altitude the number of fine particles or Aitkin nuclei is high, which means that their growth due to water vapor condensation is limited and interception by Brownian movement is likely. At high altitudes no Aitkin nuclei are present anymore; instead, a number of negative ions become the site of condensation of water vapor. Because their number is quite low, the droplets they form are large, and cause increased retention through interception and impaction.

Only at a certain altitude, (marked in Figure 6 by the letter H), will the retention of particles and water vapor in the pulmonary tracts be minimal-possbly a plus point from the point of view of health.

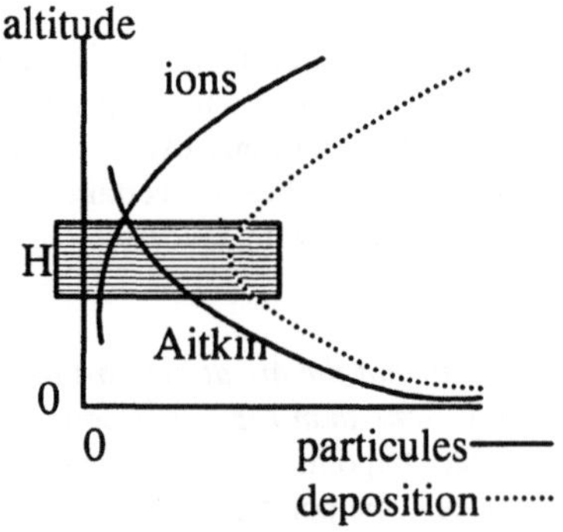

Figure 6 See text

That the presence of a great number of negative ions at high altitudes does not have a favorable effect becomes clear with the following statement:
The sensitivity of people to mountain illness is variable; this is also the case with the ability to adapt. The first symptoms are sensed in general at 3000 m and the limit of adaptation over a long period is situated at 5500 m. The most simple and effective remedy is to climb down as soon as possible. At high altitudes the oxygen level in the blood is reduced, and the flow of blood towards the brain intensified causing an increased arterial tension. The consequences are headaches and behavioral problems. Liquid accumulates in the membranes of the pulmonary aveoles (see Figure 7), later on invading the alveoles themselves and so blocking the transfer of oxygen, finally leading to choking (Beall, 1994).

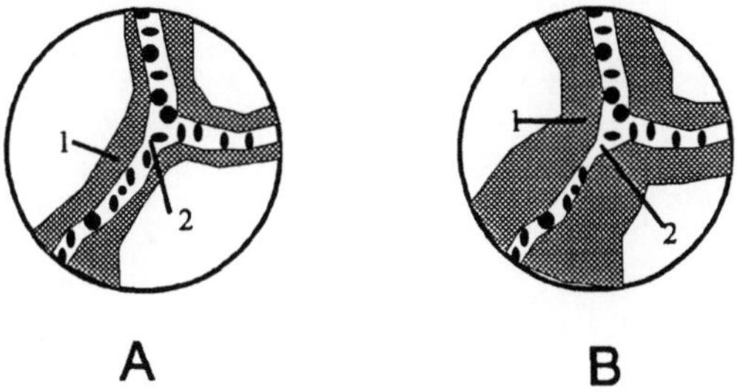

Figure 7
See also text. 1 bloodvessel, 2 liquid

CONCLUSION

It is clear that the Cartesian approach leads to a strong reduction and even elimination of a great number of generalities often proposed concerning the climate at high altitudes. Additionally, it often offers unexpected and exciting ways of understanding phenomena and leads, as a consequence, to new areas of scientific investigation. We hope that this approach will have followers, because it is increasingly necessary to leave the impoverished highways that science takes today and to look for secondary roads of investigation, the only ones that lead to really new discoveries.

REFERENCES

Allan, N.J.R., pp 107, La vie dans les montagnes, les montagnes, ed Life-Time Books B.V. Amsterdam, NL 1994

Beall, 1994, Les effets de l'altitude sur l'homme, p63, Encyclopédie de la terre: es Montagnes, Time-Life Books, Amsterdam, Netherlands, 1994.

Chalmers, J.A., Atmospheric Electricity, Pergamon Press, Oxford G.B., 1967

Encyclopedia Universalis, Volume 11, p292, Paris, 1980

Hinds, W.C, Aerosol technology, Wiley & Sons, New York, USA, 1982

Israel, H., Atmospheric Electricity, Vol II, Jerusalem, 1973.

Pani, P., Bucher, P., Zhang, Z., Hertz, J.Altitude dependence of aerosol composition, European Aerosol Conference, Zürich 1-5 October 1990.

Robert, H., Ionisation, santé, vitalité, où les bienfaits des ions négatifs de l'air, Ed.

Artulen, Paris, F, 1990
Rusfin, P, Allergies, allergiques.Asthme, J. de l'Ass. Asthme, P1., Mai 1995
Sagalyn, R.C., et al. J. Atm. Terr. Phys. 5. pp 253-272, 1954
Würtz, A.D., Chimie Médicale, I Chimie inorganique, pp. 42., Masson, Paris, 1864

PANEL DISCUSSION

The last session of the Workshop was a panel discussion.
In the first part, which was chaired by Dr. Ronald Wolff, the different papers were summarized and discussed. A written document was prepared for this panel discussion by Prof. Kvetoslav Spurny and is appended to this panel discussion report. Rapporteur was Patricia Keady.
In the second part of the panel discussion the social, economic-regulatory and research implications of the latest views on particulate pollutants and environmental health were discussed. The discussion was introduced by Prof. Morton Lippmann and Dr. Roger Mc.Clellan. Prof. Othmar Preining supplied a written document, which is also appended to this panel discussion. Rapporteur was Prof. David Swift.

Jan Marijnissen

Panel Discussion on Deposition and Retention of Inhaled Particulate Matter

Chaired by Ronald Wolff
Summary Report by Patricia B. Keady

A provocative discussion between workshop participants concentrated on issues concerning changing regulations, the health effects of ultrafine particles, and particle deposition in susceptible populations. The U.S. Environmental Protection Agency (EPA) is considering changes to the National Ambient Air Quality Standard (NAAQS). With the intent to better protect human health, the EPA may propose a new particulate matter (PM) standard based on mass measurements with an upper size cut in the range from 1 to 3 μm aerodynamic diameter. This would render the mass of the fine fraction of airborne particles. The most likely size cut is 2.5 μm ($PM_{2.5}$). The current standard requires particulate mass measurements <10 μm (PM_{10}).

Recent research suggests that particles much smaller than this (< 0.1 μm) have even more dramatic effect on human health. These so-called "ultrafine" particles may have a greater toxicity than predicted by the mass concentration. Measurements of number concentration, may prove to be a more useful measurement than mass. Unfortunately, these measurements require a new set of monitoring instruments necessitating significant new investment and training. Lippmann expressed concern that the monitoring network may not be able to make such measurements reliably.

Most workshop participants agreed that the current PM_{10} measurement is not a good scientific definition of the aerosol, either chemically or physically. McClellan cautioned that we should not confuse what we measure with what we inhale. We can measure TSP, PM_{10}, $PM_{2.5}$, or ultrafines, but individuals inhale the same thing independent of what we measure. No matter where our size cuts are, it does not change what is really there. He suggests that we need to look at more than size; we need also to look at the chemical nature of the particles.

Lippmann alleged that we do not know very much about ultrafine particle exposures, so the epidemiological data cannot support the claim that they are more highly correlated with health effects. He wonders if having a mass standard is the right approach. We have reduced the level of airborne particulate mass over the years, but may not have reduced the hazard. It may be that gas-phase conversion products cause most of the mortality, but the data is very limited. The chemistry of the fine and coarse particles can be quite different. The dichotomous sampler was never codified as an approved method, so there is little data on this.

Stöber commented that we should also discriminate between soluble and insoluble particles because they will behave quite differently once deposited. Roos added that the electrical properties of the particles will affect their deposition patterns and influence their behavior in the atmosphere.

Marijnissen cautioned that if new standards go to $PM_{2.5}$, we may get unintended results. As we clean out big particles, we may create more small particles because we have removed the scrubbing effect of the big particles. It is dangerous to set new standards, but vitally important to take more measurements. Lippmann recalled recent studies by Wichmann's group in the former East Germany. They found that the health response has not changed as much as expected with a reduction in aerosol mass.

The general consensus was that we need to take more data and measure particle number in addition to mass. To be widely accepted, measurements must be easy and inexpensive. At the very least, we should measure the aerosol in two fractions, $<10\mu m$ and $<2.5\mu m$, so we can build a base of measurements that can be compared to epidemiological data. A single measurement is no longer adequate for regulatory purposes. We will not see significant amounts of data if it is not part of a regulation. Lippmann mentioned that the EPA draft criteria document states that the volume (mass) of < 0.1-μm particles is closely proportional to number concentration. It is not correlated with < 2.5-μm particles. So it would be highly desirable if we could use a 0.1-μm cut and measure also the mass of the ultrafines.

Swift concluded by asserting that due to our choice in measurements, we assume we have causation based on what we have measured. We are getting better at measuring particle size while still mixing everything together chemically. Maybe our focus is wrong.

The discussion then turned toward the deposition of particles in susceptible populations. McClellen stated that the data we have is mainly on healthy normals. We have very limited data on those at most risk. These are the people who are most vulnerable and should be studied more closely. Wolff commented that studying susceptible populations is quite difficult in clinical tests -- maybe ethically impossible. Often you get diminished responses. Lippmann defended current research practice saying that studying a healthy normal reduces variability and gives you better control. We need to find suitable animal models that will be a surrogate for sensitive segments of the human population.

Panel Discussion

Social, Economic-Regulatory and Research Implication of the Latest Developments in the Atmospheric Particle/Human Lung System"

Rapporteur: David L. Swift, Ph.D.

Written documents by Dr. Kvetoslav Spurny and Prof. Othmar reining were submitted for this panel discussion and are appended to his report. The panel discussion was introduced by Prof. Otmar reining who introduced the two major discussants, Prof. Morton ippmann of NYU and Dr. Roger McClellan of CIIT.

Prof. Lippmann presented an overview of the history of the egulation of airborne particulate matter. He began by pointing out he secular changes in the airborne TSP values in the city of ittsburgh, PA starting in the period just after WWII. A significant eduction in TSP and general "smokiness" has occurred in Pittsburgh, s in other industrial cities of the northeastern US. The reduction f the values in these cities has been accomplished resulting from a umber of factors, not the least of which was the change in the fossil uel used domestically and in industry from coal to natural gas.

This lowering of particulate matter were done under local, ather than national, ordinances. In addition to the introduction of atural gas as a fuel substance, the coal and oil used was of ecreasing sulfur content beginning in the 1960's. Thus by the time f the first national air quality standards, a significant reduction n "gross particulate material" had already been accomplished at a ost which subsequently was seen as modest in comparison to the costs equired to meet later standards.

In 1971, the first National Ambient Air Quality Standards (NAAQS) ere established by federal law; these standards covered three classes f pollutants, viz., particulate matter, oxidant vapors, and acid ases. The specific surrogates adopted for regulation were total uspended particles (TSP), SO_2, CO, NO_x, and O_3. The law provided for rimary standards (human health related) and secondary standards ecological and "welfare" (i.e., visibility) related]. No provision as made in the law for considering the economic impact of steps equired to meet the standards.

Periodic re-evaluation (every 5 years) of the basis for the tandards was specified in the law, but the USEPA did not make these deadlines" for the first re-evaluation. In 1987, the particulate tandard was changed from TSP to PM-10, reflecting the belief that nly particles which can reach the intrathoracic (tracheo-bronchial nd pulmonary) region of the human lung should be considered with respect to health effects. Other revisions of the Clean Air Act of 1987 included the definition of a class of pollutants denoted as "Air Toxics," which were to be controlled differently than the NAAQS pollutants.

Presently the PM-10 standard is being reconsidered in view of a series of epidemiological studies reported in the literature which claim morbidity and mortality effects at very low levels of PM-10. Under court order resulting from a suit brought against the EPA by the American Lung Association (and others), the PM-10 criteria document (reviewing the existing data on health and other effects of airborne PM-10) is being "fast-tracked" in order to meet the letter of the law contrary to past experience.

Several issues were brought out by Prof. Lippmann, including the fact that short term epidemiological studies are difficult to carry out because the law calls for PM-10 to be monitored every 6th day, rather than daily. As a result, the number of epidemiological studies carried out on particulate exposure and resulting health effects are limited. Likewise, controlled human chamber studies are limited by the fact that the sensitive sub-populations (elderly, very young, and the respiratory ill) are not able to participate for practical and ethical reasons. As a result, there has been much reliance on animal studies, with their attended difficulties of differences in particle deposition, defense mechanisms, and life span.

It was pointed out by Prof. Lippmann that most animal exposure and human chamber studies have been performed with primary particulates (i.e., those generated at source as particles) as opposed to secondary aerosols (those formed in the atmosphere by gas-to-particle reactions or altered by particle-vapor interactions in the atmosphere following releases as distinct particles and vapors. Thus if one hypothesizes that such secondary particles are important to health, the existing data related to human health is sparse.

Several other issues were raised regarding the economics of ambient air quality control, including the provision in current law for carrying out an economic impact analysis of any regulatory requirement. This is contrary to the original intent of the Clean Air Act, which specifically ignored the cost of changes to comply with NAAQS and Air Toxics standards. The current provision of federal law calling for economic impact contains two components, the cost of past compliance efforts and the expected cost of proposed changes in federal law. This requirement may prove to be difficult in the case of community air pollutants.

The historical overview by Prof. Lippmann was followed by remarks by Dr. McClellan, in which he began by comparing the situation in particulate air pollutants with that of inhaled radioisotopes. The latter is based on the same considerations of the deposition, clearance, and lung retention of inhaled particles, but has more clearly understood and defined dose relationships.

Dr. McClellan then discussed the concept of attributable risk, ·inting out that there are a number of factors which are thought to as significant as particulate air pollutants in leading to ·rbidity and mortality and represent important confounders in tempts to relate the level of PM-10 or other health related measures airborne particulate matter. These include smoking status, socio-onomic status, ambient and indoor dwelling temperatures, and posure to other gases and vapors which also represent confounders.

Serious consideration should be given to the possibility of sing new particulate standards on the above mentioned idemiological studies in view of the relatively low relative ditional risk associated with such small changes (c.a. 5%) in the bient PM-10 level. The benefits of adopting lower standards may be fset by the economic costs of achieving such reductions and the ssible human health and welfare impacts of such provisions.

Thus, Dr. McClellan suggested a cautious approach to the issue of vising downward the particulate standards, as some have urged the A to do. Even though the original Clean Air Act specifically oided considering the economic impact of achieving particle andards, such impacts should be estimated as a part of the decision out the future PM-10 standards.

The further issue of adopting addition standards, such as PM-2.5 PM-1 was considered by both Dr. McClellan and Prof. Lippmann. Both pressed an opinion that PM-10 should not be eliminated in favor of e other standards, but that sampling using the PM-2.5 or PM-1 proach should be undertaken in order to provide additional data for e evaluation of health effects.

It was realized that neither PM-2.5 nor PM-1 were ideal to parate the "coarse" fraction (including wind blown dust) from the ine" fraction which is composed primarily of secondary particles sulting from gas-to-particle reactions or growth of ultrafine rticles.

Finally, Prof. Preining discussed the issue of meeting European andards in terms of controlling carbon release, stating that he lieved that such controls would be both difficult and costly.

APPENDIX 1 TO PANEL DISCUSSION,

ATMOSPHERIC PARTICULATE POLLUTANTS AND ENVIRONMENTAL HEALTH
Physico-chemical Standpoint

K.R. Spurny
Aerosol Chemist, Eichenweg 6, 57392 Schmallenberg, Germany

The fine fraction of the atmospheric anthropogenic aerosol (AAA) is an aerodispersed system of solid and liquid particles with aerodynamic sizes between about 0.01 and 2.5 μm, having different particle forms and chemical compositions. The chemical composition as well as the lung deposition efficiency and toxicity are size dependent.

Newest epidemiological studies have shown positive correlations between the fine fraction of the AAA and different health effects at concentration levels, which were lower than the existing Air Quality Standards (AQS).

The existing definition of the AAA (air particulate, PM-10) neglects the fact, that this air particulate fraction (particle sizes with aerodynamic diameters less than 10 μm) is bimodal and consists of two aerodispersed components: condensation and dispersion aerosol. It neglects further the particle size depending chemical composition. The fine fraction of the AAA is associated with the most toxic, inorganic and organic particulate air pollutants ("Lung Toxic Mode Fraction").

The aerosol physics and chemistry are able to contribute efficiently to the future solution and regulation of this problem. The AAA should be sampled in two fractions: PM<2.5 and PM 2.5-10 (aerodynamic particle diameters less than 2.5 μm and in the range between 2.5 and 10 μm). Methods and equipment for such sampling are available. Special markers of the fraction PM 2.5 could be detected (Monitoring) in situ, on line and in real time; e.g. PAH and carbon black (soot). Methods and equipment for such monitoring are available. Also free radical activities of such fine aerosol samples could be determined.

New AQS (annual means), based on newest epidemiological knowledge have to be discussed and proposed; e.g. the range of 10 $\mu g/m^3$ for PM-10, less than 5 $\mu g/m^3$ for PM < 2.5, the range of 3 $\mu g/m^3$ for soot aerosols, less than 50 ng/m^3 for PAH (and/or less than 5 ng/m^3 for BaP)

Last but not least, the existing knowledge on the field of mathematical and physico-chemical deposition and toxicity models could be probably applied for the health risk estimations of different AAA.

APPENDIX 2 TO PANEL DISCUSSION,

Othmar Preining, Clean Air Commission, Austrian Academy of Sciences

Remarks to:
Social, Economic, Regulatory and Research Implications of the latest Developments in the Atmospheric Particle/Human Lung System

The atmosphere is an aerosol (PM), there are natural and anthropogenic sources. PM is inhaled and particles are deposited in the respiratory system. What are the consequences for public health? What is the situation in Europe? What should be done and what is achievable under present socio-economic conditions?
We know there is some epidemiological evidence: PM influences public health negatively (with special emphasis on inhalable particles). It is easy (and already done in air pollution networks) to collect and measure gravimetrically PM (all sizes and chemical components) or PM 10, occasionally also PM 2.5. Inhalable are the smaller particles, PM 2.5. Most affected are urban areas. Typical values for the UK are given in the following table (1).

Component	$\mu g/m^3$	Found in fine fraction %	Fraction of total %
Total PM	35	-	-
PM 10	25 - 30	~ 60	-
Soluble ions			~50
Sulfate	2 - 10	~ 85	-
Nitrate	2 - 10	~ 75	-
Chloride	1 - 3	~ 10	-
Ammonium	2 - 6	> 95	-
H^+	0.01 - 0.1	~ 100	-
Carbonaceous			~23
EC	3	~ 80	-
OC	5	~ 80	-
Minerals	15	~ 5	~23
Metals	2 (lead 0.1)	~ 30 (lead 90)	~ 4
Toxic organic micropollutants			
Dioxins	~ 4 pg/m^3	most	trace
PCB's	~ 1 ng/m^3	most	trace
PAH's	~150 ng/m^3	most	trace

Conclusion:
- The coarse fraction (of PM 10) is mostly natural (40%)
- The fine fraction (of PM 10) is mostly anthropogenic (60%)
- The fine to coarse ratio is to some degree dependent on the local situation.
- Soluble ions, elemental and organic carbon (including toxic organic micropollutants) are in the fine fraction.
- In case the fine to coarse ratio is approximately constant PM 10 would be sufficient to evaluate the hazards from urban particulate air pollution.

Research needs: Extended survey of fine to coarse ratio at many different locations and under different weather conditions.

Of special concern are the carbonaceous fractions (TC, EC, OC) or simply soot. In Europe, the EC baseline concentrations in industrial areas are around 3-4 $\mu g/m^3$; however, in some German cities (Düsseldorf, Essen) considerably higher concentrations are observed. The variation of daily averages there is between 3-30 $\mu g/m^3$ for EC and 4-50 $\mu g/m^3$ for TC. This has to be compared with the German "Federal Clean Air Act" (Bundes-Immissionsschutzgesetz, BImSchG, Article 40, Section 2): Soot measurements are required and a standard for the measurement procedures is given. The concentration standard, effective since July 1, 1995, is for EC 14 $\mu g/m^3$ (2).
Public opinion considers air pollution and smog as most serious issues (together with acid rain and water pollution) (3).
There is considerable need for R&D for assessment of exposure to air pollution. Instruments are now available but studies are expensive and time consuming, especially for large populations; measurements and models have to be combined, terminology in Europe should be harmonized taking into account the translation into different languages (4) and the different interpretations of identical terms in different countries.
In many parts of Europe (including Austria) the development of irrational fears especially concerning dioxin can be observed. The interested and engaged scientific community should avoid to support directly or indirectly this development. The most important issue for achieving any solution in pollution problems is the confidence of the public and of decisions-makers in the rational approach by science. This confidence is essential to find support for very expensive measures in case these are necessary. **The media** is interested in dramatizing statements and this may be contraproductive for environmental issues. Reasons for this are: environmental problems are not discussed sufficiently nor openly in full complexity in our societies. The present situation needs equally environmentalists, economists, scientists, as well as political and spiritual leaders who discuss and find **together** policies for an acceptable future.

References

(1) Roy M. Harrison, M. Jones, The Chemical composition of airborne particles in the UK atmosphere. The Science of the Total Environment, 168 (1995) 195-214

(2) Hans-Ulrich Pfeffer, Jürgen Friesel, Gereon Elbers, Reinhold Beier, Klemens Ellermann, Air pollution monitoring in street canyons in North Rhine-Westphalia, Germany. The Science of the Total Environment, 169 (1995) 7-15

(3) David E. Bloom, International Public Opinion on the Environment. Science, 269 (21 July, 1995) 354-358

(4) WHO, Regional Office for Europe, Methodology for assessment of exposure to environmental factors in application to epidemiological studies. The Science of the Total Environment 168 (1995) 93-100

KEYWORD INDEX

The page number refers to the beginning of the paper in which the keywords occur.

LIST OF AUTHORS

FSC
www.fsc.org
MIX
Papier aus verantwortungsvollen Quellen
Paper from responsible sources
FSC® C105338

If you have any concerns about our products,
you can contact us on
ProductSafety@springernature.com

In case Publisher is established outside the EU,
the EU authorized representative is:
Springer Nature Customer Service Center GmbH
Europaplatz 3, 69115 Heidelberg, Germany

Printed by Libri Plureos GmbH
in Hamburg, Germany